True Visions

E.H.L. Aarts J.L. Encarnação
(Eds.)

True Visions

The Emergence of Ambient Intelligence

With 153 Figures, 5 in color and 5 Tables

Prof. Dr. Emile Aarts
Philips Research Laboratoties
High Tech Campus 36, 5656 AE Eindhoven, The Netherlands
E-mail: emile.aarts@philips.com

Professor Dr.-Ing. Dr. h.c. mult., Dr. e.h., Hon. Prof. mult. José Encarnação
Institut für Graphische Datenverarbeitung, Fraunhofer Gesellschaft
Fraunhoferstr. 5, 64283 Darmstadt, Germany
E-mail: jkl@igd.fraunhofer.de

Library of Congress Control Number: 2006923233

ISBN-10 3-540-28972-0 Springer Berlin Heidelberg New York
ISBN-13 978-3-540-28972-2 Springer Berlin Heidelberg New York

Springer is a part of Springer Science+Business Media.

springer.com

Printed in the Netherlands

Typesetting: Data prepared by the Author and by SPI Publisher Services
Cover design: *design & production* GmbH, Heidelberg

Printed on acid-free paper SPIN 11556046 57/3100/SPI 5 4 3 2 1 0

Foreword

by

Prof. Dr.-Ing. habil. Prof. e.h. Dr. h.c. mult. Hans-Jörg Bullinger,

President of the Fraunhofer-Gesellschaft,

Corporate Management and Research

The realization of *Ambient Intelligence* (AmI) requires the integration of different scientific topics, such as human–computer interaction, middleware and agent technologies, as well as aspects of virtual reality and artificial intelligence, into one single cooperative vision. The generation of new products and devices also requires the enhancement of hardware-oriented branches of research, such as embedded systems, sensor technologies, or awareness and presence technologies.

To face this challenge, the Fraunhofer-Gesellschaft – the largest organization for applied research in Europe, with roughly 80 research units at over 40 different locations throughout Germany – bundled its research competencies and research efforts and identified 12 innovation topics that will define the future of research over the next years. Intelligent products and smart environments are one major focus of these efforts. Here, intelligent products and smart environments refer to the allocation of information in an individualized manner and the cooperation of embedded devices and networked intelligent services. The environment will adaptively and autonomously adjust itself according to the user's needs and goals and will help to gain information, to establish communication, to guarantee safety, to ensure physical well-being, and to deliver entertainment and education. Furthermore, the Fraunhofer-Gesellschaft seeks not only to support humans, but also machinery, vehicles, or animals, as smart players within AmI scenarios. The realization of *Ambient Intelligence*, however, means more than the support of smart players within smart environment scenarios. *Ambient Intelligence* will overcome the device- and function-oriented interaction paradigm and will establish a user-centric approach. Instead of offering the multiplicity of diverse functions that overwhelm the user, the environment will proactively support the smart player in reaching his goals.

At present, various Fraunhofer Institutes – together with external and internal partners – are involved in many projects dealing with the realization of *Ambient Intelligence*. The focus is on the application and development of innovative sensor technologies and self-organizing sensor networks, and on the development and enhancement of middleware technologies to support ad-hoc cooperation of distributed device ensembles. Technologies that enhance the communication and interaction possibilities of their users in order to ensure the participation of aging citizens within the information society and to maintain their social connectivity are also a major focus of Fraunhofer research activities. Innovative technologies for context management, the recognition and interpretation of the situation will be researched and enhanced to develop context-aware solutions that guarantee the assistance of the smart player in the most appropriate fashion. The Fraunhofer-Gesellschaft identified several AmI scenarios, which define the guideline for future research. Within the scenario "health assistance", smart environments of the future will be developed that assist in particular the elderly. Novel sensor networks will be able to measure and interpret the activities of daily life. Consequently, the elderly will feel safe and secure as they are supported in living independently. Intelligent environments for logistics and production will overcome the centralistic approaches of today's production and logistic solutions and will thus open up new possibilities within logistics, transportation, and traffic. Finally, scenarios and solutions for intelligent travel will support the user in getting along in foreign environments and places. Context-aware and adaptive technologies will assist him in finding his destination or locations and occasions of interest.

More than 20 Fraunhofer Institutes with different background competencies are working together so that these visions will come true. Together, they will invest more than 20 million Euros in the realization and development of three demonstration platforms in order to demonstrate the AmI concepts, as well as the feasibility of the approaches proposed by the Fraunhofer-Gesellschaft. Health assistance, self-organizing logistic solutions, and intelligent travel support will each play a major role in the Fraunhofer vision of *Ambient Intelligence.*

This book is an amazing opportunity to illustrate the innovative ideas of *Ambient Intelligence.* It offers an excellent overview over the different branches of research that have to be integrated in order to realize ambitious AmI solutions. The contributions to this book range from the realization of smart environments to mobile computing, from computing to software platforms, and on to the application of context-aware technologies. Cultural issues of *Ambient Intelligence*, user support by novel user interfaces and reports from experience settings are also the focus of important chapters of this book. The contributions are all written by leading experts with a proven expertise in the area. Therefore, the content of this book is at the highest level of technical quality. The reader not only gets an excellent overview but also detailed knowledge about the relevant key components of *Ambient Intelligence* and related enabling technologies which we estimate to be the most important research subjects in the realization of the vision of *Ambient Intelligence.*

Darmstadt, March 2006 *Hans-Jörg Bullinger*

Foreword

by

Dr. Rick Harwig,

Chief Technology Officer of Philips

Since the gebinning of recorded history substantial evidence has accumulted for the proposition that technological innovation has contributed in a major way to the advancement of mankind. Evidently, groundbreaking inventions such as the incandescent lamp, the telephone, and the television have changed the lives of many people in the world. Looking at more recent times the development of the Internet can certainly be counted as one of the most influential and disruptive technologies boosting the information and communication society through global connectivity and personalized services. So, anyone who has harbored serious doubts that technology is the key to modern society's innovation are put to rest by these and many other technological innovations that have provided ordinary people with support and comfort over time. Obviously, each age dictates its own characteristic socio-economical problems and issues. The development of the rapidly growing new economies such as China, India, and the Latin American countries set the pace for technological innovation in the classical western economies of Europe, Japan, and the United States. The proliferation of digital personalized media, the sustainable development of our society, and the care for the aging population are the contemporary fields that call for disruptive innovation. Interestingly, the role of technology in this respect is drastically changing, as it can no longer be seen as the only driving force of innovation. Moreover, existing innovation processes are subject to major overhaul due to the increased self-awareness of people and the complexity imposed by the large diversity of stakeholders that are involved. Through its unique way of combining elements from technology, society, and business into a unifying multi-dimensional approach to innovation, *Ambient Intelligence* can be seen as one of the most interesting developments of present times.

To Philips as a company *Ambient Intelligence* has been quite influential in introducing a common view on the development of future electronic products and services. In the world of *Ambient Intelligence* few things operate on a stand-alone basis. Lighting, sound, vision, domestic appliances and personal healthcare products and services all communicate and cooperate with one another to improve the total user experience. Spearheaded by Philips Research, the vision of *Ambient Intelligence* has brought a common and tangible focus to a variety of endeavors, encouraging many in our organization to see how their activities fit into the overall concept of an environment that is sensitive and responsive to the presence of people and sympathetic to their needs. The *Philips Active Ambilight Television*^{TM} and the *Philips Ambient Experience*^{TM} are two examples of products that resulted from the AmI vision and were successfully launched in the market. As user-system interaction research is recognized as one of the key factors in the realization of *Ambient Intelligence*, Philips Research established its HomeLab in 2002 as a fully functional house equipped with an extensive observation system that offers behavioural scientists a unique instrument for studying human behaviour in context to support the company to deliver on its band promise known as Sense and Simplicity. The AmI vision has also enabled Philips Research to establish new collaborations and focussed partnerships with other strong players in the field such as the Fraunhofer Gesellaschaft, IBM, INRIA, IMEC, MIT, and Thomson Multimedia.

Along with the build up of the vision for Philips, a parallel track was followed which was aimed at positioning the vision as an open initiative for the advancement of the innovation in information and communication technology in Europe. During a series of workgroups organized by ISTAG (Information Society and Technology Advisory Group) which serves as an influential advisory board to the European Community, the vision of *Ambient Intelligence* was adopted as the leading theme for the sixth framework on IST research in Europe. This is a major achievement in the development of a united Europe because it provides the European Union with a unique instrument that can be used to set novel competitive landmarks in relevant innovation domains such as High-Tech systems, Creative Industries, Healthcare, and Mobility. It is my true conviction that major business innovations in these domains can only be achieved in the years to come by building an open and collaborative atmosphere that allows both large and small enterprises to develop jointly new products and services from within an eco-system of collaborative efforts in which no major dominant players are active. Moreover, disruptive and renewing developments leading to breakaway and market innovations often take place at the crossroads of well-established domains, and *Ambient Intelligence* has proved to be quite helpful in developing such cross border domains because of its inherent multi-disciplinary nature.

Finally, there is the issue of outreach, which calls for direct participation of society in the development and realization of the vision. As *Ambient Intelligence* implies for people that technology will be invisibly integrated into their environment, this puts them into the foreground. So, they need to be involved early on, so as to become aware of the potential of this novel development. Never before in the history of modern society have we as technologists been faced with such a challenge. We can no longer hide inside our laboratories and research new technologies with the conviction that what we are producing is exactly what people are waiting for. We need to open up and involve society. We have to tell and educate ordinary people about our innovation endeavors with the aim of involving them preferably as co-creators or at least as co-users. This book *True Visions* can play a role in this respect because it brings together in a unique way all aspects relevant to the development and roll-out of the AmI vision.

So, after its first five years of steady development we can safely state that the AmI vision has reached a well-recognized status of maturity. Both inside and outside Philips the build-up of the vision needs to continue and we need to demonstrate its added value to society. This calls for a united approach within an open innovation spirit that empowers those working on the realization of *Ambient Intelligence*, and convinces those who are not directly involved that this vision is going to be something that will provide added value.

Eindhoven, March 2006 *Rick Harwig*

Preface

The vision of *Ambient Intelligence* (AmI) provides a paradigm for user-centric computing for the first two decades of the 21st century. Building on early developments in *Ubiquitous Computing and Pervasive Computing*, *Ambient Intelligence* opens a new venue based on the design of smart electronic environments that are sensitive and responsive to the presence of people. AmI environments consists of many embedded electronic devices that enable users to interact with their environment in a natural way, thus providing them ubiquitous access to content, information, and services in a context aware, personalized, and (pre)responsive way, thereby enhancing productivity, healthcare, well-being, expressiveness, and creativity.

The salient novel aspect of *Ambient Intelligence* is the incorporation of the physical world into the interaction between human being and computing devices. This incorporation can be achieved by a massive embedding of intelligent computing devices. *Ambient Intelligence* aims at taking the integration of computing devices to a maximum by involving any physical object in the interaction between people and their environments. The perceived user value is the improved productivity, well-being and self-expression through the enhanced user-system interaction. Large scale embedding of electronic devices, ubiquitous communication, networked intelligence, and natural user interfaces are the main technological issues posing major research questions for the realization of the AmI vision. No matter how important and challenging these technological aspects may be, the true merits of the AmI success will depend on the take up it receives from ordinary people in their every-day lives.

Evidently, the focus of the AmI paradigm is not only on the physical integration of electronics, but also on the creation and generation of enhanced experiences. As a consequence thereof many disciplines need to be combined to achieve the required insights that may lead to the ultimate user experience. Obviously, integration is the magic word, and this in more than one respect. Video, audio, vision, and lighting need to be integrated into a novel approach to multimedia applications. Physical and virtual worlds need to be integrated into a single end-user reality. Physical objects and devices need to be integrated into a seamlessly interoperable environment. Users need to be integrated into content production chains to set up their own activities in creative industries. Finally, the digital and the physical worlds need to be fully integrated. It all comes down to the challenge of mastering the experience through advanced interaction concepts. This implies that many scientific disciplines come together to constitute a novel interdisciplinary science and technology building on the individual achievements made in mathematics, computing science, electrical engineering, industrial engineering, and human factors.

True Visions is one of the first books that brings together in a single volume all the constituent elements originating from the various disciplines that matter to the development of *Ambient Intelligence.* The broadness of the spectrum of activities is amazing and we as editors where astonished by the amount of relevant items that were brought up by the authors in their individual chapters. All authors are experts in their own specific domains and without exception they have succeeded in positioning their domain of expertise within the greater endeavour called *Ambient Intelligence.* We have chosen the title *True Visions: The Emergence of Ambient Intelligence* to emphasize that the development of *Ambient Intelligence* is still in an early phase. It all started with a vision of a new digital world, providing us with novel opportunities to become more productive, express ourselves better, and improve our well-being. We have now left the stage of articulating and evangelizing the vision, in order to enter the stage of its realization, which is a tedious and very challenging effort. True Visions reports on the achievements made so far, and we are confident that it contains a wealth of information for the interested reader. We also believe, however, that anyone who has digested its contents will share our opinion that True Visions reveals that the integral realization of *Ambient Intelligence* in daily life still requires a major effort.

Eindhoven, Darmstadt
March 2006

Emile Aarts
José Encarnação

Contents

List of Contributors

Anne Trefethen
Deputy Director e-Science Core
Programme – DTI/EPSRC
Polaris House
North Star Avenue
UK-Swindon SN2 1ET
UK
+44 1793 444284
Anne.Trefethen@epsrc.ac.uk

Anton Nijholt
Universiteit Twente
Postbus 217
NL-7500 AE Enschede
The Netherlands
+31 53 4893686
anijholt@cs.utwente.nl

Berit Svendsen
Telenor ASA
P.O. Box 1 D7d
N-1331 Fornebu
Norway
+47 810 77 000
berit.svendsen@telenor.com

Bernt Schiele
Department of Computer Science
ETH Zürich
Haldeneggsteig 4
CH-8092 Zuerich
Switzerland
+41 1 632 0668
schiele@inf.ethz.ch

Berry Eggen
Technische Universiteit Eindhoven
Den Dolech 2 (HG 2.54)
NL-5612 AZ Eindhoven
The Netherlands
+31 40 247 5227
j.h.eggen@tue.nl

Boris de Ruyter
Philips Research
High Tech Campus 37
NL-5656 AE Eindhoven
The Netherlands
+31 40 27 44982
boris.de.ruyter@philips.com

Christine Kallmayer
Fraunhofer IZM
Gustav Meyer Allee 25
D-13355 Berlin
Germany
+49 30 46403 228
kallmayer@izm.fraunhofer.de

David De Roure
Department of Electronics and
Computer Science,

Southampton University
Highfield Campus – Hartley Avenue
SO17 1BJ Southampton, UK
+44 23 8059 2418
dder@ecs.soton.ac.uk

Dick Broer
Philips Research
High Tech Campus 4
NL-5656 AE Eindhoven
The Netherlands
+31 40 27 42746
dick.broer@philips.com

Dirk Heylen
Human Media Interaction
Computer Science – University of Twente
P.O. Box 217
7500 AE Enschede, The Netherlands
+31 53 4893745
d.k.j.heylen@ewi.utwente.nl

Ebba Thora Hvannberg
Department of Computer Science,
University of Iceland
Hjardarhaga 2-6
IS-107 Reykjavik
Iceland
+354 525 4702
ebba@hi.is

Emile Aarts
Philips Research
High Tech Campus 5
NL-5656 AE Eindhoven
The Netherlands
+31 40 27 43203
emile.aarts@philips.com

Henk van Houten
Philips Research
High Tech Campus 4
NL-5656 AE Eindhoven
The Netherlands
+31 40 27 42769
henk.van.houten@philips.com

Herbert Reichl
Fraunhofer IZM
Gustav Meyer Allee 25
D-13355 Berlin
Germany
+49 30 46403 122
reichl@izm.fraunhofer.de

Hugo De Man
IMEC
Kapeldreef 75
B-3001 Leuven
Belgium
+32 16 281201
hugo.deman@imec.be

Ingrid Moerman
IMEC vzw – Ghent University – IBBT
Department of Information Technology (INTEC)
Gaston Crommenlaan 8 bus 201
B-9050 Gent
Belgium
+32 9 33 14 925
ingrid.moerman@intec.ugent.be

Jaap den Toonder
Philips Research
High Tech Campus 4
NL-5656 AE Eindhoven
The Netherlands
+31 40 27 46508
jaap.den.toonder@philips.com

James Crowley
INRIA Rhône-Alpes
655 Avenu de l'Europe
F-38330 Montbonnot-St. Martin
France
+33 4 76 61 52 10
crowley@imag.fr

Jean-Claude Burgelman
Institute for Prospective Technological Studies
Edificio Expo, Calle Inca Garcilaso s/n
E-41092 Seville
Spain
+34 954 48 84 96
jean-claude.burgelman@cec.eu.int

Joelle Coutaz
Laboratoire CLIPS, Université Joseph Fourier
385 Ave de la Bibliothèque
F-38041 Grenoble
France
+33 4 76 51 48 54
joelle.coutaz@imag.fr

José Encarnação
Fraunhofer Institute for Computer Graphics
Fraunhoferstraße 5
D-64283 Darmstadt
Germany
+49 6151 155130
jle@igd.fhg.de

Jürgen Wolf
Fraunhofer IZM
Gustav Meyer Allee 25
D-13355 Berlin
Germany
+49 30 46403 606
wolf@izm.fraunhofer.de

Martin Ouwerkerk
Philips Research
High Tech Campus 4
NL-5656 AE Eindhoven
The Netherlands
+31 40 27 42677
martin.ouwerkerk@philips.com

Nikolaos Georgantas
INRIA-Rocquencourt
Batiment 8, INRIA – UR de Rocquencourt
Domaine de Voluceau,
Rocquencourt – BP 105
F-78 153 Le Chesnay Cedex,
France
+33 1 39 63 51 37
nikolaos.georgantas@inria.fr

Paola Inverardi
Dipartimento di Informatica
Universita' dell'Aquila
via Vetoio 1 – Coppito
I-67010 L'Aquila
Italy
+39 0862 43 3127
inverard@di.univaq.it

Patrick Reignier
Laboratoire GRAVIR, INRIA Rhone Alpes
655 Avenu de l'Europe
38334 St. Ismier
France
+33 4 76 61 54 11
patrick.reignier@inrialpes.fr

Paul Lagasse
University Gent
Sint-Pieternieuwstraat 41
B-9000 Gent
Belgium
+32 9 2643315
paul.lagasse@intec.rug.ac.be

Paul van der Sluis
Philips Research
High Tech Campus 4
NL-5656 AE Eindhoven
The Netherlands
+31 40 27 42343
paul.van.der.sluis@philips.com

Philip Treleaven
Department of Computer Science,
University College London
Gower Street
UK-London WC1E 6BT,UK
+44 207 6797288
p.treleaven@cs.ucl.ac.uk

Privender Saini
Philips Research
High Tech Campus 37
NL-5656 AE Eindhoven
The Netherlands
+31 40 27 44342
privender.saini@philips.com

Rainer Wasinger
German Research Center for AI (DFKI)
Stuhlsatzenhausweg 3
D-66123 Saarbruecken
Germany
+49 681 302 3393
Rainer.Wasinger@dfki.de

Rifat Hikmet
Philips Research
Prof. Holstlaan 4
NL-5656 AA Eindhoven
The Netherlands
+31 40 27 43449
rifat.hikmet@philips.com

Rudy Lauwereins
IMEC
Kapeldreef 75
B-3001 Leuven
Belgium
+32 16 281244
rudy.lauwereins@imec.be

Ruud Balkenende
Philips Research
High Tech Campus 4
NL-5656 AE Eindhoven
The Netherlands
+31 40 27 42577
ruud.balkenende@philips.com

Stefano Marzano
Philips Design
Emmasingel 24
NL-5611 AS Eindhoven
The Netherlands
+31 40 27 59032
stefano.marzano@philips.com

Stephen Emmott
University College LondonDepartment of Computing
Gower Street
UK-London WC1E 6BT
United Kingdom
+44 20 7679 7288
semmott@microsoft.com

Stephen Klink
Philips Research
High Tech Campus 4
NL-5656 AE Eindhoven
The Netherlands
+31 40 27 43182
steve.klink@philips.com

Steven Kyffin
Philips Design
Emmasingel 24
NL-5611 AS Eindhoven
The Netherlands
+31 40 27 59195
steven.kyffin@philips.com

Thomas Kirste
Department of Computer Science,
University of Rostock
Albert-Einstein-Strasse 21
D-18059 Rostock
Germany
thomas.kirste@uni-rostock.de

Tony Hey
Department of Electronics and Computer Science
University of Southampton
Highfield Campus – Hartley Avenue
SO17 1BJ Southampton UK
+44 1793 444022
tony.hey@epsrc.ac.uk

Torsten Linz
Fraunhofer IZM
Gustav Meyer Allee 25
D-13355 Berlin
Germany
+49 30 46403 670
linz@izm.fraunhofer.de

Valérie Issarny
INRIA-Rocqencourt
Batiment 8, INRIA – UR de Rocquencourt
Domaine de Voluceau
Rocquencourt – BP 105
F-78 153 Le Chesnay Cedex
France
+33 1 39 63 57 17
valerie.issarny@inria.fr

Wolfgang Wahlster
German Research Center for AI
Stuhlsatzenhausweg 3
D-66123 Saarbruecken
Germany
+49 681 302 5252 or 5251 or 5080
wahlster@dfki.de

Yves Punie
Institute for Prospective Technological Studies
Edificio Expo, Calle Inca
Garcilaso s/n
E-41092, Seville
Spain
+34 95 448 82 29 (direct)
yves.punie@cec.eu.int

1

Into Ambient Intelligence

E. Aarts and J. Encarnação

"Ambient Intelligence is about everyday technology that makes sense"

1.1 The Vision

Ambient Intelligence (AmI) refers to electronic environments that are sensitive and responsive to the presence of people. The paradigm relates to a vision for digital systems in the years 2010–2020 that was developed in the late 1990s, and which has become quite influential in the development of new concepts for information processing, combining multidisciplinary fields including electrical engineering, computer science, industrial design, user interfaces, and cognitive sciences.

In an AmI world, devices operate collectively using information and intelligence that is hidden in the network connecting the devices. Lighting, sound, vision, domestic appliance, and personal health care products all cooperate seamlessly with one another to improve the total user experience through the support of natural and intuitive user interfaces.

The AmI paradigm provides the basis for new models of technological innovation within a multidimensional society. The essential enabling factor of the AmI vision is provided by the fact that current technological developments will enable the integration of electronics into the environment, thus enabling the actors, i.e., people and object to interact with their environment in a seamless, trustworthy, and natural manner. In addition, the past years reveal a growing interest in the role of information and communication technology to support people's lives, and this not only refers to productivity but also to health care, well-being, leisure, and creativity. A major issue in this respect is given by the growing awareness that novel products such as devices and services should meet elementary user requirements such as usefulness and simplicity. So, it is generally believed that novel technologies should not increase functional complexity, but merely should contribute to the development of *easy to use* and *simple to experience* products. Obviously, this statement has a broad endorsement by a wide community of both designers and engineers, but reality reveals that it is hard to achieve in practise, and that novel approaches, as may be provided by the AmI vision, are needed to make it work.

The notion *ambience* in Ambient Intelligence refers to the environment and reflects the need for an embedding of technology in a way that it becomes nonobtrusively integrated into everyday objects. The notion *intelligence* reflects that the digital surroundings exhibit specific forms of social interaction, i.e., the environments should be able to recognize the people that live in it, adapt themselves to them, learn from their behavior, and possibly act upon their behalf. This leads to the following list of salient features of Ambient Intelligence into.

- *Integration* through large-scale embedding of electronics into the environment
- *Context-awareness* through user, location, and situation identification
- *Personalization* through interface and service adjustment
- *Adaptation* through learning
- *Anticipation* through reasoning

Evidently, the new paradigm is aimed at improving the quality of peoples' lives by creating the desired atmosphere and functionality via intelligent, personalized, interconnected systems and services. However simple this requirement may seem, its true realization is for the time being not within our reach. To bring this ideology closer to its realization, substantial investigation into integration technology, natural interaction concepts, and human behavior is needed, and this is the aim of Ambient Intelligence.

1.2 Trends and Opportunities

At the occasion of the 50th anniversary of the *Association of Computing Machinery* in 1997, computer scientists from all over the world were asked for their opinion about the next 50 years of computing (Denning and Metcalfe 1997). Their reaction was strikingly consistent in the sense that they all envisioned a world consisting of distributed computing devices that were surrounding people in a nonobtrusive way. *Ubiquitous computing* is one of the early paradigms based on this vision. It was introduced by Weiser (1991) who proposes a computer infrastructure that succeeds the mobile computing infrastructure and situates a world in which it is possible to have access to any source of information at any place at any point in time by any person. Such a world can be conceived by a huge distributed network consisting of thousands of interconnected embedded systems that surround the user and satisfy her needs for information, communication, navigation, and entertainment.

Evidently, the AmI vision closely follows the early developments in ubiquitous computing and it should not be seen as a revolutionary disruption but merely as a natural evolution caused by a number of factors related to worldwide developments which took place at the end of the last century, and that paved the way for a novel concept in the development of electronic systems and the way people operate them. These factors can be classified into three categories, which relate to developments in technology, global connectivity, and socioeconomic aspects. Below, we briefly elaborate on each of these in more detail.

1.2.1 "More Moore" and "More than Moore"

An interesting frame of reference for our discussion on technology is provided by the developments in semiconductor industry. It is generally known and accepted that developments in this domain follow Moore's law (Noyce 1977), which states that the integration density of systems on silicon doubles every 18 months. This law seems to hold a self-fulfilling prophecy because the computer industry follows this trend for already four decades. Moreover, other characteristic quantities of information processing systems, such as communication bandwidth, storage capacity, and cost per bit of input–output communication, seem to follow similar rules. Advances in Moore's law show the following three lines of development. 1D-Moore is the one-dimensional continuation of the classical Moore's law into the submicron domain of microelectronics, resulting in small and powerful integrated circuits that can be produced at low cost. 2D-Moore is the development of two-dimensional large-area electronic circuitry at extremely low cost, possibly using other technologies than silicon such as polymer-electronics. 3D-Moore refers to the development of ultrahigh functional three-dimensional circuitry consisting of microelectronic mechanical systems (MEMS) that integrate sensor, actuator, computing, and communication functions into a single nanoelectronics system.

Following the lines of thought imposed by the different developments of Moore's law one may conclude that the design and manufacturing of electronic devices has reached a level of miniaturization which allows the integration of electronic systems for processing, communication, storage display, and access into any possible physical object like clothes, furniture, cars, and homes, thus making people's environments smart.

1.2.2 Weaving the Web

What had started in the 1980s as an early attempt to connect the scientific world through a network supporting the exchange of scientific documents and results among researchers had developed by the end of the past century into a truly world-wide network allowing not only researchers but also ordinary people to have access to digital information in the broad sense (Berners-Lee et al. 1999). Access percentages in the western world started to exceed 50% and it soon became obvious that the Internet would grow into a truly ubiquitous access network. Furthermore, the Internet had reached a point in its development were it became evident that the long proclaimed convergence of mobile computing, personal computing, and consumer electronics would become effective. Nowadays this promise has become reality as many Internet providers offer what they call *triple play*, which is the use of (mobile) telephony, television, and date access over the Internet through a single subscription, and the next step will be the development of the seamless handover between these services among different wirelessly connected devices. In addition to these network developments there were also major developments in the way data were stored in and retrieved from the network. So, people started to develop means not only to store raw data on servers and terminals, but also to generate metadata on the raw data that could provide additional information and support the retrieval process. These developments led to the introduction of the *semantic Web*, which uses all sorts of ontologies among data to enrich the Web leading to the development of high-level applications such as *context aware media browsers* (Berners-Lee et al. 2001).

The developments of the Internet brought a next step within reach, which could lead to a major paradigm shift in the way intelligence was generated among electronic devices. So far the only assumption had been to generate intelligence through complex software programs executed on powerful high-cost terminal devices. Now it would be conceivable to design systems consisting of simple low-cost terminal devices that could fetch intelligence from the network to which they are connected leaving the rendering of the information as the only task to be carried out by the terminal device.

1.2.3 The Experience Economy

By the end of the past century, socioeconomic investigations revealed that a next wave of business development was emerging based on mass customization leading to a new economic order, from which an answer could be obtained on the question whether Ambient Intelligence could contribute to the development of new business and greater wealth to all people in the world. Pine and Gilmore (1999) describe in their compelling bestseller a new economy, which they call the *experience economy*. They position this economy as the fourth major wave following the classical economies known as the commodity, the goods, and the service economy. The general belief of the experience economy is that people are willing to spend money on having experiences, and from certain enterprises such as the holiday economy, one indeed may conclude that this might be very well true. A salient property of an experience is given by the fact that it can feel real, irrespective of whether it has been generated by a real or a virtual cause; what counts is the belly feeling. Personal reminiscences that bring back good old feelings are nice examples of such experiences. Richard Florida (2001) takes the ideas of the experience economy a step further by sketching a world of urban centers in which people create a living by working as artists in the way they create new products and services. These so-called *creative industries* fully exploit the concept of experience design and provide a new economy supplying local-for-local markets.

Prahalad and Ramaswamy (2004) use similar concepts as in the experience economy to propose a new way of value creation for the 21st century based on the co-creation and development of novel goods and services that satisfy the greater needs of customers.

All these novel socioeconomic ideas open up major possibilities for making money in markets that exploit Ambient Intelligent technology, thus providing the necessary economical foundation for the development of Ambient Intelligence.

1.2.4 Advances in Design

In addition to the tree key factors mentioned above, the late 1990s also showed a general development that was following up on the profound desire to have more things that were simply useful, and to move users or people in general into the center of our activities. In other words, the information society had resulted into an overload of products and services for which the user benefits were unclear. There was a call to design things that were easy to understand and simple in their use. Norman (1993); Negroponte (1995) and Winograd (1996) are all examples of designers who were fiercely opposed against the existing means and concepts for user interfaces and man–machine interaction, and who were desperately in search of novel paradigms that would make the interaction with electronic systems more natural. It was generally believed that the social character of the user interface would be determined by the extent to which the system complies with the intuition and habits of its users and this led to a number of groundbreaking insights. One of these developments is that of a novel research area which is called *affective computing*, and which is characterized by a multidisciplinary approach to man–machine interaction that combines different methods from psychology and computer science (Picard 1997). Another angle is provided by the approach followed by Reeves and Nass (1996) who state in their *Media Equation* that the interaction between man and machine should be based on the very same concepts as the interaction between humans is based, i.e., it should be intuitive, multimodal, and based on emotion. Clearly, this conjecture is simple in nature but it has proved to provide many designers with some foothold in the development of their products. Loosely related to this is the concept of *user-centered design* (Norman 1993; Beyer and Holtzblatt 2002), which is applied by many interaction designers. It states that the user is to be placed in the center of the design activity, which follows a number of consecutive design cycles in each of which the designer reevaluates her concept and its realization through specific user evaluations and tests.

The development of the AmI vision has been largely influenced by these design developments because they clearly articulated the need for a user-centric approach to the design of novel electronic environments. They also contributed to the understanding that the use of Ambient Intelligent environments can only be measured by the user benefits that are ultimately achieved by them, irrespective of their intricacy and sophistication.

1.3 A Brief History of Ambient Intelligence

In this section, we briefly review some of the AmI developments. Although Ambient Intelligence is still young, its development over the past years shows a number of interesting landmarks. Obviously, the origin of Ambient Intelligence is tightly coupled to the groundbreaking developments in computing science imposed by the developments in ubiquitous computing. Its unique identity, however, is imposed by the fact that the AmI vision is to a large extent a user-centric industrial vision with a major impact on scientific research.

1.3.1 Early Developments at Philips

The notion of Ambient Intelligence was developed in 1998 in a series of internal workshops that were commissioned by the board of management of the Philips company. The workshops were aimed at investigating a number of different scenarios that would lead a high-volume consumer electronic industry from the current world which was called *fragmented with features* into a world near 2020 with fully integrated user-friendly devices supporting ubiquitous information, communication, and entertainment. Palo Alto Ventures, a US management consultancy company, acted as the facilitator (Zelka 1998). The first public presentation on Ambient Intelligence was given at the Digital Living Room Conference 1999 by Roel Pieper who at that time was a member of the board of management of Philips Electronics responsible for consumer electronics equipment (AmI Keynote 1999).

The first official publication that mentions the notion “Ambient Intelligence” appeared in a Dutch IT journal (Aarts and Appelo 1999) and emphasized the importance of the early work of the late Mark Weiser who already for more than ten years was working on a new concept for mobile computing which he called ubiquitous computing (Weiser 1991).

The AmI vision has also been used by Philips Research to establish new and promising collaborations with other strong players in the field. In 1999, Philips Research joined the Oxygen alliance; an international consortium of industrial partners that collaborated within the context of the MIT Oxygen project (Dertouzos 1999, 2001). The Oxygen project is a joint effort of the MIT Computer Science Laboratory and the Artificial Intelligence Laboratory, and it aimed at developing the technology for the computer of the 21st century. It allows multimodal controlled handheld communication units to connect through environmental units to a broadband communication network, thus supporting ubiquitous information access and communication.

In the mean time the vision grew mature. In 2000, first serious plans were launched to build an advanced laboratory that could be used to conduct feasibility and usability studies in Ambient Intelligence. After two years of designing and building, HomeLab was eventually opened on 24 April 2002 (Aarts and Eggen, 2002), and the opening event officially marked the start of a new research program on Ambient Intelligence at Philips Research (de Ruyter 2003). In 2003, Philips published *The New Everyday* (Aarts and Marzano 2003). The book contained over 100 contributions on Ambient Intelligence on a broad range of topics ranging from materials science up to marketing and business models. Most of the contributions are from Philips authors, but about ten of them are from renowned specialists emphasizing various aspects related to Ambient Intelligence ranging from promising new applications to critical remarks that warn for the possible societal disorientation that might result from Ambient Intelligence. *The New Everyday* can be seen as the first in a series of activities undertaken by Philips to open the vision by sharing the ideas Philips had developed so far with a broader audience.

1.3.2 Opening up the Vision

Along with the build up of the vision for Philips, a parallel track was developed which was aimed at positioning the vision as an open initiative for the advancement of the innovation in information and communication technology in Europe. Following the advice of the Information Society and Technology Advisory Group (ISTAG) (ISTAG 2001) issued in 2001, the European Commission used the vision for the launch of their sixth framework (FP6) in IST, Information, Society, and Technology, with a subsidiary budget of 3.7 billion euros. The influence of the European Commission has been crucial for the development of the vision and it is hardly conceivable that the paradigm could have grown in the strong way it did without the Commission's support. The reason for this is obvious. It soon had become evident that a single company, however big, could not turn a vision as broad as Ambient Intelligence into reality. A neutral and influential party was needed to bring the different stakeholders together and facilitate and manage the development processes. As a result of the many initiatives undertaken by the European Commission the development of the AmI vision really began to gain traction. These initiatives resulted in the start of numerous research projects that were aimed at the development and realization of the vision. Last year's IST Conference in Den Hague featured results of more than 30 major projects on a large diversity of applications including personal health care, consumer electronics, logistics and transportation, and e-mobility (IST 2004).

At the same time, the vision was recognized as one of the leading themes in computing science by the Association for Computing Machinery (ACM), and as a result thereof a book chapter on Ambient Intelligence was invited to the book *The Invisible Future* (Aarts et al. 2002). The book was published at the occasion of the ACM1 conference, which was aimed at providing the electrical engineering and computer science community of the world with new insights into the future of computing at large. In addition to the chapter on Ambient Intelligence, the book contains a wealth of contributions from various renowned scientists in the world expressing their vision on a variety of subjects ranging from computer hardware and programming up to health, education, and societal issues.

During the past years several major international initiatives were started. Fraunhofer Gesellschaft originated several activities in a variety of domains including multimedia, microsystems design and augmented spaces. Their *InHaus* project (inHaus, online), which is similar to Philips' *HomeLab* a research facility that supports investigation into feasibility and usability aspects of Ambient Intelligence, can be seen as a first approach to user-centric design and engineering. MIT started an AmI research group at their Media Lab with a special emphasis on research in personal health care. More than ten 5 million euros or more research programs on Ambient Intelligence were started at national levels in a large variety of countries including Canada, Spain, France, and The Netherlands. A novel European subsidiary instrument has been announced under the name Experience and Application Research Centers, which is aimed at the financial support of research facilities that conduct research into user behavior for the purpose of user-centric design and co-creation of novel products and services inspired by the vision of Ambient Intelligence.

Over the years, the European Symposium on Ambient Intelligence (EUSAI) has grown into the most interesting event for the exchange of novel ideas in Ambient Intelligence (Aarts et al. 2003; Markopoulos et al. 2004). In addition to EUSAI, which is devoted to Ambient Intelligence only, many large key conferences have started to set up special events highlighting developments in Ambient Intelligence that are of interest to their audience. Examples are DATE03, CHI2004, INTERACT2004, CIRA2005, and INDIN2005.

In the mean time, several books have been published which deal with different aspects of Ambient Intelligence. (Mukherjee et al. 2005) discuss physics and hardware related aspects including microelectronic systems, large-area electronics, and MEMS design. Weber, et al. (2005) present a collection of edited chapters on a variety of technological aspects including connectivity and low-power design. Basten, et al. (2003) discuss challenges in embedded systems design related to Ambient Intelligence. Verhaegh, et al. (2003) present a collection of edited chapters on the design and analysis of Ambient Intelligent algorithms. Riva et al. (2005) present a collection of edited book chapters on sociocultural aspects of Ambient Intelligence.

1.3.3 Where Are We Headed?

Over the past years, the awareness has grown that the classical type of industrial research facility can no longer provide the technological innovation required to drive the world's economical development. For more than half a century industrial research laboratories applied the policy that hiring the best possible people and stimulating them to generate as much as possible intellectual property rights would provide the most effective way to technological innovation. Many of the existing industrial research laboratories consider the protection of their knowledge as quite important and consequently the nature of these laboratories reflects that of a closed facility. More recently, new models for industrial research were proposed that followed the developments of the networked knowledge economy. Based on the belief that tapping into as many as possible bright people can develop more innovative ideas, industrial research has widened its scope to become more collaborative and open-minded.

Ambient Intelligence has proved to be quite instrumental in the realization of open innovation. The broadness of the vision allows many different parties to contribute from within their own specific angles. Looking at Europe we see again a number of interesting initiatives that are aimed at international collaboration. The *AmI@Work* (online) group combines parties from both the public and the private domains to provide a forum for the discussion of the use of Ambient Intelligence to improve productivity and support health care and well-being. The European Technology Platform for the development of embedded systems called *Artemis* uses the AmI vision to define their strategic research agenda. In addition to the many activities by the European Commission, there are also private initiatives that are aimed at international collaboration. As an example we mention AIR&D, the Ambient Intelligence Research and Development consortium in which INRIA, Fraunhofer Gesellschaft, Philips, and Thompson jointly conduct precompetitive research in Ambient Intelligence. These are just a few examples of a large variety of novel and promising initiatives that use the AmI vision to support open innovation.

So after five years of successful developments, one may wonder which direction Ambient Intelligence will take from here. For the longer term this question is hard to answer because its answer will be strongly influenced by the many external factors that determine the success, and hence the lifetime of a vision such as Ambient Intelligence. There are two key factors for success that can be mentioned. The first one is given by the extent to which Ambient Intelligence can drive innovation processes that turn into business successes, and the second one is determined by the length of time over which researchers will stay inspired by the vision and are willing to contribute to its realization. For the shorter term, i.e., the next one or two years, however, the answer is obvious and determined by output that will become available from the many initiatives that have been defined for this period of time. All in all the forthcoming years promise to become very inspiring for the development of Ambient Intelligence with many new challenging initiatives that carry the promise to contribute substantially to the fruitful realization of the AmI vision.

1.4 Realizing Ambient Intelligence

Ambient Intelligence will make a substantial contribution to science, as well as to the economy, if its realization contributes noticeably to human well-being. The cycle of technology becoming increasingly difficult and inscrutable in its usage must be broken. Up to now, it has been the user's responsibility to manage her personal environment, to operate and control the various appliances and devices that are available for her support. It is obvious that the more the technology is available and the more options there are, the greater the challenge of mastering your everyday environment of not getting lost in an abundance of possibilities. Failing to address this challenge adequately simply results in technology that becomes inoperable and thus effectively useless.

1.4.1 For the Well-Being

It seems to be contradictory, but the more options and functionalities a bundle of devices offer to their users, the more functions remained unused. But the reason is simple: Technology as it exists today forces its metaphors upon the user. The responsibility of finding a strategy that combines all the functions that are offered by the environment equipped with smart devices and functionalities is shifted to the user. Consequently, it is possible for the user to be more occupied with finding strategies and functions than she is with her actual goals.

Stated more emphatically: with increasing technology, the user is in danger of losing sight of her goals or of changing her goals, because she is not able to find the appropriate combination of strategies for device functions. The user is forced to pay more attention to complete lists of functions than to her actual goal. Instruction manuals for today's devices are a good example of the metaphor change that took place. These manuals explain menu-based selection lists or parameter adjustments. But the instructions do not explain the possible user goals, which can be achieved by using the corresponding device. Furthermore, which goals could be achieved, if devices were interconnected, is generally not mentioned at all.

In summary, technological development has shifted in the last few years from goal orientation to function orientation. The development effort concentrated more on the extension of the pool of functions than on the user and her goals and her well-being. The most important task of Ambient Intelligence is to reverse this metaphor change back to the user's goal orientation (see Fig. 1.1). Technology has to meet the wants of the user and not vice versa.

Through Ambient Intelligence, the environment gains the capability to take over mechanical and monotonous control tasks – as well as stressful feature selections and combinations – from the user and manage appliance activities on her behalf. To do this, the environment's full assistive potential must be mobilized for the user, tailored to her individual goals and needs. Realizing this, the user becomes an active part of her environment; she will be

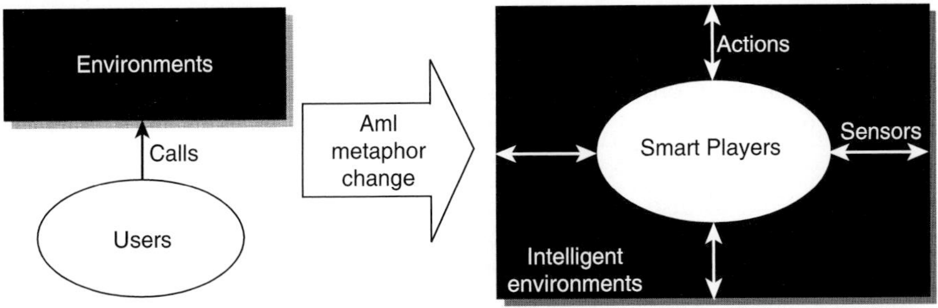

Fig. 1.1. The metaphor change of Ambient Intelligence from function orientation to goal orientation effects that a user becomes a smart player within a proactive intelligent environment

more than only a user who is trying to reach her goals by using the available environment technologies. The user becomes a *smart player*, who is proactive assisted by her environment. The term smart player indicates that the interaction of the user and the environment changes from unidirectional to bidirectional. Thus, it expresses the awareness of the environment in respect of the smart player's needs and goals (in the following the terms *user* and *smart player* are used as synonyms).

Consequently, Ambient Intelligence extends the technical foundation that was laid by former initiatives like *ubiquitous computing* and *pervasive computing* respectively. These technologies triggered the diffusion of information technology into various appliances and objects of the everyday life. But now, Ambient Intelligence has to guarantee that those smart devices behave reasonably and that they unburden – instead of burden – the user. This means, the approach of the former initiatives, which is more technology-oriented (innovations by *technology-push*), must be replaced by a more user- and scenario-oriented approach, respectively (innovations by *user-pull*). Consequently, Ambient Intelligence enables the smart player to concentrate on what she ultimately wants to achieve: her actual goals.

1.4.2 Reorientation

The AmI vision is aimed at creating a reactive and sensible (one can also say "intelligent") environment for the smart player's needs and well-being, the basic requirements are the following:

- The environment (and its devices) must be aware of the smart player's current situation, her interaction within her environment and its own current state (and possible changes in this state).
- In addition, the environment must be able to interpret those occurrences into user goals and, accordingly, into possible reactions that enable a cooperative, proactive support for the user.

- In a final step, the environment must be able to translate the interpreted goals into strategies that can be fulfilled by the environment's devices and functionalities in order to adapt itself to the smart player's needs.

This cycle can be called the *Principle Workflow Cycle* of Ambient Intelligence. Regarding each step of the workflow cycle, Ambient Intelligence forces the scientific – and also the industrial – community to think about different challenges and to find appropriate solutions for them. Some of the arising questions should be mentioned at this point:

- "*How does the user interact with her environment and what does she wish to express?*"
  Of course, if input from the user is necessary, the environment should speak the language of the user. The user must be allowed to interact in a natural way, by means of voice or gestures. But there is more to it than that. To assist the user in an effective way, Ambient Intelligence affects a transition from a function-oriented interaction with devices to a goal-oriented interaction with device systems. The user should be able to express goal states rather than select functions of devices. It means that the interaction metaphor has to change from the device-oriented vocabulary to a more user-oriented vocabulary. The environment must speak the language of the user and not vice versa. Consequently, Ambient Intelligence demands a transition from conventional unimodal, menu-based dialog structures to polymodal, conversational dialog structures that assist the user in defining her goals rather than selecting function calls.
- "*How should an ensemble of devices interoperate?*"
  It cannot be expected that the user is able or even willing to orchestrate multiple – invisible!? – devices in order to fulfill her needs. Particularly, if the user is not aware of each individual device and its functionalities. Consequently, device ensembles that define the environment have to provide methods of self-organization. That means the interoperability of devices must be guaranteed, but also their extensibility and reliability. Ambient Intelligence must dismantle the handcrafted design and implementation of component and device ensembles as it is done today and must replace them with new methods of self-organized ad hoc device cooperation. That means the technology must move from an accidental collection of independent devices to a system that acts as a coherent ensemble.
- "*How should an ensemble react to environmental changes and user interactions?*"

The environment, in which the user is located, should do the obvious. That means in technical terms that the environment variables that determine the actual state of the environment should change reasonably in order to fulfill the user's goals. Obviously, the environment – that is defined by its devices – should not demand any input from the user if her intentions can be inferred from the given context. Consequently, to realize the vision of Ambient Intelligence, devices and components are needed that behave context aware and, above all, reasonably. Device ensembles should do what is obvious for the user and what corresponds to the personal preferences of the user and the current state of the environment.

In summary, the AmI vision – to create reactive environments to reasonably assist the smart player according to the given context and her preferences – will cause some technological reorientations. That will involve, for example, self-organizing component and device ensembles (Encarnação and Kirste 2005), but also innovative multimodal interaction technologies, machine readable and understandable ontologies, visualization technologies, and also the integration of Artificial Intelligence into context-aware solutions. The usage of technology will realign from pure function orientation to being smarter player- and goal oriented.

1.4.3 Impact Through Integration

The paradigm shift from function to goal orientation that is affected by the vision of Ambient Intelligence will create a stronger link and cooperation between different technologies and also activate new interdisciplinary research activities for many different branches of research. Science and industry will also be motivated to purposefully work for the well-being and the needs of humans.

Therefore, Ambient Intelligence must integrate scientific subjects like human–computer interaction, middleware and agent technologies, or virtual reality and different disciplines of computer science, such as context management or Artificial Intelligence, into one cooperative vision. Furthermore, more hardware-oriented branches of research, such as nanotechnologies or embedded systems, mobile technologies or sensor technologies, awareness and presence technologies, must be homogenously integrated to create new devices, products, and systems, but also new kinds of content, integrated and personalized services and applications, in order to pursue the goal of realizing the AmI ideas (see Fig. 1.2).

In addition, Ambient Intelligence will make it necessary for different research fields to cooperate. This means new kinds of multidisciplinarity have to be established in order to make the integration of different technologies possible, as well as to integrate nontechnological disciplines like psychology, social science, or medical science. Because the concepts of Ambient Intelligence may not be limited to a certain culture domain the applicability of AmI solutions and implementations across different cultures has to be verified and specialization toward the specific features unique to individual cultures has to be provided. Consequently, multiculturality will have a strong impact on the realization of AmI scenarios.

Besides multidisciplinarity and multiculturality, interoperability must not be forgotten. If the interoperation and the cooperation of devices and components that are developed by different companies or vendors in different nations

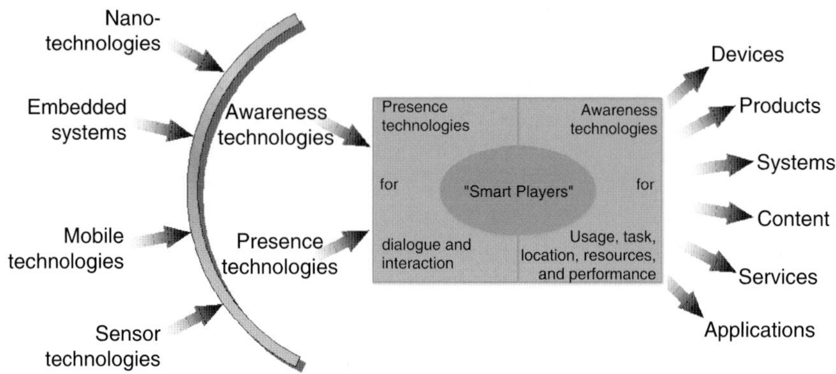

Fig. 1.2. The enhancement of different technologies and their application in AmI scenarios in an integrated fashion will lead to new products and systems

were to be guaranteed, AmI scenarios could be assembled ad hoc from different stand-alone devices that autonomously configure themselves into a system that acts as a coherent ensemble.

Thus, Ambient Intelligence burdens those who want to make all of this happen with strong requirements, because it demands the cooperation of multiple research branches and the consideration of the needs and characteristics of different cultures, as well as (technical) interoperability. Only if all requirements are accomplished AmI technologies will be eventually accepted by the majority of the users and result in both scientific and economic success.

1.4.4 Turning Vision into Reality

Obviously, in order to realize Ambient Intelligence, sizable advances in a multitude of research fields are necessary, not only in the field of software engineering but also in interdisciplinary fields like human–computer interaction. For the implementation of Ambient Intelligence, the following two research tracks must be taken in parallel and in close cooperation.

- The further development of certain technologies (e.g., the enhancement of centralized agent platforms to self-organizing decentralized component ensembles, the construction of intelligent scenarios and rooms – not in a hand-crafted way, but in such a way that devices configure themselves autonomously according to the given smart player goals, or the enhancement of poly-modal dialog technologies).
- The development and implementation of AmI scenarios under strict consideration of the smart player's needs her preferences and her cultural demands.

The requirements of the smart players must directly affect the development the adjustment of AmI technology and vice versa the enhancement of technology may only happen under consideration of the smart player's needs.

The Fraunhofer Gesellschaft, the largest organization for applied research in Europe, bundles its efforts towards the realization of Ambient Intelligence in an appropriate initiative under the leadership of the Institute for Computer Graphics in Darmstadt (Fh-IGD) and the Institute for Open Communication Systems in Berlin (Fh-FOKUS). First, some scenarios are analyzed to examine the requirements and the needs of users while interacting in their home, in their car, or in their office. Cars could recommend points of interest in the nearby surroundings or could provide the user with her favorite music. The car could also manage incoming telephone calls in accordance with the current driving situation, thus preventing the driver from engaging in dangerous driving behavior.

The intelligent house organizes the shopping according to the preferences of the residents, but also under consideration of their personal date planner (in case there are guests to be entertained). Particularly the needs of the elderly receive special consideration. Loudspeakers and microphones in the room where the resident stays manage telephone calls. The house contacts the relatives or the physician in case the person's vital functions experience a change for the worse. However, not only humans, but also animals or even work pieces in production can profit from the developments in Ambient Intelligence. Animals and work pieces can also be the smart player in AmI scenarios. By means of self-organization, work pieces will find the most appropriate machine that corresponds best to its task schedule for instance. Thus, Ambient Intelligence will open up the way for the self-organizing factory and new kinds of logistics.

The result of the detailed specification of possible AmI scenarios is a complete definition of the affected environment variables and their values, a complete pool of possible user and environmental goals, as well as their implication on environmental changes. Finally, possible conflicts (between devices and/or components) are identified and appropriate conflict resolution mechanisms are found.

Those results form the basis of the definition of the AmI workflow, which defines a reasonable workflow cycle from (user) interaction via goal detection to environmental changes. The results also form the basis for the definition of the AmI application space, which defines the basic functionalities that have to be provided to realize AmI scenarios.

Afterwards, detailed requirements concerning the AmI technologies could be specified. Certainly sensor and monitoring technologies, self-organizing middleware technologies, and multimodal interaction technologies will have to be developed or adjusted. Some requirements concerning machine-readable ontologies or inference and strategy engines will also emerge.

The last step in making a huge step towards Ambient Intelligence is to implement the proposed scenarios by assembling the developed technologies. The AmI initiative of the *INI-GraphicsNet* (AmI@INI, online) that was established in 2004 forms the basis of the described research approach. In addition the Fraunhofer Institute for Computer Graphics is involved in several projects that are researching certain AmI issues. The project DynAMITE, (online) for example (an abbreviation of DYNAmic Adaptive Multimodal IT Ensembles) researches the topology of AmI component ensembles as well as implements a self-organizing middleware for the composition of device ensembles (Hellenschmidt and Kirste 2004a,b). In order to manipulate devices – also in foreign environments – the *Personal Environment Controller* (PECo, online) provides a three-dimensional visualization of the room to allow the user to interact directly with her environment. Thus new kinds of interaction for direct manipulation instead are investigated.

In a nutshell: Ambient Intelligence could become reality, if researchers from different research fields share the same vision of a future where all of us are surrounded by intelligent electronic environments that are sensitive and responsive to our needs and if those researchers cooperate to define a common reference model that builds the basis for the development of new technologies and new assistive proactive scenarios.

1.5 Ambient Intelligence Becomes a Success

Ambient Intelligence implies a seamless environment of computing, advanced networking technology and specific interfaces where the smart player is assisted by reactive environments according to the given context, and her personal needs and preferences. Because Ambient Intelligence will have a strong impact on science as well as on economy, it represents a long-term vision for the EU Information Society Technologies Research program. To be successful, researchers across multiple disciplines like computer science, engineering but also social science and psychology will have to cooperate. This cooperation will be the key for future technical innovations and for the development of effective applications. But Ambient Intelligence becomes a success only if the needs of the smart players are in the focus of the cooperative research activities. Consequently, the vision of Ambient Intelligence creates a strong link between interdisciplinary research fields, leading to new innovative technologies and applications, and finally strengthening science and economy.

One of the major new things in Ambient Intelligence is given by the development to involve humans in the design and creation of novel concepts and products. This call for application and experience research extends far beyond the classical approaches to user-centered design in terms of manner and the extent in which the end-users are involved. Most classical design methods adopt a more-or-less static approach in which functional user requirements are first elucidated through use-cases or target-group studies and subsequently transformed into design requirements. Ambient Intelligence calls for an approach in which users are an integral part of the design and operation of their environment, thus enabling them to live and work in a surrounding that stimulates their productivity, creativity, and well-being in an active way.

2

Information, Society and Technology

J.-C. Burgelman and Y. Punie

"Ambient Intelligence is more than an enhanced Internet, a smart phone, an interactive television, or a combination of them."

2.1 Introduction

Ambient Intelligence is fundamentally a European concept for a future information society where intelligent interfaces enable people and devices to interact with each other and with the environment in real time and pro-actively. Technology operates in the background while computing capabilities are everywhere, connected and always available. This intelligent environment is aware of the specific characteristics of human presence and preferences, takes care of explicit and implicit needs and is capable of responding intelligently to spoken or gestured indications of desire. It even engages in intelligent dialogue. Central to the AmI concept is "human-centred computing", "user-friendliness", user empowerment and the support of human interaction (ISTAG 2001; Aarts et al. 2002). Ambient devices and services for work, health, comfort and sanity will need to function in a seamless, unobtrusive and often invisible way. Ambient Intelligence flags the idea of machines that become really active, that think for us but only – and this is crucial – when people want it and only on their conditions.

That is why the abbreviation of Ambient Intelligence as AmI is used – it should signal a move beyond concepts such as user-friendliness into a servitude to people in a way users can never realise on their own. Ambient Intelligence is therefore a human-centric approach to next generation Information and Communication Technology (ICT), and by being so responds to some fundamental European values. The humanistic dimension of technology is becoming an increasingly important driver in the information society. This clarifies to a certain extent why a concept that only emerged in mainstream technology discourses at the end of last century (ISTAG 1999) – and that was built upon the notion of ubiquitous computing as coined by Weiser (1991), a computer scientist at the Palo Alto Research Center (Xerox Parc) – is already significantly widespread. It might explain why the number of hits that are generated by Google on the term "Ambient Intelligence" is significant: 440,000 hits in

April 2006 (against 38,500 in November 2004). The words Ambient Intelligence without quotes almost returned 500,000 links (against $> 200,000$ in November 2004).

Ambient Intelligence is an emergent property (ISTAG 2003a) at the core of the Information, Society and Technology (IST) priority of the sixth RTD Framework Programme (FP6) of the European Union (EU), following the work of ISTAG and of other consultative procedures organised by the European Commission. The IST thematic priority was set up to contribute to realise the so-called Lisbon goals, i.e., "to become the most competitive and dynamic knowledge-based economy in the world capable of sustainable economic growth with more and better jobs and greater social cohesion" (EC 2002). But also outside the main EU instrument for RTD funding, there are considerable efforts in Europe dedicated towards realising the building blocks for Ambient Intelligence. There is for example the ITEA consortium at the pan-European level and also national programmes exist, such as the UK Equator one (Equator, online; ITEA, online). As a result, the AmI vision has gained momentum within the IST community, where it helps to focus resources on a common project in Europe possibly leading to greater economies of scale while avoiding fragmentation and duplication of efforts (Punie 2005).

The development of AmI may signal new encounters between technology and society; encounters that are different compared to the past. AmI encounters are supposed to be friendly, pleasant, useful and seamless, but probably – like all Schumpeterian innovation ("creative destruction") – the encounters will also be surprising because the new always entails uncertainty. That is why this article alludes to Steven Spielberg's classical science-fiction movie "Close Encounters of a Third Kind" (Dirks 1996) where people have close encounters with UFOs and aliens, but in a benevolent and enlightening way. Although obviously Ambient Intelligence is nothing extra-terrestrial, the parallel with one of the film's poster slogans "We are not alone" can be drawn, since Ambient Intelligence is planned to support and help people and to enhance their everyday life experiences. This is the "creative" part of "creative destruction", i.e., the changes in our way of living and working Ambient Intelligence will bring about.

Ambient Intelligence indeed may represent a new paradigm for citizens, administrations, governance and business. Radical social transformations are expected from its implementation. Although the AmI vision is not a panacea for social and societal problems, it could offer innovative ways to address the fundamental socio-economic challenges that Europe will be facing during the coming years, such as the increase of its customers and citizens, the population and increased mobility (ISTAG 2003a).

We argue that as the AmI vision is currently being constructed in ways that enable new directions for Europe, it has the unique potential to bring something new in terms of the information society we are building. Firstly, because the AmI concept is different compared to earlier technology visions and, secondly, because Ambient Intelligence offers opportunities to deal with Europe's specific challenges in ways that are "Euro-friendly".

2.2 The AmI Vision: A Different Encounter Between Technology and Society

As we know from Schumpeter, all major technological innovations or paradigms are ambiguous: old and well-established forms of doing things are challenged or destructed whilst at the same time new but uncertain ways of doing things emerge. The challenge for developers and policy makers is then to make this destruction creative, leading to better ways of doing things. Ambient Intelligence has the potential to become a major force of creative destruction because three interrelated shifts occur: (1) shifts in the "logic" of ICT and ICT usage, (2) shifts in EU RTD and (3) shifts in designing Ambient Intelligence.

2.2.1 Shifts from Technology to Usage

The AmI vision implies several related shifts in IT and ICT. A first technology shift concerns the way computing is organised. Computing systems have changed from mainframe computing (1960–1980) to personal computing (1980–1990), and are evolving from embedded and multiple computing (2000 onwards) towards invisible computing (2010 onwards). A second shift is related to the move towards distributed and peer-to-peer computing whereby resources (e.g., computation power or content) are networked, decentralised and shared amongst peers in a location independent way. A third shift concerns communication processes: from people talking to people, to people interacting with machines, to machines/devices/software agents talking to each other and to people. A fourth important change is the one which presumes that interfacing with computing capabilities will become natural and intuitive, in contrast with current Graphical User Interfaces (GUI) and past keyboard-based input systems. What Ambient Intelligence offers is a combination and integration of these different shifts, and by doing so, it adds a new and significant layer to the current evolution in ICTs (ITEA 2005).

These changes in the way ICTs are functioning go hand in hand with progress in the diffusion and acceptance of past and current ICTs. It should not be forgotten that businesses, homes and individual lives have embraced many technological devices and services during the last decades, although there were also notable failures such as Videotext in the 1980s. In the last 10 years mobile telephony in Europe has grown into a market of over 300 million users. In many European countries, mobile telephony penetration rates are above 70% of households. Short Message Service (SMS) has also given a considerable boost to the mobile services market in recent years. Residential Internet access rose to 43% in the EU15 in June 2003 versus 28% in October 2000 (Eurobarometer 2002; SEC 2005). The Internet has become huge and it is in the meantime already changing due to the roll out of broadband networks and always-on connections. In the EU15 in July 2004, roughly 80% of the people could be reached by broadband, although only 8% were subscribers (EITO online). A reason for the latter observation is that Internet chat and e-mail rather require always-on connection while peer-to-peer computing also requires high bandwidth. The convergence between broadband and wireless networks might stimulate however the demand for new converging multimedia services and for broadband access. Another driver is related to the fact that digital storage is becoming very cheap, hence the emergence of MP3 players and USB keys. These and many more new technologies are being introduced and taken-up by millions of consumers in Europe.

2.2.2 Shifts in EU RTD Framework Programmes

Ambient Intelligence reflects in a certain way that views have changed on how to develop and support new ISTs. The focus has shifted from linear and deterministic models of technological change to a socially and culturally constructed process of innovation that features interactivity, and dynamic and continuous change in an interplay between research, development, design, marketing and the uses of new technology. The top down approaches characteristic for the 1980s have in the 1990s slowly been supplemented by bottom up approaches (OECD 1992; Smits and Kuhlmann 2004). Ambient Intelligence builds upon this change by systematically combining and integrating both top down and bottom up approaches. This means that Ambient Intelligence is certainly not only about users but also about ISTs, but it stresses the interaction between both. Ambient Intelligence therefore stands for an integrated and systematic approach to innovation. It starts from needs and looks for technical solutions rather than looking for where techniques can be matched with needs.

This can be illustrated with different European RTD Framework Programmes (FP) (Cordis Acts, online; Cordis Esprit, online; Cordis FP5, online; Cordis Telematics, online). ICT research has a long history within these programmes, starting with ESPRIT (European Strategic Programme for Research and Development) in 1984 (FP1) which was on ITs, in particular on the emerging information infrastructure. ESPRIT continued in the successive FPs and was in the meantime joined by RACE (Research and development in Advanced Communications technologies in Europe) and TELEMATICS, which is a merger of the more application-oriented programmes DELTA, DRIVE and AIM. The major focus of these ICT programmes moved gradually from hardware-based infrastructure initiatives towards software-based applications and services.

This was consolidated in FP5 (1999–2002), where these different programmes were merged under the common denominator of a "user-friendly information society". Although de facto the programme was more about developing useful and relevant applications for the information society, this move was important in terms of providing a common approach to IST innovation which is not only based on technological performance but also on awareness of the importance of user needs and user relevance. Users were defined mainly in terms of businesses, companies, industries but also the public sector and ultimately, the wider public as end-users were to be taken into account.

The IST programme within FP6 (2002–2006) builds further upon this trend by offering more focus and integration, the latter two also being the biggest priority in terms of budget and effort in FP6 ("Focussing and Integrating EU Research" [Cordis FP6, oline]). Furthermore, the IST programme has five thematic priorities. These are: (1) applied IST research addressing major societal and economic challenges, (2) communication, computing and software technologies, (3) components and micro-systems, (4) knowledge and interface technologies and (5) Future and Emerging Technologies (FETs). Ambient Intelligence is not one of these five priorities but given its horizontal nature, it is an authoritative guiding principle for most European IST research (Gago Panel Report 2005).

The central role of Ambient Intelligence in FP6 thus flags a double change in approaching IST innovation in Europe, first by offering a common vision on the future of the information society in Europe and second by the substance of the vision, i.e., an integrated and systematic approach to IST innovation. The latter issue will be elaborated further in the next section.

2.2.3 Designing Ambient Intelligence is Designing Social Structures

The integrated and systemic nature of Ambient Intelligence means that it represents a step beyond the current concept of a user-friendly information society (ISTAG 2001). It signals a move beyond user-friendliness and "usability" concerns to favour relations between people and their intelligent environment that are seamless, intuitive, natural and humanistic. Ambient Intelligence thus stands for a "people-friendly information society".

Usability research tends to objectify the relationship between people (as "users") and technologies. It deals with concrete human–machine interactions, based on functional translations of user requirements into the design of new artefacts. These are mainly routed within the traditions of behavioural science and computer engineering, but increasingly, efforts are undertaken to bridge both worlds and to take users seriously, especially within the fields of Human Computer Interaction (HCI), Computer Mediated Communication (CMC) and Computer Supported Cooperative Work (CSCW) efforts are.

With Ambient Intelligence, human–machine interaction is not just about a simple relation between an individual user and an individual artefact. It goes beyond the usual focus on individual users, especially when taking into account that AmI products and services will be intelligent, adaptable and networked, in contrast with stand-alone products. As Tuomi (2003) argues, machines are to be seen as media that connect systems of social activity. This means that designing a product actually means designing the structures for social interaction. Designing Ambient Intelligence consequently means designing social structures.

This is partly acknowledged in one of the latest reports of ISTAG (2003a; 2004a) where the concept of EARCs (Experience and Application Research Centres) is proposed as a new approach to prototyping necessary for the successful development of AmI products and services. Functional, technical, social, economic and cultural requirements of systems gathered from users and stakeholders need to be put at the centre of the development process, revisited through design, implementation, checking and testing. Experience prototyping can be used to understand user experiences and their contexts, explore and evaluate new designs and communicate ideas to designers and stakeholders. This should "allow people to live in their own future" and should bring AmI research closer to the needs of citizens and businesses (ISTAG 2003a).

Such a view on IST innovation also has implications for the way the relationship between technology and society is conceived. Visions of the future of technology in society tend to be shaped by what the technologies have to offer. They often suffer from technological determinism (Marvin 1988; Flichy 1995). Every time a new technology pops up, revolutionary social changes are promised and promoted. They only look at what is technologically feasible and ignore the socio-economic context and user dynamics that are shaping the innovation process as well (Burgelman 2000). But Ambient Intelligence claims to be different, i.e., "human-centred".

Right from the start, the AmI vision explicitly focused on people, not on technologies. People need to benefit from services and applications supported by new technologies in the background and they need to be given the lead in the way systems, services and interfaces are implemented. The four scenarios that were developed in the 2001 ISTAG report also emphasised a key feature of Ambient Intelligence, which is that the technologies should be fully adapted to human needs and cognitions. According to ISTAG (2001), the social and political aspects of Ambient Intelligence will be very important for its development. A series of necessary characteristics that will permit the eventual societal acceptance of Ambient Intelligence were identified. These are carefully balanced between technological determinism and social reductionism. At the level of discourse, this is what makes Ambient Intelligence already different from earlier technology visions (Punie 2005).

2.3 IST, the European Social Model and the Lisbon Objectives

Ambient Intelligence as a concept has already gone through an evolution from when it started in 1999 to where it evolved now. This has to do with moving from a technology-based vision to increasingly interconnecting it with the specificities and challenges Europe is facing. The stronger and better this interconnecting is made, the more chances Ambient Intelligence has to really make a difference.

The so-called European model emerged in Europe after the Second World War, under the umbrella of what later became labelled as the welfare society (Calabrese and Burgelman 1999). The European model covers many different policy areas (e.g., health, social protection, welfare, education) but also contains a set of common values that are based on four principles: (1) growth to enable full employment, (2) solidarity, (3) equal opportunities and (4) sustainability. There are of course differences between European countries in how these principles are applied, but overall this synthesises the main characteristics of the European model.

Important is that this current European model is being challenged in the mid-term to long-term future by new developments. These developments can be summarised under the headings of enlargement, ageing population and global competition. First, the May 2004 enlargement has raised the EU population with 20% to more than 450 million people while it only increased its GDP by 4.5%. It indicates that socio-economic disparities across the EU are becoming wider. This, moreover, is not only the case across the Union (i.e., between countries) but also between regions. Social cohesion is therefore so high on the agenda of the enlarged Union. Second, there are, however, also similarities between the EU15 and New Member States, especially in terms of demographics: old and new Europe faces the same demographic challenges. And this explains why the future of our health care systems, pensions and active employment is core concern too (European Commission 2004a). Third, global competition is the rule for almost all sectors of our economies rather than the exception. It demands for high flexibility and mobility in the organisation of labour and living in order to be competitive.

Demographic and social trends, such as individualism, diversity, mobility and the choice of personal life styles, all affect the structure of groups and communities and the ways we live and work. Mobile phones, for instance, are enablers of lifestyles that are increasingly individual and mobile. Household structures (family size and composition) are changing too, with a decline of traditional nuclear families and an increase of dual income households and single parent/single person households (Gavigan et al. 1999; Ducatel et al. 2000).

All the changes affect the four pillars of the European model while at the same time current ICTs and future AmI environments might affect them too (Clements et al. 2004). This becomes obvious from the following arguments:

- ISTs are essential for growth and employment. Firstly because the IST sector is a growth sector per se. The ICT equipment and services sector on its own right have grown from 4% of EU GDP in the early 1990s to around 8% and accounted for 6% of employment in the EU in 2000. R&D investment in the ICT sector accounts for 18% of overall EU spending in R&D (OECD 2002; COM_2004/757). The latter is also the case when looking specifically at the top 500 private R&D-investing firms in the EU (European Commission 2004b). Secondly, ISTs are also important for growth and employment because of their indirect impact. They are central to stimulating productivity and improving competitiveness. Between 1995 and 2000, 40% of the productivity growth in the EU was due to ICT (COM 2004).

- ISTs are essential for solidarity and cohesion because of their fundamental character to bridge the limitations of time and space. ISTs can bring people, regions and countries that are socio-economically disparate closer together. This is especially relevant now that the EU has enlarged with ten New Member States from regions that lag behind in terms of economic prosperity. Also better governance and smarter health can be realised via IST.
- ISTs are essential to develop and maintain equal opportunities in a knowledge society that is increasingly based on digital networks and electronic communications. Issues such as the digital divide, e-learning but also digital identities and privacy are key for developing a future Europe in terms of providing equal opportunities and chances to all. This is based on the idea that IST are designed and used to serve people and not vice versa.
- ISTs are essential to sustainability. There are significant opportunities for improving environmental sustainability through ICTs in terms of, for instance, rationalising energy management in housing (or other facilities), of more efficient and more safe transport (passenger and freight) and of enabling a product-to-service shift across the economy. There are rebound effects that need to be taken into account, but ISTs are expected to have a beneficial impact on environmental sustainability (Rodriguez et al. 2004).

As a result, IST provides a systemic technology as they touch upon the foundations of the post–World War II European society. It follows from this that a successful information society policy should be a holistic policy that takes into account technological, economic, political and socio-cultural issues.

This also explains implicitly why ISTs are regarded as crucial for realising the Lisbon objectives of a competitive and dynamic knowledge-based economy with sustainable economic growth, more and better jobs and greater social cohesion. ISTs are seen as key contributors to realising the Lisbon goals (EC 2002COM 2004). In the next section, some of the key social drives and IST applications to that end are described.

2.4 Foresight in IST in Europe

The discussion on the developments in (IST) in Europe is largely centred on societal questions. Europe has reached a consensus on the requirement that novel technological developments should be inspired by needful applications that support functional use within society in the large. More specifically, this implies to a large extent that novel IST developments should be driven by social rather than by technological factors.

2.4.1 Social Drivers

Factors that shape the speed of ICT progress can be technological, economic, social, or political and they are often interrelated. An overview of some of the major drivers, trends and challenges for the future information society is provided by the FISTERA (Foresight on IST in the European Research Area) thematic network. FISTERA is an EC-funded network that aims to understand the key factors driving IST in a future Europe. In its review of the major scenario studies and foresights of the emerging information society, FISTERA has noted that especially the social drivers for future ISTs have remained particularly stable during the last years. The network has identified 15 key social drivers for IST R&D development up until 2010 and beyond. These are presented in Box 2.1.

Box 2.1 Social drivers for future IST

- Aging population and implications for health applications
- The maintenance of languages, cultures and life styles in an enlarged Europe
- Using novel ways of community learning and knowledge sharing
- Increasing demand for personal mobility
- The demand for improved public services
- Increasing requirements for personal privacy and trust
- Assuring ICT service security and robustness
- Complying with increasing "bottom line" ethical requests
- Bridging the digital divide
- Building ICT-related skills allowing social innovation (supporting ICT use and employing ICTs)
- Increasing demand for system integration and interconnection
- Ongoing globalisation of services and business
- Enhanced awareness of environmental issues and sustainable growth
- ICT-based applications for enhanced security (Compañó et al. 2004).

As areas to work on for the future, most foresights on IST in Europe list health care, ageing population, transport and mobility, education, governmental services, leisure and changing social relationships, including cultural diversity and migration. As drivers, these topics have changed very little over the past few years indicating that what drives socially the demand for ICT at the macro level is more or less constant. It provides a sound foundation for the creation of the European knowledge society. The only exception is "security". It became a major concern after the September 11, 2001 terrorist attack and ICT is now increasingly considered to provide indispensable tools for private and public security and defence (Compañó et al. 2004).

2.4.2 Promising IST Applications

FISTERA results provide an account of what the most important future IST application areas are, according to national foresight studies and a recent

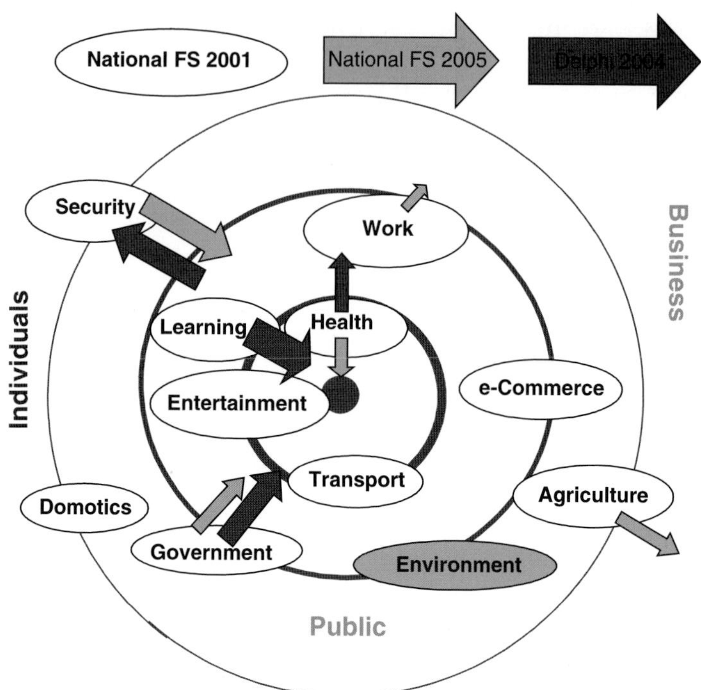

Fig. 2.1. Promising IST application fields (FISTERA 2005)

Delphi exercise (Popper and Miles 2004). Interestingly, the results show how some of the priorities have changed during the last years, by comparing the results of foresight researches undertaken in 2000/2001 with more recent ones and with a recent Delphi exercise done by FISTERA. The results are shown in Fig. 2.1. Importance is measured in terms of the appearance of the application fields. The most important application fields, i.e., the ones that are most frequently mentioned and/or most prominently visible in the different foresight exercises, appear in the centre of the graph, around the red dot and the first bold circle. As the centre of gravity moves to the periphery, the application fields are regarded as less important.

The 2000/2001 national foresight studies indicate that the major application areas for IST seem to related to, firstly, health care, learning, entertainment and transport. Second order importance goes to governmental services, e-commerce and work. The outer circle of importance consists of agriculture, security and domotics. It might be surprising that security is perceived to be less important, but that might be the result of the fact that until September 11, 2001, foresight studies treated security as an aspect of individual technologies, e.g., security of IT from virus attacks, security of the food chain from diseases like BSE, security from natural disasters like flooding, etc. (Compañó et al. 2004).

The green arrows highlight the changes according to the more recent foresights. Health becomes the single most important domain. Government applications move to first order of importance. Security becomes now more prominent, after September 11. Environmental concerns and quality of life emerge as new topics, at the expense of work-related and agriculture-related IST applications.

An online Delphi executed in 2004 with 413 experts confirmed on the one hand the importance of learning and egovernment but decreases the central role of health on the other hand. Also, security applications are seen as less prominent, as is shown by the red arrows in Fig. 2.1. In the next section, four of the most promising IST application fields are further developed within the context of the opportunities offered by Ambient Intelligence.

2.5 AmI Innovation in Europe

The vision of an intelligent environment based on the convergence between ubiquitous computing, ubiquitous communication and intuitive, intelligent interfaces has the potential to respond to these challenges and social drivers in a "European way", i.e., in a way that respects the basic European areas of consensus mentioned above. More provocative in this respect is the book of Jeremy Rifkin where the European Dream is contrasted against the American one.

The European Dream emphasises community relationships over individual autonomy, cultural diversity over assimilation, quality of life over accumulation of wealth, sustainable development over unlimited material growth, deep play over unrelenting toil, universal human rights and the rights of nature over property rights and global cooperation over the unilateral exercise of power (Rifkin 2004).

In addition to the central role of Ambient Intelligence for the key areas of growth, employment and competitiveness, it promises to take on board values such as inclusiveness, diversity, quality of life and sustainability in addressing these key areas. Four cases are now developed to illustrate how Ambient Intelligence could address socio-economic realities and challenges in an innovative and European way. The cases are linked to the most promising IST application fields as mentioned above: (1) sharing of knowledge, learning and experiences, (2) Ambient Intelligence in health care, (3) Ambient Intelligence in eGovernance and (4) Ambient Intelligence and biometrics identification, building upon the potential of ICT for security.

2.5.1 Sharing of Knowledge, Learning and Experiences

There is a potential for Ambient Intelligence to play an increasingly significant role in social learning and the exchange of knowledge, particularly now that network infrastructure and network access are becoming ubiquitous. The challenge would thus be mainly on realising smart content, i.e., content that is produced and controlled by the users themselves and that adapts itself to changing contexts and situations, and on enhancing and improving content manipulation, storage, archiving and retrieving technologies.

An example of the potential of Ambient Intelligence to support spontaneous learning and to establish a "collective learning memory" is described in the ISTAG Scenarios on Ambient Intelligence in 2010 (ISTAG 2001). One of the scenarios was "Annette and Solomon". It describes a meeting of an environmental studies group that is led by a human mentor but facilitated by an "Ambient" knowing the personal preferences and characteristics of the participants (real and virtual). The scenario implies significant technical developments such as high "emotional bandwidth" for shared presence and visualisation technologies, and breakthroughs in computer supported pedagogic techniques. But it also presents a challenging social vision of Ambient Intelligence in the service of fostering community life through shared interests. The current popularity of chatting, weblogs and peer-to-peer computing indicates there is an interest in digitally enhanced sharing of (rich) content.

Ambient Intelligence could prove to be relevant for such a purpose, as it will able to integrate and communicate tacit, context-dependent knowledge more easily than current-day technologies can. Social learning might be facilitated in such an AmI environment since it can bring people from different backgrounds and different contexts closer together. The intelligent environment would facilitate the sharing of experiences by making the necessary translations (Van Bavel et al. 2004). A first step towards this environment would be provided by linguistic translations, as described by ISTAG (2004b) as a "multilingual companion" that makes multilingual and cross-lingual information access and communication virtually automatic. Such a companion would be extremely useful in enlarged Europe but the grand challenge would be to also encompass Europe's cultural diversity (and thus not only linguistic diversity).

Digital content for entertainment, culture and leisure that is shared amongst people would require, however, a re-assessment of digital copyrights. This needs to be seen in the context of what can be described as a Virtual Residence (Beslay and Punie 2002), a common space for digital content that is shared amongst family members and peers. This could not only give rise to new and unexpected (grassroots) uses of content but also create opportunities for new business and/or revenue models that are based on usage rather than on ownership of, for instance, a physical copy. In the same way as for instance someone can step into your car and listen to "your" music, one should be able to invite others to access and enjoy – perhaps temporarily – digital content. A major challenge for the realisation of this will consist of the management and control of access rights to these digital assets.

2.5.2 Health Care

Health care is a priority in Europe and is central in the European social model. It will become probably an even more important priority in an enlarged EU, where considerable regional and national differences in health care systems exist and where the population is ageing. The social challenge is to keep the costs of healthcare systems under control while at the same time maintaining a high quality service. eHealth applications are expected to contribute to addressing this challenge by reducing costs and by delivering better and more efficient health care. More specific advantages are easy access to various medical experts and to efficiently exercised second/third opinions, early diagnosis, improved disease tracking and prevention measures and better record keeping.

But Ambient Intelligence in the context of eHealth could be more than maintaining and improving existing health care systems. It could support a paradigm shift in health care delivery by focussing on the autonomous citizen (i.e., proactive with respect to her/his own health, enabled to self-care, seeking services for prevention and disease management and aware of lifestyles) and independent living (i.e., living autonomously and safely as long as wished).

Today, medical files are primarily managed by medical institutions and not by the patients themselves. In the future, the user will need to get more control over this because medical information will not only be gathered via existing institutionalised forms (e.g., anamnesis, hospital check-up, etc.) but also via direct sensor-based monitoring, at different levels: in the body (implants), embedded in clothes (smart fabrics, so-called eWearables) and at dedicated places such as the smart home. All this monitoring of (health) information needs to be managed (e.g., medical files), secured and protected against, for instance, unauthorised access. The use of interoperable systems will be crucial for the exploitation of collected data in the context of the European mobility of citizens and patients. An additional benefit relates to biomedical research (e.g., by using data-mining in monitored data) provided that privacy protective rules are taken into account.

Moreover, AmI-based health care applications or Ambient Care (Cabrera et al. 2004) would be based on a holistic view of human "wellness" (physical health is one component, but not the only one). Moreover, focusing on prevention rather than on cure is more inclusive with regard to socially or physically vulnerable groups (dependent elderly and disabled, children, etc.). ISTAG (2003) envisages a "Ambient Care System" that is responsive and proactive, that places the user in control of their health care management, including communication with professional careers, friends, family and the wider community. Ambient Intelligence would for example help older and disabled people who remain in their own homes for longer by providing them and their carers with increased safety and reassurance, and supporting treatment, rehabilitation and care. Moreover, this would not necessary lead to loss of personal contacts and social interactions at the expense of virtual and remote (medical) encounters. There are also new opportunities for sociability offered by Ambient Intelligence. (Cabrera and Rodríguez 2004) describe for instance a scenario whereby AmI-based assistant technologies enable elderly people not only to live autonomously for longer but also to retire elsewhere while staying connected at the same time.

2.5.3 eGovernance

eGovernment has become an explicit component of both public sector reform and information society initiatives (e.g., eEurope). ICTs are seen as crucial instruments for the modernisation of public services in terms of increasing efficiency and of providing better services to citizens and companies, and ultimately, to strengthen democracy. The social changes Europe is facing (cf., infra) combined with the potential of Ambient Intelligence could however provide a unique opportunity for re-thinking the role of public services in society. A workshop organised by the Institute for Prospective Technological Studies in March 2004 focussed on developing a balanced vision of what eGovernment in the EU would look like in 2010, taking into account Ambient Intelligence (Centeno et al. 2004).

The vision that emerged from the workshop defines eGovernment in the EU in the next decade as an enabler for better government in its broadest sense. It places eGovernment at the core of public management modernisation and reform that not only pursues cost-efficiency and effectiveness but also the creation of public value. The latter is a broad term that fits within the European model as described above and that encompasses the various democratic, social, economic, environmental and governance roles of governments. Concrete examples of these roles are the provision of public administration and public services (health, education, social care); the development, implementation and evaluation of policies and regulations; the management of public finances; the guarantee of democratic political processes, gender equality, social inclusion and personal security; and the management of environmental sustainability and sustainable development.

Four key components would constitute the vision: more user-centric, more knowledge-based, more distributed and more networked. More user-centric means that the needs of citizens and businesses will guide the delivery of eGovernment services rather than the specific demands of people as consumers of public services. This would constitute a shift towards giving the user more control over eGovernment services and ultimately towards empowering users in the process of democratic participation. Emphasising the role of knowledge in government is nothing new but the diffusion of ICTs, the development of the knowledge society and the potential of Ambient Intelligence enable to revitalise the discussion on the role of knowledge in government. More knowledge-based government implies a shift from providing information towards more efficient creation, management and use of knowledge in interaction with citizens and businesses in order to create public value. This means that governments become more flexible and adaptive to changing and diverse environments and needs. More distributed eGovernment opens up the possibilities for a stronger involvement of intermediaries (private, social and public partners) in the delivery of public services and in the exercise of democratic governance. Governments will need to better understand the potential of these actors, in order to develop more innovative and longer term collaborative models and partnerships with them. Finally, there are several trends in public administrations in Europe towards the development of a networked eGovernment, which will require strong co-ordination and collaboration among all actors (also citizens and businesses). Networked eGovernment is crucial for knowledge creation, sharing and dissemination, and for the creation of public value (Centeno et al. 2004).

Ambient Intelligence would offer new models to delivery of government that are mobile, always available, anywhere and via any device. Specific but not exclusively for the eGovernment domain is that these services need to be trusted, secure and, above all, need to respect identity and privacy of citizens and businesses. Many services can be delivered based on an authentication that not necessarily identifies people or that contains lots of personal and private data. Since Ambient Intelligence does enlarge considerably the possibilities for surveillance, people could refuse to live in an AmI environment where governments but also companies and other people know too much about each other. The challenge here is to find a right balance between protecting privacy and providing secure and useful services (ISTAG 2001; Clements et al. 2003). In the next section, the issue of electronic identification is further elaborated.

2.5.4 Biometric Identification

An enlarged EU will be increasingly characterised by a high level of mobility, not only of people but also of devices and goods. In combination with the characteristic of modern society to become networked and digital, a strong need for electronic identification emerges. Technologies such as RFID and biometrics provide new tools for more reliable identification. They also establish connections between the real and digital world. This is crucial for future AmI environments where the real and the virtual will become closely intertwined and maybe even merged. Identification technologies are necessary access points to AmI environments. Without automatic and seamless identification, Ambient Intelligence will not function, e.g., adapting the environment to person needs. Biometric technologies are an important technology for reliable and seamless identification, quite apart from the present-day security concerns as a result of September 11, 2001.

Biometric identification is a technique that uses biometric features such as fingerprint, face, iris, voice and signature to identify human beings. Biometric features are deemed "unique" although some are less "distinct" than others and thus less useful for automated identification purposes. Biometric technologies can provide more convenience and security to the processes of (automatic) authentication and identification. Identity authentication consists of verifying that people are who they claim to be while identification is focussed on discovering the identity of unknown people. Access to physical premises can be done just by speaking to a microphone or looking through a camera that are located next to the door. For digital access, biometrics can replace the use of many different and complicated passwords. In contrast with passwords that can be forgotten or keys that can be lost, biometrics are, in theory, always available (Maghiros et al. 2005).

Biometric technologies are high on the political agenda, both in Europe and abroad, as a response to the September 11 terrorist attacks and the concerns about threats to global security. There are already many experiments with biometric-based border control at airports and passports will increasingly contain biometric data in the future. It can be expected that once the public becomes accustomed to using biometrics at the borders, their use in commercial and other civil applications will follow. This "diffusion effect" signals the possibility of increased acceptance of biometric identification as a result of governments' initiatives to use biometric identification. Such a diffusion of biometrics could be an important enabler for the realisation of Ambient Intelligence provided it is done in ways that respect European values, traditions and legal frameworks. Biometrics can, for instance, also protect privacy because authentication can also be done without necessarily revealing a person's identity. Submitting a registered fingerprint can just be enough to get access to a service. Providing it is recognised that biometrics are never 100% accurate, that the necessary fallback procedures are foreseen and that the purpose of biometric applications are clearly defined, opportunities for a European way of deriving maximum benefit from the deployment of biometrics could be available (Maghiros et al. 2005).

2.6 Conclusions

Ambient Intelligence will be everywhere, anytime, always on, but on demand and thus only when needed and under control of the person. The concept is, in a certain sense, not new or revolutionary since the bridging of time and space has been on the agenda of IST research for many years. The need for IST to be user-friendly, embedded and unobtrusive has also been raised for several years already. More recently, the "service" notion of IST is more and more highlighted, i.e., the notion of technology in the service of mankind in contrast with people needing to adapt to the technologies.

What is really new is that Ambient Intelligence aims at integrating all these features in a way that it allows the IST community to offer Euro-specific responses to the main societal, social and economic challenges of Europe. At the core of the AmI concept is the ambition to offer world-class services and competitiveness whilst at the same time caring for well-being and diversity. In this article, four cases are described to illustrate this but many more and many different application fields are to be developed.

Ambient Intelligence is therefore more a new conceptual approach to IST innovation or even a paradigm than a well-defined set of technologies or social practices. For Europe, and the EU in particular, Ambient Intelligence is also a unique concept to promote a way of making sense of IST research in Europe that is based on European strengths and values and that is not driven by military and/or defence-related inspirations. It is not just anecdotic but rather significant to note that in French, the acronym “AmI” means “friend”, hence the title of this article: an encounter between technology and society of a different kind.

3

Ambient Culture

S. Marzano

"The house of tomorrow will look more like the house of yesterday than the house of today."

3.1 Prelude

Let us start with the following observations:

- We are currently entering a period in which many of our traditional certainties will be challenged. For example: What does it mean to be human? What makes an experience real? What is ethical behavior? Where is the borderline between the natural and the artificial?
- Five hundred years ago, Leonardo da Vinci and Michelangelo were engaged in a great enterprise to find out more about the human body, the human spirit, and its environment, initiating a new evolutionary path for the mind. Today, we are about to embark on a project to extend at least some human qualities to machines. In doing so, we will not only be holding up a mirror to ourselves, but we shall also have to live with the results of our creative efforts. It is therefore vital that we get it right today.
- When the Ancient Greeks and Romans established the basic form of our present democracies, they laid down rules to govern the relationships between members of the community so that a balance of interests might be maintained. In the same way, we now have to lay down the rules of the new community, in which an entire sub-community of proactive devices joins humans.

3.2 Questions, Questions, Questions

Artificial Intelligence, Ambient Intelligence... they are terms that trigger the imagination, and, at the same time, call up all sorts of questions in people's minds. What is it? What will it do? How will it do it? What will it look like? How intelligent will it be? Can I keep control, or will it take over? When will it be available? And do I want it, anyway? All very human questions, which come down to "What will be the relationship between this intelligence and me?"

Businesses will also be asking themselves questions. Will Ambient Intelligence find a favorable response deep down in people? Will it meet fundamental needs and desires? Will it be the next big opportunity for our business? Is it something we should get involved in? How will institutions, governments, and authorities stand in relation to Ambient Intelligence? How will it be regulated? Will they encourage it and provide the necessary infrastructures?

In this chapter, I wish to posit – and suggest answers to – a number of questions relating to the cultural issues raised by Ambient Intelligence. First, "What will (or should) Ambient Intelligence do?" Then, "How will it do it?". That is, what experience should it deliver? And consequently, what are the practical implications for the businesses that seek to supply that experience?

3.3 What Should Ambient Intelligence Do?

First, what should Ambient Intelligence do? Obviously, we want it to improve the quality of people's lives, and I believe it can do that. However, technology is, in itself, a force for neither good nor bad. Whether it works positively or negatively depends on what we decide to do with it, because not everything that is possible with technology is actually desirable. It is therefore essential that we make the right choices with respect to Ambient Intelligence.

We can only do that if we agree on what quality of life and what sort of developed world we would like to see. There is a growing consensus that to achieve sustainability we need to achieve a good balance among three factors, sometimes referred to as the three P's: People, Profit, and Planet. This concept is changing social, political, and business agendas. In developing Ambient Intelligence, we would do well to consider how we can take due account of these factors, and allow them to guide our work. I wish to suggest some ways in which we can do that, specifically as they affect the cultural aspects of Ambient Intelligence.

Reformulating the three P's to apply more specifically to Ambient Intelligence, we need to take account of the following three factors. First, fundamental human drives: what we need and what we want out of life. This is the People aspect. The second is the constraints that our physical and social environments place on us. This is the Planet part. And, finally, there is the Profit part, i.e., the potential of Ambient Intelligence for generating economic growth, profit, and wealth.

3.3.1 Fundamental Human Drives

Let us start by looking at the first factor, fundamental human drives. I do not believe that these have changed significantly since we first lived in caves.

We have a reliable guide to them in the many patterns of behavior that anthropologists have found repeated – in different places, at different times – over millennia.

Exteriorization of Powers

Ever since human beings began to paint images on cave walls, scratch lines on sticks, or chip stone tools, we have been amplifying our mental and physical powers by exteriorizing them. The French anthropologist Leroi Gourhan (1993) has described the emergence of these activities as the most significant turning point in the history of humanity. They allowed us to make the intangible tangible. Gradually, from that point, we began to exploit everything around us to improve our lives and expand our powers. Today, the means may be different, but the goal of our activities is the same.

Drive Towards Ultimate Freedom

But why are we driven in this way? Simplifying, I suggest it is because we want to survive, to attain the highest possible levels of comfort and freedom, and to make sense of the world. So, from finding food and protecting ourselves from the elements and other animals, we started to climb Maslow's hierarchy of human needs (Maslow 1943, 1954, 1971), becoming interested in our physical comfort. And our most basic instincts have been driving us to climb higher and higher ever since. In fact, those instincts do not recognize any limits to our aspirations. We want to be everywhere, to do everything, and to know everything. Simply surviving is not enough: we want to become invincible, immortal, and essentially demi-gods – at all costs and as our top priority.

This deep-seated human longing is widely reflected in myths and legends, and popular culture: in the Faust story, for instance, and in Superman, and Bionic Woman. It is also reflected in many religions, where gods or goddesses are often seen as all-powerful and in essentially human form. And – particularly relevant to us – throughout history this instinct to find ever greater comfort, power, knowledge, and freedom has also been the main driving force behind technological innovation.

Maximum Comfort, Minimum Effort

Being able to do, know, and experience things is not enough. We want it all with maximum comfort and minimum effort. For example, we like flying in a plane because it helps us to be everywhere. But it is far from ideal. We would prefer to fly like a bird, or be beamed around, as in StarTrek. And in the home, we want technology, but we do not really want to sacrifice valuable space for it. On mobility and Ambient Intelligence, see Stoop (2003). This desire to have devices that amplify our powers, but do not hinder us or clutter our lives in the process can be seen as the driver of increasing miniaturization. Many devices have already made the transition from big static objects to small objects that we can carry around on our bodies: clocks are now wrist watches, and more recently phones and audio systems have reached the stage of becoming worn on the body. Potentially, many of the devices that we have created to exteriorize and expand our powers will make the journey back inside us and become effectively re-interiorized; see also Marzano (1998).

3.3.2 Social and Physical Constraints

In pursuing these timeless human goals, however, we find there are a number of constraints imposed on us by our social and physical environment.

Using Intelligent Systems

Some of the main elements in our surroundings that we, as human beings, have tried to exploit in pursuing our instinct, have been intelligent systems. We enslaved or subjugated other nations, and we persuaded or forced others to work or fight for us, even sacrificing them to the gods to help us control Nature. We domesticated wild animals and cultivated wild plants. And today, we cultivate microbes, manipulate molecules, and are just starting to modify our own genetic code.

Civilization

However, amplification of our powers thanks to technology has led to greater contact with others and therefore to greater socialization. And the more we get to know about other people, the more we realize that we all share the same aspirations. In the past, that meant, for example, that if we were considering enslaving our neighbors, they were probably thinking of enslaving us. A status quo was reached and we gradually developed rules of good neighborliness – a civilized culture – that put constraints on how far we could go in exploiting other people to achieve our own goals. As the concept of who our neighbors were expanded (from family, to tribe, to nation, to the whole human race) so the idea of slavery and, more generally, the idea of achieving one's goals at the expense of another became less acceptable. Gradually, the emphasis shifted to achieving fulfillment through self-generated personal growth.

Sustainability

Today, we are also beginning to constrain our exploitation of animals, and we are becoming concerned about the natural environment (not only close to home but also globally), including the genetic modification of crops and the cloning of animals and humans. All this concern is not just altruism: we are also inspired by what has gone wrong in the past, and by the fear of what might go wrong in the future. As a result, we are much more cautious today about exploiting new intelligent systems. In effect, we are rapidly developing a global civilized culture based on sustainability, an appreciation that the world and its inhabitants are ultimately interdependent. It is a culture that goes beyond environmental issues to include questions of social responsibility and economic viability.

The abolition of slavery, the rise of democracy, and the emancipation of many groups in society have led to the development of new technological devices to perform tasks formerly carried out by slaves and servants, resulting in today's sophisticated mechanical and electronic devices that amplify our abilities, and increase our comfort and freedom of movement. Now, although we had to give up using organic intelligent systems because of social and ethical concerns, the use of ambient, artificial intelligent systems can be seen as the next natural step, because they will help us get closer to achieving our objective, but in a civilized way. This is increasingly important, because, the more imbalances are put right, the more impatient people become to see the rest also put right. Ambient Intelligence can help accelerate the process.

3.3.3 Potential for Growth, Profit, and Wealth

We now have one factor left to consider: the P for Profit. How comfortably does Ambient Intelligence sit with commerce? Is it a viable business proposition? If, as I have suggested, it is in line with basic human ambitions, then ultimately it can hardly fail. However, companies will need to pursue it as they would any other business, i.e., seeing to differentiate themselves effectively from the competition. As companies come closer together in terms of their technologies, they are finding it harder to differentiate themselves on the basis of technology alone, and will therefore need to find other ways. For example, they might seek to ensure that their Ambient Intelligence properly respects the P for Planet, including not only environmental, but also social and economic issues. As is now well known, such corporate environmental and social responsibility is likely to become an increasingly important differentiator in the next few years. Ambient Intelligence is obviously largely invisible (although there will be a "something" in which it is embedded), and for companies used to differentiating themselves in terms of appearance, setting themselves apart from the competition will be a challenge. Each company must find its own way of establishing its "personality," perhaps through a distinctive interface, or through the way its Ambient Intelligence "behaves."

3.4 How Will Ambient Intelligence Do What It Does?

This brings me to my next question. How will Ambient Intelligence do what it does? What would be the ideal "form"? What would be the ideal Ambient Intelligence experience? From our research at Philips Design, in collaboration with a network of specialist institutes, we have isolated a number of general principles, as well as specific processes and tools for generating and validating ideas about this – in the home, on the person, at work, in public places, and on the move.

3.4.1 Relevant, Meaningful, Understandable

First of all, it has become clear that Ambient Intelligence should have a form and experience that users perceive as being relevant, meaningful, and understandable. They want to feel that it fits in with what they want to do in their specific situation, and how they want to do it in their specific culture. They want it to be significant to them in a deeper sense. And they want to be able to interpret it clearly and easily.

Relevant

What is relevant? Certain objects in our surroundings are timeless and indispensable: elements like tables, chairs, walls, floors, ceilings, and clothes exist in some form in almost all cultures. Although subject to superficial variations, they have remained essentially unchanged for millennia. They have proved themselves to be relevant. But our research has also shown that the relevance of such objects may differ from place to place and from culture to culture. For us, in Western Europe, for instance, chairs are more important than carpets, but the reverse is the case in Arab cultures. And, in the case of clothing, the relevance of garments is clearly affected by culture. Contrast the baseball cap with the burkha, for instance.

Meaningful

What is perceived as meaningful also differs from culture to culture. In Japan, for instance, memories and ancestors are very important: even the smallest flat will have room for a shrine. In Italy, the *present* family is culturally very significant, so family gatherings are highly valued. The British like their privacy. In Islamic culture, offering hospitality is very important.

Understandable

Ambient Intelligence is almost by definition largely invisible to the user. Interaction between the user and the intelligence is also required. But interacting with the invisible and intangible is not something that comes easily to human beings. We are used to decoding the world through physical objects and signs. We therefore need to make sure that we use items in the environment that are meaningful and have iconic value to serve as interfaces. These can then develop further and form new archetypes for interfaces with Ambient Intelligence.

3.5 Some Examples

Below, we discuss three examples of AmI design projects that were carried out at Philips Design with the aim to explore and develop concepts for a new culture resulting from Ambient Intelligence.

3.5.1 La Casa Prossima Futura

An example of what I mean can be seen in our project *La Casa Prossima Futura* (Marzano 1999): the house of the near future. The project explored the idea that many of the modules in a home system would be unobtrusive helpers, more or less invisible, either hidden completely out of sight or incorporated into objects that have always belonged in the home – and will always belong in the home – like tables and chairs, cupboards, or lamps. Other modules would form works of art in their own right, like pictures on the wall or sculptures – things that people have always been proud to show off in their homes. Today's black boxes are nowhere to be seen, and the timeless elements are backstage. The result is a much richer experience – more fun than in the home of the past or the home of today.

One set of concepts in the project centered on the activities and rituals involved in preparing and eating food. For a leisurely breakfast in bed in twenty-first-century-style, for instance, we came up with a wooden breakfast tray with magnetic metal contacts integrated into the tray, cups and plates to ensure that they do not slide about. Though cool to the touch themselves, these contacts also provide power to the crockery, keeping coffee and croissants warm, or orange juice and cereals cool. The tray has a removable touch screen, for reading online newspapers and checking e-mail.

For the kitchen, we developed display screens that resemble sturdy terra-cotta oven-dishes, and loudspeakers that could be integrated into the tiling: these devices are easy to wipe clean and have no buttons or knobs to catch the dirt. To make cooking slightly less stressful, we also conceived a washable intelligent apron, with an integrated power circuit and a built-in microphone. This will allow the cook to operate kitchen appliances hands-free: voice-activated, it can be used to turn down a hotplate or recall a recipe on the screen while one's hands are busy. Alternatively, it can be used to call in help from the family or tell them lunch is ready. For optimum food storage, we designed microclimates to keep any food in perfect condition. They can keep a meal warm on the table or can be used to store – and present – delicate cheeses or homemade cakes. Chips are clicked onto the microclimate to tell it which food is being stored. Sensors in the top of the glass dome keep track of both temperature and humidity.

Finally, for dining, we developed a cordless group-cooking tool to enhance the social aspects of eating. It can be used for all sorts of cooking at table – fondue, raclette, tabletop barbecuing, or grilling. The halogen heating elements create soft patterns of light to suit the mood of the evening. The tool is powered by induction, through an integrated power circuit woven into the washable linen tablecloth, while the cloth itself remains cool. Special ceramic plates can be used to keep the food warm during the meal, giving both guests and hosts time to enjoy the food as well as the company. Ambient sound is provided by the ceramic audio speakers, which are both functional and decorative.

We also developed concepts for other parts of the house, such as a Home Medical Box for the bathroom, which would function as an interactive medical encyclopedia, with in-depth explanations and simulations. It also provides access to the family doctor via a video link, enabling guidance in the use of a number of simple diagnostic tools, such as a stethoscope microphone, an electronic thermometer, and a blood-pressure wristband.

Other concepts have focused on sleeping rituals. Nebula, for instance, allows people to create their own unique ambience in their bedroom with a projector that can be personalized to meet user needs. Contained in an elegant vase at the bottom of the bed, this projector projects images onto the ceiling. The images can be personalized and manipulated by the users as they lie in bed, either by 'programming' it using a special token, or by moving about on the bed.

La Casa Prossima Futura demonstrated convincingly that the house of the future will actually look more like the house of yesterday than the house of today. The functions that are contained in black or gray boxes today will tomorrow be embedded into the infrastructure or into all sorts of traditional-looking objects.

3.5.2 New Nomads

Already today most people carry a number of technological devices around with them: a mobile phone, an MP3 player or CD-player, a palm-top computer, and so on. The more such tools become an essential part of our lives, the more we will want to enjoy the same sort of connectivity and digital support we get at home or at the office, but without the bulky equipment which it normally tends to entail. Again, manufacturers have a choice here: they can either focus on the technology, and condemn people to wearing the equivalent of black boxes; or they can embed the technology in such a way that it still makes it possible for people to do what they have always done with their clothing, namely to convey messages of sexual attraction, authority, or membership of a social group. This latter approach is the one we explored in a project we called *New Nomads* (Marzano 2000).

This project built on our initial exploration of wearable electronics, made in the late 1990s as part of the *Vision of the Future* project (Marzano 1995). Bringing together a multidisciplinary team of designers, scientists, and engineers, it took a fresh approach to the future of lifestyle and fashion. The key challenge was to integrate into clothing an electronic infrastructure to which intelligent objects could be attached.

This meant that we first needed to develop conductive textiles that would enable data, power, and audio to be conveyed around a garment through a "body area network." We also needed to consider how people might engage with such a body area network. We therefore looked at the ritual of dressing: how, by combining personal and cultural aesthetic preferences, people "assemble" a personal identity through their dress. *New Nomads* extended this metaphor of "assembly" to understand how people might use electronic functionalities to create their own personal portable electronic environment, one that would allow them to make a statement about themselves and their preferred lifestyle.

Which groups in society would be the early adopters of wearable electronics? We looked at a spectrum of likely subcultures, from clubbers and active sportspeople to urban commuters and gamers, and developed a number of concepts that our research led us to believe would interest them.

For sportspeople, for instance, we envisaged high-performance sportswear incorporating not only personal audio and communication technology, but also biometric sensors for monitoring pulse, blood pressure, body temperature, respiration, and other vital signs. The resulting data, we anticipated, would be analyzed and displayed immediately so that the wearer could adjust their training appropriately.

Another concept we came up with was an audio jacket: an expressive item of street wear for young people, providing them with personal downloadable sound entertainment on the move. All the integrated audio devices were controlled via a simple control panel, which is located on the sleeve for easy access. On the back of the jacket, an electro-luminescent digital display showed the output of an embedded spectrum analyzer, the light emitted reflecting the frequencies of the music being listened to. For children, we designed a coat incorporating GPS technology (as in mobile phones) to pinpoint the wearer's location, and a Webcam, giving parents peace of mind and enabling children to play exciting outdoor games safely.

Finally, we envisaged a "relaxation kimono," a garment with a conductive spine embroidered on the back. The conductive threads spread an electrostatic charge, giving the wearer a soothing, tingling sensation.

3.5.3 Living Memory

In *La Casa Prossima Futura*, we explored how Ambient Intelligence might function within the home; in *New Nomads*, we looked at how it might function within a single garment. Will Ambient Intelligence also have a role to play beyond the home or the person, in the community at large?

In a project for the European Union called *Living Memory*, carried out collaboratively with a number of universities and other companies, we explored how communities develop online and offline, and how people interact within them. With the aim of combating the growing fragmentation of local communities, we were essentially looking at how the connections within a modern village-like community actually develop and operate.

To that end, we set up a network in one part of Edinburgh, a neighborhood that formed a reasonably self-contained social community. The network infrastructure itself was hybrid in nature, combining physical locations and a closed computer network. Beforehand, we studied how people normally operated within this community: the places they went to, the people they met there, what they did, and so on. We then developed appropriate interfaces that could be placed at key locations (shops, libraries, cafés, bus stops, and so on) so that people could enter all sorts of information, ideas, comments, questions – anything they liked – into a database. This database was designed to mature, so that people could continue to access material that was used a lot, while material that was only accessed rarely would gradually "sink down," and be buried in the depths, in a manner analogous to what happens in a human memory. In this way, the system allows people to use modern technology to maintain – or even create – the same sort of community feeling that once existed in villages. Even though the people involved today are probably more spread out than in a traditional village, lead more complex lives, and keep different hours, thanks to the *Living Memory* network they can still keep in touch with each other and with the community as a whole.

One of the interfaces we came up with was a café table. The tabletop incorporates a touch-screen. If, after accessing information via the screen, users want to take that information away with them, they simply insert a solid-state token, download the information to it, remove it, and take it with them.

Community networks of this sort could be useful not only in consolidating neighborhoods, but also other types of communities, both large and small. Within companies, for instance, they could help to promote a sense of community among staff, many of whom work from home, or at widely separated branch offices. Schools, colleges, or churches could benefit similarly. And in the case of families, such a network could help to keep distant family members in touch, while within the home it could serve as a sort of digital cross between the refrigerator door, the scribbled note on the table, the family conference, and the answering machine or voice mail.

3.6 How Do We Create the "Right" Ambient Intelligence – the Relevant Hypothesis for a Desirable Future?

So how do we actually set about developing a concept proposition in practice? We have to focus on satisfying universal and timeless human needs, but we also need to do it in ways that people will find relevant and meaningful *within their own timeframe and culture*. We also need to make sure that any ideas we suggest are technologically feasible, economically viable, and in line with environmental and social sustainability.

At Philips Design, we are doing this by continuously collecting information about large-scale social trends expected to develop over the next five to seven years and that will affect the way people act and feel. We also conduct field research into short-term trends – the latest developments in areas such as street-life, movies, video games, websites, lifestyle, fashion, and cars, up to a maximum of two years ahead. Similarly, we monitor and plot the course of emerging technological developments and developments in the business world.

3.6.1 Imagineering

Based on this information, we use a process we call Strategic Futures to "imagineer" or create scenarios or hypotheses about what everyday life may be like in the future. We first identify what is possible, i.e., what benefits we could theoretically provide within the next five to seven years, given present and emerging technologies. We then apply our findings concerning trends and social developments to identify precisely which of those possible benefits people are most likely to see as relevant to their lives. For instance, in Western society today, there is a growing interest in sustainability, in discovery, and in networking: our assumption is that projects that take these or other current interests into account are more likely to succeed than those that ignore them.

In this process of scenario-building we make use of "personas," fully rounded-out descriptions of individuals whom we envisage living in the future world, affected by the various social trends and developments we see ahead. By imagining these personas going about their day-to-day lives tomorrow, we are able to explore future needs, wishes, and expectations in greater depth. After its completion, each scenario is validated by experts, opinion leaders, and a network of research institutes using the Delphi method.

The next stage brings "real" products one step closer. In a series of brainstorming sessions, we envisage products or services that would provide the required benefits in ways that would be acceptable within the future temporal and social context. The resulting product concepts (often quite numerous) are filtered, using criteria of technological feasibility, relevance to Philips and its strategic direction, size of market potential, and so on. Realistic mock-ups are then made of the concepts that have withstood the filtering process, and short video clips are made of people using them, in order to bring them to life.

Having envisaged the product or service that provides the required benefits in the required way, we also look ahead to see what sort of customer interface will be most appropriate for presenting the product, system, or service to consumers. For instance, will we want to sell the product through our traditional channels, or will a different channel be more appropriate or effective?

Finally, we need to consider what competences will be required to produce these products and services: if we do not already possess them, will we obtain them by joining up with a suitable partner, or will we want to acquire or develop them ourselves?

3.6.2 Communication

The next stage is to generate feedback on these ideas. For this purpose, we stage public exhibitions, media events, and debates – as was the case with *La Casa Prossima Futura*, for instance. In this way, we share our ideas with the general public and the feedback they give us enables us to identify areas for future research, to formulate clearer hypotheses about "preferable futures," and to draw up roadmaps for the development of competences and capabilities.

McLuhan and Fiore (1967) once remarked that we drive into the future using only our rear-view mirror. As I have tried to show, our Strategic Futures methodology allows us to overcome some of the problems of relying solely on historical data in two ways. First, we identify directions in anthropological developments that have such long terms so as to be effectively timeless, and may even reflect innate human properties (e.g., the desire to be everywhere, know everything, and be able to do everything, with the maximum of ease). Second, we identify emergent phenomena (socio-cultural, technological, and economic), and use these as the basis of our projections, rather than relying on already fully established (and therefore potentially moribund) phenomena.

By presenting our ideas to a wide audience through exhibitions and communication events, our Strategic Futures methodology not only provides us with feedback on the suitability of our ideas and on the best way of approaching consumers with them, it also actively involves people in setting the direction of change. It does this by implanting in consumers' minds what neuroscientist Ingvar (1985) has called "memories of the future": expectations that people use actively (though subconsciously) to prime their observations and to trigger aspirations. Both he and the American neurophysiologist Calvin (1995) found that people who think ahead and formulate an idea of what they want to happen are better placed to recognize signs relevant to those ideas. Thinking about potential future developments seems to open your mind to receive them; not thinking about them tends to close your mind to them.

Incidentally, this approach also has interesting implications for the design process itself. By making our design ideas so tangible and explicit, we as designers are able to work together in a more focused way, and are better able to discuss alternatives with each other.

3.6.3 Getting Real

Up to this point, we have been talking about showing people mock-ups and films of possible uses. Now, having filtered the various ideas down to only those that the public and the experts think are most promising, we develop working prototypes, to test the technology, the design, the materials, and the way people use them. We also want to find out to what extent people would prefer to have the technology on show, or hidden away.

This stage also gives an opportunity to test out appropriate new materials that would allow us to produce the shapes required, and create an appropriate look and feel. We need to experiment with modes of interaction, exploring the sorts of relationships people would like to have with the product. Our prototypes take all these factors into account. They also look good and are in tune with the times, because we have to communicate with people in the visual language they understand. We then closely observe different people using them in natural settings.

Finally – especially in Ambient Intelligence products, which operate together as a system – we need to understand how people will interact with them as a system. Can people easily understand how the various component parts work together? What sort of interface will they prefer to use? Which aspects of the system will they interact with most often? And which least often? These and many more practical questions need to be explored at this stage through experimentation and observation.

3.6.4 Into Production

The result of all the observation and experimentation is a working prototype that has survived all the filters. At that point, the decision may be taken to put it into production. We then need to be sure that the new product is going to be perceived as fashionable and attractive, in terms of its appearance and positioning. This is where our research into short-term trends comes in. This research (which we call Culture Scan) enables us to give the product the right color, shape, or interface to have maximum appeal when it enters the market.

3.7 Business Issues

Let us now turn to the issue of generating business with Ambient Intelligence. Of all the relevant issues that may be discussed in this respect we single out two items that are of special interest to the development of Ambient Intelligence because of their social implication, i.e., partnerships and the bottom of the pyramid.

3.7.1 Partnerships

The embedding of intelligence in the environment means that artifacts produced by many traditional areas of business will finally become smart. Digital technology will become a core part of many industries in which it is currently seen as peripheral. This development implies a new model of the relationship between electronics and other areas of business, one which places digital technology at the center. This new model has both practical and more fundamental consequences.

First, on a purely practical level, companies will need to find partners who can provide the right complementary capabilities. Then, having found such partners, they will need to work out appropriate ways of working together, since each industry has its own cultures and codes. The new partners will also need to share the same values and affinities, so that consumers find the partnership, and the propositions that it develops, credible and therefore acceptable. Part of that credibility will also depend on how the partnership relates to the consumer. The two parties have to work out a new form of joint customer relations. In short, industries will need to develop jointly a totally new competence in building good relationships with complementary partners.

3.7.2 Bottom of the Pyramid

AmI partnerships will not be immune to general developments in the business world, such as globalization and the increasing attention being paid to corporate social responsibility. How will Ambient Intelligence fit into these emerging scenarios? For example, as Prahalad (2004) has recently argued, there is an enormous market at the "bottom of the pyramid" – people in developing countries, with somewhat different needs and severely limited resources. We will clearly need to consider how Ambient Intelligence might be applied in a culturally appropriate way to serve these potential consumers. And we will need to do this not only because it may be profitable, but also because it is right to apply the knowledge we acquire in developing Ambient Intelligence to help redress the balance of welfare between the advanced and less advanced countries. Companies need to work together to facilitate the creation of wealth, of buying power. It is also important that this wealth is distributed fairly around the world, and that companies seek to improve access to the new systems for people everywhere, not just in advanced Western societies.

As we pay greater attention to less developed markets, more and more products and services will need to be specifically targeted: solutions will need to be tailored to local aspirations. The global market is not uniform and standardized, it is one for which a wide variety of customized, highly local solutions can be proposed on the basis of globalized technologies. The new partnership business model I mentioned above is also likely to be very relevant to new, emerging markets: its flexibility allows globally operating industries to partner with local infrastructure, local service providers, and local government to provide services and new propositions directly relevant to different locations.

An important aspect of Ambient Intelligence is the linking of home networks with external networks: this could be exploited equally well both in advanced and less advanced markets. Take web-based education, for instance. In the advanced industrialized world, it is an interesting alternative to bricks-and-mortar education. In the remoter parts of Brazil or Africa, however, where schools are relatively few and students live far apart, it could be a relatively inexpensive way of raising educational standards quickly, and allowing these societies to leapfrog up to date. In advanced markets, the same technology could also provide a service to older people in the form of online health monitoring or home diagnostics, perhaps combined with a shopping ordering and delivery service. In this way, by combining new technologies with new business models, Ambient Intelligence may enable the development of highly versatile solutions to meet diverse needs. We will need imagination and creativity to spot all the opportunities that most surely exist.

3.8 Deeper Issues

Let us now turn to consider briefly certain other issues of a more abstract kind that, as we develop Ambient Intelligence, we shall need to confront at some point.

3.8.1 Educating Our Intelligent Objects

Very soon, familiar objects that used to be static and unintelligent will become "subjects" – active and intelligent actors in our environment. Ambient Intelligence is about creating new types of relationships, not only between people and their active objects, but also among those objects themselves. We will need to lay down the ground rules that govern these relationships.

Consider a smart home in which various products are activated as we arrive home from work. The message device reports on who has called, the audio system plays our favorite music, the cooking center tells us that dinner is ready, and promptly at 6 O'clock the TV flicks on automatically for the early evening news. If you have had a hard day at the office or a tiring journey, the last thing you want is to be greeted with a lot of noise, as all these devices compete for our attention. So we need to educate our Ambient Intelligence systems: or rather, we need to build good, social behavior into them, along with the ability to learn. In doing that we will essentially be defining the culture of Ambient Intelligence.

Ambient Intelligence products will be the modern-day equivalent of trusted butlers, maids, and valets. In the past, if you wanted your butler or valet to serve you well, you had to train them to do things the way you wanted them to do, quietly and efficiently. In our Ambient Intelligence world, our new "servants" will need to be educated in the morality, etiquette, and culture of the society they are entering. They will need to be socialized; their personalities will need to be shaped. In that sense, they are no different from, say, new employees, who need to learn new tasks and a new culture, or babies, who need to learn an incredible range of skills before we accept them as "full" members of society.

But we are patient with babies: we give them a couple of decades to achieve an acceptable level of socialization. We will not be so patient with objects. We will want them to come into our lives almost fully socialized. They will need to be ready to use, grown up, and sociable, from day one. For more discussion of this issue, see Marzano (1998).

3.8.2 The Next Challenge

At the moment, we are only envisaging the first phase of Ambient Intelligence, in which intelligent devices will behave in a predictable way, performing actions selected from a closed set that has been programmed in at the design phase. However, in due course, we will move on to a second phase, with objects that can develop behavior that has not been explicitly programmed in, but which arises through the interaction of various parameters.

Over time, an Ambient Intelligence of this sort will grow up, as it were, and become gradually customized to the requirements of the user. Users may be able to invest in "training" for their Ambient Intelligence, which in turn will mean that they will almost "care for" it in the way good parents care for their children and good employers care for their employees. People will be active protagonists and co-creators of their environment, rather than passive end-users; see also Andrews (2003) and Andrews et al. (2003) for more discussion. We will need to be prepared to deal with the cultural implications of this development.

We have already started to explore this issue in a project we have called *Open Tools*. We envisaged "tools" that were "open" in that they were initially relatively "undefined": users would be able to largely specify and develop the functionalities of these tools as they wished. The tools would evolve over time, adapting and shaping themselves to the user's habits and needs, and becoming increasingly specialized as they did so. The project explored the impact such tools would have on users, on current mass-customization paradigms and on the relationship between brands and their customers.

Our open tools were conceived as "service units," physical products connected to larger systems providing easy access to functionality, services, and content, as well as to other products and peripherals. Containing a range of latent functions, bounded only by a set of rules, they would allow users to negotiate and unfold their potential in users' own ways and at their own pace. An open tool would "remember" how you had used it before, and use this memory to decide how best to link into your own pool of digital functions, data, and services. They would link to digital resources, to each other and to other devices when required or desired.

One of these open tools was called OpenDesk. It offered a touch–display work surface for manipulating digital media. The desk was "open" to software and was able to connect with other devices and open tools to extend its capabilities. This and other proposed open tools were "tested" against a six-month scenario, to show how they could evolve, and become appropriately specialized and personalized.

3.8.3 Where Is the Boundary?

Another cultural issue we will need to consider at some point will be the desirability of Ambient Intelligence being incorporated into an even more intimate ambience into our own bodies. As we have seen above, we are already incorporating Ambient Intelligence into clothing; and people are quite happy to have a pacemaker built into their bodies. But this is just the beginning, it seems. In 2002, a British professor (Kevin Warwick of the University of Reading) took things a step further by having a chip implanted into his wrist linked to the median nerve, which operates the muscles of the hand (Warwick 2004). At Rush University Medical Center in Chicago, silicon chips are being implanted in the eye of sufferers of macular degeneration of the retina (Rush, online), while work on implanting electrodes into the brain is also being carried out at Emory University in Atlanta (Emory, online), the aim being to help people who have been almost totally paralyzed to communicate through electrical brain impulses, detected by an electrode and interpreted by a computer.

Such implants have a clear medical justification. But how long will it be before we accept the implantation of chips for non-medical reasons? Attitudes to the body are already changing. Body piercing, tattoos, and cosmetic surgery are much more common than a generation ago. And a US company, Applied Digital Solutions, recently (October 2004) received FDA approval for the medical use of its VeriChip implant (ADSX, online). No larger than a grain of rice, this chip can contain essential medical information. Health care providers can scan the chip to access this information when necessary.

If this sort of product finds widespread public acceptance, will we have crossed an important boundary? Where will people draw the line between the organic and the inorganic, the real and the artificial? And how will that affect how we view and treat our Ambient Intelligence systems ... and each other?

3.8.4 Which Reality Is Real?

A less obvious, but equally fundamental issue that awaits us is an ontological one, about the nature of existence itself, or at least how we perceive it. McLuhan (1964) famously said that the medium was the message, and that we were becoming more interested in television, for instance, than reality. The French sociologist Baudrillard (1988, 1995) of the Sorbonne believes we have already passed beyond this point. He argues that the traditional relationship between media and reality is being reversed. Increasingly, the media are no longer seen as just reflecting or representing reality. They constitute a new, hyper-reality that is felt to be even more real than "real reality." The fact that we call semi-staged shows like *Big Brother* "reality TV" probably says more about what people think of as "real" than we suspect.

Will we get so used to interacting with our Ambient Intelligence that it will affect the way we interact with real people? Today we already can read press reports of workers preferring to convene with colleagues by e-mail rather than engage in a face-to-face interaction. If, through Ambient Intelligence, we come to experience more of the real world through technology rather than directly through our senses, are these indirect experiences less valid? Is hyper-reality less valid than physical reality? Where can we draw the boundary between physical reality and imagination?

We may not want to engage in deep philosophical discussions like this every day, but, at some point and in some form, these are issues we will need to confront.

3.9 How to Measure Intelligence?

For now, let us come back to earth with a more practical topic. Communicating with consumers about something as intangible as Ambient Intelligence may prove to be difficult. We are used to quantifying electrical power in terms of watts, and motor power in terms of horsepower, but how can we talk about Ambient Intelligence – in terms of its IQ? Above, I compared Ambient Intelligence systems to a staff of butlers, maids, and valets. I suggest that the metaphor of the butler may prove to be useful in helping us talk meaningfully about Ambient Intelligence with consumers.

Some time ago, an Italian anthropologist attempted to compare the standard of living of a freeman in Ancient Greece with someone in Western society in the 1980s. The problem was to find a common measure. Based on the fact that a freeman in Ancient Greece was allowed to own 8 slaves, and taking into account what work the slaves did, the researcher managed to work out that today, with all our domestic appliances and conveniences, we have the equivalent of 36 slaves!

Perhaps we should use the butler as a unit of measure of convenience and ease in Ambient Intelligence. It is, after all, hardly different from measuring motor power in terms of horses. When the term "horsepower" was first coined, it must surely have called up a wonderful picture of a carriage drawn by an enormous team of invisible horses. Who would not want a carriage like that? So perhaps we can soon describe a top-of-the-range Ambient Intelligence system as "a 120-butler system." People can boast: "I have 55 butlers," and real estate agents can praise houses as "A highly desirable townhouse: 6 bedrooms, 112 butlers."

3.10 The Culture of Ambient Intelligence – Human Culture in the Broadest Sense

To conclude, Ambient Intelligence is more than just a question of embedding technology into objects. It involves human culture in its broadest sense: universal desires; complex social relationships; different value systems; individual likes and dislikes; the sustainability of economic and natural ecosystems; and codes of ethics, conduct, and communication, both in civil society and in business. The more we understand all these aspects, the better we will be able to find satisfactory answers to all our questions, shape our propositions, and succeed in providing people with an Ambient Intelligence with which they feel comfortable and which they can welcome into their lives as something that truly improves their quality of life.

4

Smart Materials

D.J. Broer, H. van Houten, M. Ouwerkerk, J.M.J. den Toonder, P. van der Sluis, S.I. Klink, R.A.M. Hikmet, and R. Balkenende

"Ambient Intelligence will be built around cautious and pro-active equipment, unobtrusively hidden in the environment and only exhibiting itself when there is a need for its function."

4.1 Introduction

Building hardware that fit within the philosophy of Ambient Intelligence often requires access to responsive materials. For this purpose responsive materials are defined as materials that change appearance or shape as a function of an external stimulus. That may be as much as the way a person experiences the look of a material in its interaction with light related to addressable properties such as absorption (colour, brightness, transmission) and scattering (velvet tones, metallic, transmission). But it can also be related to other properties to trigger a person's perception of its ambient, such as surface roughness, odour release, acoustic reflection, etc. The driving force for the change in the material's property can be an electrical voltage or current as a response to a stimulus from a sensor. In that case the sensor picks up a signal from the environment it is in. Examples are the presence of a person, change in conditions such as temperature, humidity, ambient light, the use of audio-visual equipment, etc. But the response of the materials can also be more autonomous where it changes properties as a result of the dynamically changing environmental conditions without the intervention of an additional sensor. When integrated in commodities as furniture and electronic equipment or even in the walls of a building or in the interior or exterior of vehicles, one can speak of so-called smart-skins or electronic skins which draw much attention in the world of military equipment and in programs such as Ambient Intelligence.

The technologies applied to create smart materials are diverse and include, among many others, self-organizing, conducting and chiral polymers, and similar organic materials, liquid crystals, biotechnology, ceramics, piezoelectrics, photonics, microsensors, shape memory alloys (SMAs) and polymers, and the broad world of nanotechnology. In general the requirement is that one should be able to produce them on large surface areas despite the fact that their mechanisms are based on sub-micron structure control even going down to a molecular level. To enable this production of ultra-small feature sizes on large and various types of substrates often the combination of top–down and bottom–up technologies is proclaimed. For instance, the use of photolithography (top–down) and self-organization as exists within liquid-crystalline media (bottom–up) proves to be very valuable to create field-activated or autonomous responding elements.

This chapter will discuss a number of developments, both within Philips Research Laboratories as elsewhere, related to smart or intelligent materials. Examples are thermochromic skins, electronic skins/paintable displays, polymers with a mechanical response, switchable mirrors and ambient lighting devices.

4.2 Chromogenic Materials

Chromogenic materials change their colour when subjected to a change in the environment. This is caused by a reversible change of some intrinsic properties of the material. In this respect these materials are different from systems based on liquid crystals or dispersed particles, where the variation in properties is brought about by a change in the orientation of the constituents. Depending on the external stimulus, chromogenic materials are referred to as thermochromic (temperature sensitive), photochromic (light sensitive) or electrochromic (voltage sensitive).

For many applications, chromogenic materials are deposited as a thin film on a substrate. Properties of these thin chromogenic films, like transmittance or colour, change in response to, for example, a change in the ambient illumination level. In this way the properties of windows and displays can be tuned in such a way that they adapt to external conditions, for example by applying a photochromic layer or an electrochromic layer coupled with an illumination sensor. In the case of displays an interesting application is the improvement of the perceived picture quality by optimizing the trade-off between image brightness and contrast. Other applications are sunroofs and mirrors that can be tinted. The effect is then used to reduce glare or to reduce solar heat. Further, in buildings these materials can be used to save energy by tuning the reflective and absorptive properties to optimize solar control as an active replacement of the present passive reflective films and shading devices. Changeable windows can also be used for privacy purposes. Thermochromic materials have also been proposed as energy-efficient glazing. In this respect especially the semiconductor to metal transition that is observed for vanadium dioxide is attractive, as this transition leads to a huge increase in the IR-reflectivity upon heating. The switching temperature was precisely set at around 30°C by element doping. Further, vanadium dioxide has been presented as an attractive thin film material for electrical or optical switches, optical storage and laser protection. An application that is more directly interacting with the user is heat indication. This for instance can be used to indicate whether a hot surface is already safe to be touched. For these types of applications, a switching temperature of about 60°C is needed. The thermal stability of the material, however, should be much better in order to withstand much higher operating temperatures.

A large number of systems show a thermochromic response. A common example of thermochromic transitions in metal complexes is the transition between the blue tetrahedral and pink octahedral coordination of cobalt (II) when cobalt chloride is added to anhydrous ethanol and the temperature changed. Examples of thermochromic transitions in inorganic compounds include Ag_2HgI_4, vanadium dioxide and several inorganic sulphides. Further, there are thousands of organic thermochromic compounds, with well-known examples including di-beta-naphthospiropyran, poly(xylylviologen dibromide) and ETCD polydiacetylene.

As an example the vanadium dioxide system will be discussed in some more detail. Vanadium oxide shows a change in electrical and optical properties around 68°C. At this temperature a structural transformation from the monoclinic to the tetragonal rutile-type phase occurs, leading to an electrical transition from a semiconductor with a band gap of 0.7 eV to a metal (Goodenough 1971). A change in conductivity by a factor of 106 has been reported for crystalline sample. The transition temperature can be lowered by introducing small amounts of W, Mo, Nb or F. However, usually this also decreases the transmission in the visible range. The transition temperature can also be increased, by introducing elements like Ti, Fe or Al (Burkhardt et al. 1999).

The metal–semiconductor transition leads to a change in optical properties. A strong change in the reflection in the IR is found, which indicates that the material is studied extensively for the use in thermochromic windows lately. Also in the visible region, optical changes occur. The absorption of the material only changes slightly (the low temperature phase is bluish, the high temperature is more transparent), but a large change in refractive index occurs. Figure 4.1 shows the refractive indices of the material for the low and high temperature phases.

This change in refractive index of vanadium oxide can be used to create a thermochromic effect. In single crystals, repeatability of switching is limited due to mechanical stress that is induced as an effect of the volume change accompanying the phase transition. However, for thin films, and crystalline particles, good multiple switching behaviours have been reported. Thin films in addition open the possibility to strengthen the visible optical effect by incorporating the vanadium oxide layer in an optical stack. This stack would for instance consist of a substrate, coated with a highly reflective layer (the substrate can also be the high-reflectivity layer), a layer of vanadium oxide, and a layer of a dielectric material such as silica. The choice of the various materials and the thickness in which they are deposited determine the colours that are observed at low and high temperatures, respectively. An example of such a coating stack demonstrating the optical effect is shown in Fig. 4.2.

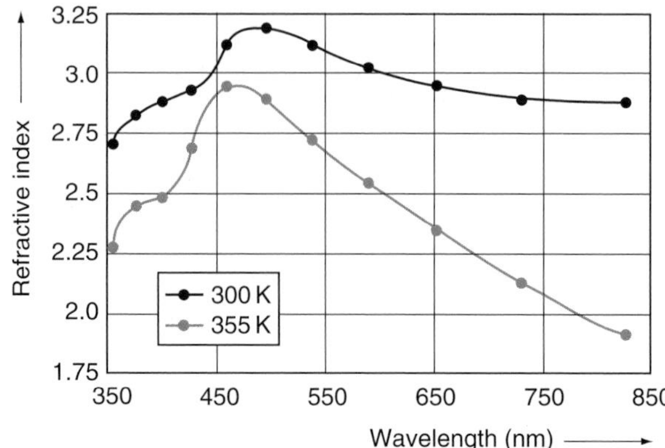

Fig. 4.1. Refractive indices for the low and high temperature phases of vanadium oxide

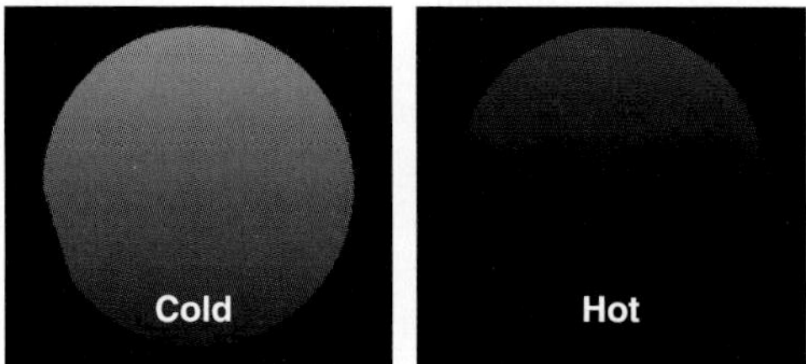

Fig. 4.2. Si with a thermochromic coating at room temperature (left) and at 80°C (right)

4.3 Thermochromic Skins Based on Liquid Crystals

Most practical organic thermochromic materials are based on liquid crystals (Brown 1983). Well known are the thermochromic materials based on cholesteric liquid crystals. They are known from for instance the simple wine or beer thermometers that can be stuck on a bottle or integrated in the label or the thermometer plaster that can be stuck on the skin (Brown 1983). Examples can be found in advertisements on the Internet (http://robbie.ebigchina.com). The principle is based on a chiral-nematic calamitic (= rod-like) liquid crystal that organizes itself in an ensemble of molecules of which the average direction describes a helix in the direction perpendicular to the molecules. Because of the molecular anisotropy, the uniaxial optical indicatrix also describes a helix into the same direction and reflection of light will occur when Bragg's conditions are met (Collings 1990). The reflection wavelength λ relates to the helicoidal pitch p through $\lambda = \bar{n} \cdot p$, where $\bar{n}$ is the average refractive index (De Gennes and Prost 1993). The bandwidth of reflection is $\Delta\lambda = \lambda \cdot \Delta n/\bar{n}$ with Δn being the difference in refractive index along the average molecular orientation and its orthogonal components. The pitch p can be modulated by temperature and as a consequence also the reflection wavelength will shift. The pitch becomes especially temperature sensitive when chiral-nematic molecules that exhibit a smectic phase at lower temperatures are selected. The smectic phase unwinds the helix when the transition is approached and the colour changes very steeply with temperature. Simple thermochromic coatings can be made by dispersing the cholesteric material in a polymer binder such as a polyurethane or by microencapsulating the liquid crystal and dispersing the capsules in a polymeric binder. These dispersions can be applied from solution as a film on a substrate and dried or cured. Best results are obtained when the films are applied on a black substrate.

A drawback of the thermochromic paints based on cholesteric materials is that the colour saturation is relatively low. This is particularly caused by the property of the cholesterics to reflect only a single handedness of circular polarized light related to its helicoidal structure. The opposite handedness is transmitted and absorbed by the substrate. This indicates that at maximum only 50% of the light is reflected unless right- and left-handed materials are combined, for example in a stack of two layers or by mixing two emulsions with opposite handedness. Another serious drawback for many applications is that the colours and the sequence in which they appear upon changing the temperature cannot be chosen freely. Related to the negative sign of $\mathrm{d}p/\mathrm{d}T$ of the cholesteric liquid crystals, the colour changes from being red at low temperatures to blue at high temperatures, just opposite to what is desired for those applications where the colour change is used for warning signal (since red is perceived as being related to a hot surface). Another imperfection is that the coating is still filled with a liquid that has a finite vapour pressure and will slowly evaporate at higher temperatures or extracted when in contact with solvents.

To overcome these deficiencies, we developed an alternative system, also based on a liquid crystal material (Heynderickx and Broer 1995). In order to make a thermally stable structure and avoid evaporation of material we selected a liquid-crystalline polymer rather than a low molar mass material. The polymer backbone is a polysiloxane, which can withstand high temperatures. The backbone is also very flexible enabling rapid changes in the molecular organization when passing a transition temperature. In its isotropic state the polymer is fully transparent. But when cooled down to its liquid-crystalline phase the polymer becomes turbid and scattering because of refractive index steps along the domain boundaries. An example of a block copolymer that performs well for this purpose is shown in Fig. 4.3, upper left. The polystyrene block provides dimensional stability by microphase separation. The siloxane block is smectic (SmA) at room temperature and exhibits a smectic to isotropic phase transition at 80°C. The occurrence of a smectic phase rather than a nematic phase enhances the scattering even more. For coating applications the polymer is dispersed in a photocurable monomer, in this case ethoxylated bisphenol-A diacrylate, that after polymerization has

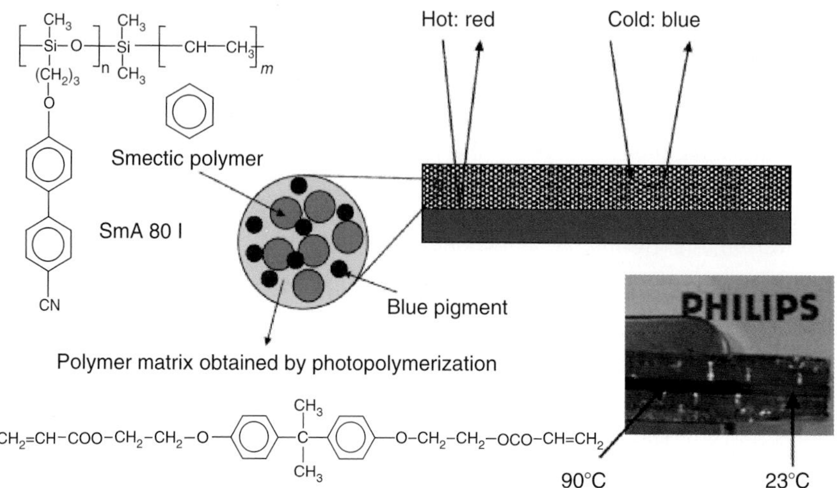

Fig. 4.3. Schematic representation of a thermochromic element based on a smectic liquid crystal

the same index as that of the isotropic copolymer. Adding a small amount of a dye or pigment to the polymer dispersion creates the colour transition. In the isotropic state of the copolymer the dye is hardly visible and the colour perception is dominated by the substrate colour (red in case of the demonstrator in Fig. 4.3). As soon as the polymer dispersion starts to scatter while cooling down the colour of the dye in the dispersion is intensified and will dominate the appearance (the blue part).

The material can be easily processed in the uncured state of the ethoxylated bisphenol-A diacrylate. Without solvent it behaves as a viscous paste that can be doctor bladed on a substrate. Coating properties are enhanced by the addition of small amounts of solvents such as xylene. After the formation of the film, usually in the order of 60 mm thick, the sample is cured by a short UV exposure that excites a dissolved photoinitiator. Thorough curing is promoted by a UV postcure at elevated temperatures above the clearing temperature of the liquid crystal polymer. The application of these materials can be functional, for example as a warning signal for hand burning temperatures, or purely decorative. Some examples are shown in Fig. 4.4.

4.4 Switchable Mirrors

In 1996 it was discovered that when a Pd-coated yttrium film is exposed to hydrogen, the metallic reflecting Y film becomes transparent (Huiberts et al. 1996). The semi-transparent thin metallic Pd layer is necessary both for protection of the underlying Y film and for the catalysis of the hydrogen uptake. Soon after it was discovered that all the lanthanides show this behaviour,

Fig. 4.4. Mock-ups of potential applications of thermochromic material for indication or decoration

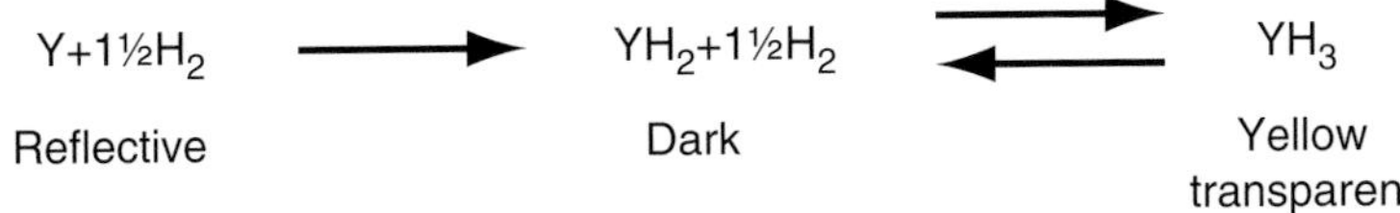

Fig. 4.5. Partially reversible hydrogen uptake by a thin Pd-coated Y film with associated optical changes

Fig. 4.6. Reversible hydrogen uptake by a thin Pd-coated $MgGdH_2$ film with associated optical changes

but unfortunately the reaction is only partly reversible and the transparent hydrides all have yellow to red colours (depending on the metal), see Fig. 4.5.

The transparent YH_3 phase could only be switched back to a dark coloured YH_2 phase and not to the reflective Y phase. It was found that this transformation could be carried out in an electrochemical cell (Notten et al. 1996). However, one year later it was discovered that alloying with Mg resulted in a material system that could be switched from a highly reflecting to a colour-neutral transparent layer in a reversible manner (Van der Sluis et al. 1997).

Switching with hydrogen gas is technologically not very appealing, but fortunately, the same optical changes could be induced in electrochemical ways, i.e. by applying a voltage over a suitable electrochemical cell. The result is a switchable mirror, which can be switched from reflecting via a black state to transparent by just applying a voltage, see Fig. 4.6.

In this case the electrochemical cell consists of two glass plates with a liquid electrolyte. Small prototypes of all-solid state electrochemical cells have been made in which the whole electrochemical cell is deposited as thin layers on one side of a glass pane (Van der Sluis and Mercier 2001). Further improvements in terms of durability and optical performance are required to exploit these devices.

Several applications of such a switchable mirror can be envisaged. The largest scale application would clearly be in architectural glass. During nights the glass would be switched to the reflective state in order to keep the temperature constant. During the day, in the absence of sun, the glass can be switched to the transparent state, so that light can enter the building. When the sun starts to shine and the light becomes unpleasant, the glass can be switched to the dark state when heat gain of the building is wanted, for example in cold climates during winter. When heat gain of the building is unwanted, the glass can be switched to the reflecting state. All these transformations can be carried out gradually. Significant energetic and comfort gains can be expected from such windows. Imagine how the looks of a city centre would change when such glazing is applied on a large scale. In a similar way switchable mirrors can be used for climate control in cars. Heat uptakes of parked cars in the sun can be decreased by 80%.

In the reflecting state the transparency of the window can be adjusted to below 10^{-5}. That means that no matter how the light is distributed (e.g. dark outside, light inside) it will not be possible to look through the glass. This privacy option is an added benefit to using switchable mirrors and can also be useful indoors. Think of a boardroom that can temporarily be made private.

More indoor applications can be envisaged once reliable solid-state mirrors become a reality. Imagine a television set with such a window in front. When the television is in use the window will be switched to the transparent state, for unobstructed viewing. When the television is switched off, the result normally is an unpleasant dark rectangular object inside the living room. By switching the window to the reflective state, the television has become a mirror reflecting the colours of the room. An added benefit here lies in the dark state. When a lot of ambient light is present the contrast of the image goes down because the ambient light is reflected off the display. A dark coating in front of the display combined with an increased brilliance will enhance the contrast significantly because light from the display goes through the coating once and ambient light goes through the coating twice. With a simple light sensor a constant contrast depending on the ambient light can be constructed.

An application for small area devices lies in handheld displays such as mobile phones and PDAs. Currently, when such a device is in use, LEDs illuminate the LCD display at the expense of valuable battery power. When a lot of ambient light is available a reflector behind the display would be enough, but that would interfere with the backlight in the normal operating mode. With a switchable mirror this can easily be accommodated.

Other applications could be found in the area of lighting control, for example to change the character of a luminary from floodlighting to spot lighting. Overall the application of switchable mirrors will be very visible in everyday life. However, this depends on the not yet available all-solid state devices.

4.5 Switchable Cholesteric Mirrors

Cholesteric liquid crystal phase, as already briefly described for the thermochromic liquid crystals, is obtained when a nematic phase is doped with chiral molecules. Chiral molecules are optically active and are known to show optical rotary dispersion in the order of $1^\circ\,\mathrm{cm}^{-1}$. However, in the cholesteric phase they induce rotation of the long axes of the liquid crystal molecules (the director n) about a helix as shown in Fig. 4.7.

Such a macromolecular arrangement leads to optical effects unique to this phase. For example, the optical rotary dispersion shows a dramatic increase and reaches values in the order of $100^\circ\,\mathrm{cm}^{-1}$. Furthermore, a band of circularly polarized light having the same sense as the cholesteric helix is reflected while the band with the opposite sense is transmitted. The upper (λ_{max}) and lower (λ_{min}) boundaries of the reflected band are $\lambda_{\mathrm{min}} = p * n_{\mathrm{o}}$ and $\lambda_{\mathrm{max}} = p * n_{\mathrm{e}}$, where p is the cholesteric pitch corresponding to the length over which the director rotates 360°, n_{e} and n_{o} are the extraordinary and the ordinary refractive indices of a uniaxially oriented phase. The reflected bandwidth $\Delta\lambda$ is given by $\Delta\lambda = \lambda_{\mathrm{min}} - \lambda_{\mathrm{max}} = p * (\underline{n}_{\mathrm{e}} - n_{\mathrm{o}})$.

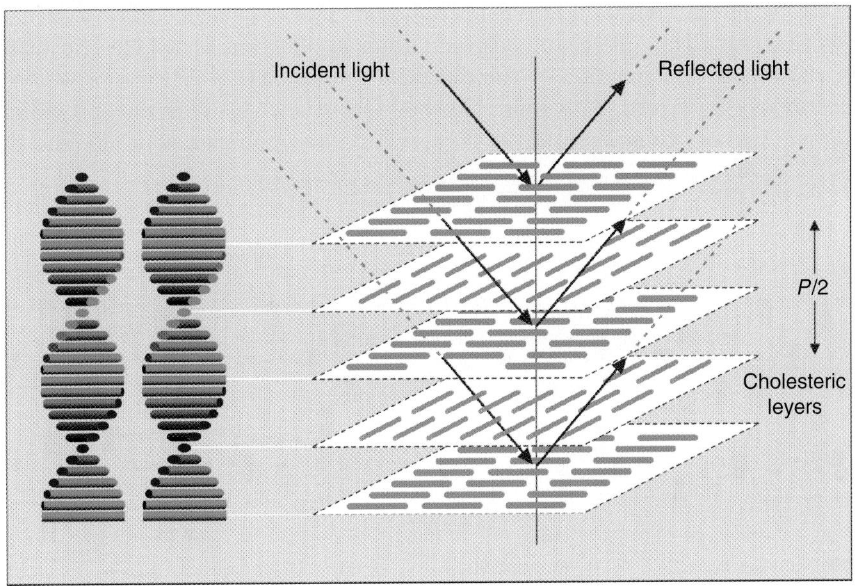

Fig. 4.7. Schematic representation of the cholesteric structure

In conventional liquid crystal cells, an orientation layer placed on top of the cell surfaces induces long-range orientation of liquid crystal molecules. Switching is induced by applying an electric field across transparent electrodes present on the cell surfaces underneath the orientation layers. Upon removal of the field, the liquid crystal molecules revert to the initial orientation state under the influence of these orientation layers. As shown in Fig. 4.7, cholesterics have a complicated helical structure and in order to obtain sufficient reflection the cell gap needs to be at least ten pitches thick. When such a structure is switched from a defect free planar orientation it is almost impossible for the system to reorient itself back to the initial state under the influence of the surface orientation layers. Instead, the cholesteric helix becomes oriented in various directions and the cell shows a scattering texture. In order to obtain fast switching a memory state needs to be built into such a cholesteric system. We tried to do this by creation of a lightly cross-linked network dispersed within the non-reactive liquid crystal molecules (Hikmet and Kemperman 1998). This is done by in-situ polymerization of a liquid crystal monoacrylate and diacrylate mixture in the presence of non-reactive liquid crystal molecules. The planar orientation of the cholesteric mixtures containing monomers with reactive groups is obtained in cells containing uniaxially rubbed polymer layers. The polymerization of the reactive molecules is induced by UV radiation freezing-in the cholesteric configuration and orientation by creating a network containing non-reactive liquid crystal molecules (anisotropic gel). The gel structure is schematically represented in Fig. 4.8.

The network in these gels consists of monoacrylate molecules forming the side-chain polymers, which are cross-linked by the diacrylates. The network is in strong interaction with non-reactive liquid crystals, which can be switched together with the side-chain polymer upon application of an electric field. In these gels, the function of the diacrylate molecules, which are present at fractions of a percent, is twofold: (1) they form the cross-links thus providing system memory function and (2) they preserve the polymer structure and its

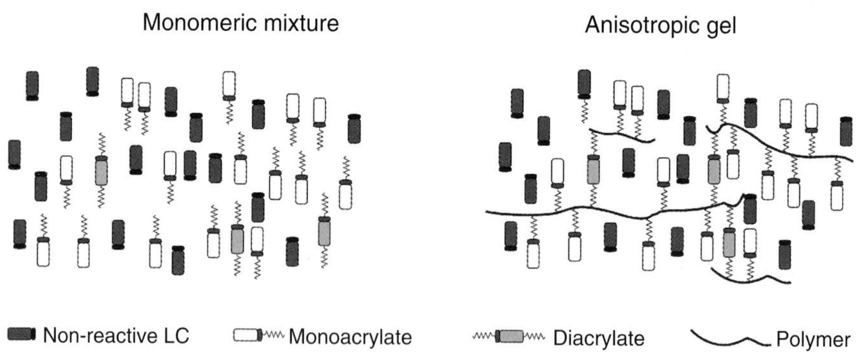

Fig. 4.8. Schematic representation of gel formation

distribution within the system preventing its diffusion. The second function is especially important in producing broadband reflectors and patterned gels described below.

In the gels, two different types of switching have been characterized. In the first mode of switching the reflection band shifts gradually to low wavelengths with increasing voltage before decreasing in magnitude and shifting back to higher wavelengths and disappearing at higher voltages. Shifting of the reflection band to lower wavelengths is a well-understood effect associated with the cholesteric layers getting tilted with respect to the incident beam of light followed by helical unwinding.

In the second mode the reflection band becomes narrower and decreases in magnitude as only the lower limit of the reflection band retains its position with increasing voltage. The fact that the lower limit of the reflection band remains the same while the upper limit decreases indicates that the cholesteric structure and the helical pitch remain the same while molecules tilt to become oriented in the direction of the applied field.

As described above, the pitch and the birefringence of the cholesteric phase determine the width of the reflection band. This means in most common cases a bandwidth of about 50–70 nm. In order to produce broadband switchable cholesteric reflectors we used excited state quenchers. Such quenchers work by capturing the energy of the excited state of the initiator, thus radical formation can be prevented. In these gels, the distribution and the size of the domains, which give rise to the broadening of the reflection band, are still not clear. One possibility is the formation of a pitch gradient as a result of light intensity variation across the cell due to absorption of UV light.

Due to the absorbance of the initiator and the additives, around 50% of the UV light is absorbed within the cell, which means a factor two difference in the intensity of light at the two surfaces of the cell. Such a pitch gradient has been observed in polymerized acrylate networks (Broer et al. 1995). In the case of studies of other gels, the band broadening occurred as the system phase separated into two regions containing liquid crystal and the liquid crystal swollen network. In the current system, inhomogeneities throughout the system are expected to contribute to the band broadening. These broadband cholesteric gels could be switched reversibly between silver coloured reflecting and non-reflecting transparent states. Upon application of the electric field, the cholesteric structure disappears and the cell becomes transparent as shown in Fig. 4.9. Upon removal of the voltage, the cell reverts to the silver coloured reflecting state very rapidly.

4.6 Electronic Skins and Paintable Displays

Liquid crystal displays (LCDs) make up for most of the displays that have been integrated in everyday commodities ranging from watches, pocket calculators to more sophisticated electronic products like mobile phones, PDAs,

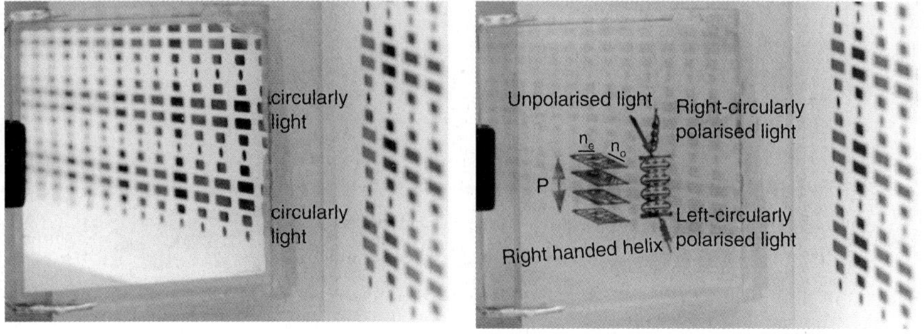

Fig. 4.9. Switching of a broadband cholesteric gel

television sets and car navigation systems. Also in the living room the traditional CRT TVs are more and more being replaced by LCD-TVs. The main reason why LCDs are so widely used is basically their thin form factor, low power consumption and the fact that there are a wide variety of LCD effects to suit a particular application, for example simple segmented reflective TN-LCDs for watches, high end active matrix IPS-based and VAN-based LCDs for monitor and TV applications and bi-stable nematic effects for electronic book applications.

One property that sometimes hinders further integration of conventional LCDs in new applications, product ranges or product designs is their restraint to flat device surfaces as imposed by their glass substrates. As a direct consequence LCDs are also very fragile. Another limitation, basically caused by their fabrication methods and related substrates handling, are their limited sizes and restrictions to rectangular shapes. The recently developed concept of Paintable LCDs by Philips Research represents a new flexible LCD manufacturing technology by which an LCD is made by the sequential coating (painting) and UV curing of a stack of tailored organic layers (Penterman et al. 2002a,b, Vogels et al. 2004). Since the Paintable LCD technology is based on a sequence of coating processes, the manufacturing process is relatively simple and fast, and can be applied to a wide variety of substrates including flexible plastic substrates. Furthermore, in theory there is no substrate size limitation for this LCD manufacturing technology. Paintable LCDs based on plastic flexible substrates exhibit the following features: they are non-rectangular, thin, flexible and robust displays. This technology will allow designers to design the display around a future product, instead of the other way around. Figure 4.10 shows some examples of the application possibilities.

The heart of a conventional LCD is a thin liquid crystal layer (typically 5 mm) sandwiched between two 0.7 mm thick glass substrates. In Paintable LCDs the liquid crystal material is confined between a substrate and a polymer sheet, with the important difference that the latter is formed during processing. This in-situ polymer sheet formation, also known as stratification, is the result of a photopolymerization-induced phase separation of a coatable

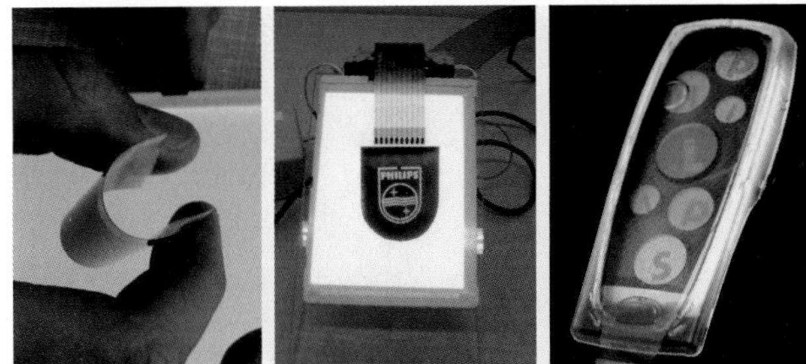

Fig. 4.10. Paintable LCD prototypes that demonstrate their unique features: Paintable LCDs are thin and flexible, can easily be cut to the required shape, and can therefore be laminated onto devices or integrated into garments

formulation of a liquid crystal (LC) material and a polymer forming material. Phase separation is among the other factors induced by the increasing incompatibility between the increasing number of polymer chains with the LC material. The direction of the phase separation (the polymer sheet is formed on top of the LC layer) has been ensured by the presence of a UV light absorbing reactive monomer in the coatable formulation (the mechanism is based on a photopolymerization reaction that converts monomers into a polymer by a (UV) light-induced decomposition of initiators into free radicals). The reaction can be controlled by modulation of the light intensity in the monomer film. In the Paintable LCD technology the direction of the phase separation has been ensured by the presence of the UV light absorbing reactive monomer in the coatable formulation, which during the UV exposure step establishes a gradient in the UV intensity across the layer. As a result, the photopolymerization predominantly takes place in the top of the layer. This induces a diffusion of monomers in the film in the upward direction where they become incorporated in the polymer top coat, and a concomitant diffusion of LC molecules in the reverse direction. Ultimately, a polymer sheet is formed on top of the LC layer.

The latest innovation in the Paintable LCD technology involves the formation of arrays of small liquid crystal-filled capsules, being the heart of the Paintable LCD, by a single UV light exposure step. A Paintable LCD owes its robustness to this multi-capsule geometry. As was mentioned above, the direction of the phase separation perpendicular to the substrate has been ensured by the presence of the UV absorber, whereas the control of the direction of the phase separation parallel to the substrate has been ensured by a local modification of the substrate with an adhesion promoter. At the position of the adhesion promoter, the polymer top coat becomes covalently attached to the alignment layer.

The processing steps of a Paintable LCD are as follows (see Fig. 4.11a–d): Via spin casting a poly(imide) alignment layer is deposited on a plastic substrate provided with IPS electrodes. Upon curing and rubbing, the alignment

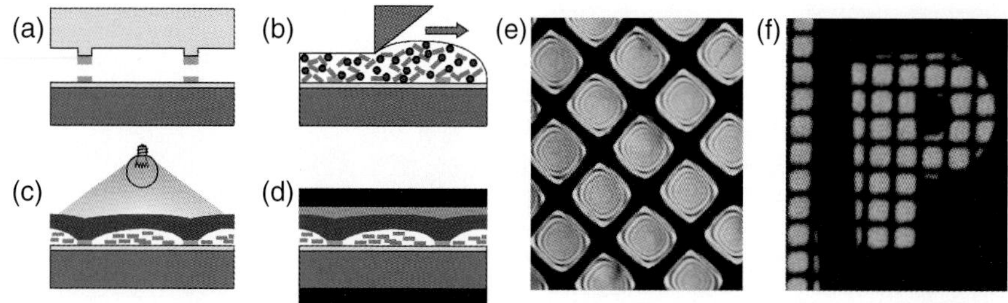

Fig. 4.11. (**a**) Via offset printing the alignment layer is modified with a grid pattern of adhesion promoter. (**b**) Subsequently, a thin film of the LC/prepolymer formulation is coated on the modified alignment layer. (**c**) A single UV exposure step creates the array of the liquid crystal-filled capsules. (**d**) The display can be completed by deposition of a planarization layer and lamination of two polarizers to the front and back of the Paintable LCD stack. (**e**). An enlarged polarization microscopy picture of the LC-filled polymer capsules (top view), the size of the capsules is $450\,\mu m \times 450\,\mu m$. (**f**) The LC material in the capsules can be switched by in-plane electrical fields originating from interdigitated electrodes that have been structured on the substrate

layer is locally modified with an adhesion promoter via an offset printing step (Fig. 4.11a). With the aid of a rubber stamp a pattern of a concentrated solution of the adhesion promoter can simply be printed. After evaporation of the solvent, the adhesion promoter lines form a grid pattern of $500\,mm \times 500\,mm$. With the doctor blade coating technique a thin film of the LC/prepolymer formulation is coated on the modified alignment layer (Fig. 4.11b), and a single UV exposure directly yields the structures depicted in Fig. 4.11c. The display can finally be completed by deposition of a UV-curable acrylate coating for planarization, and lamination of two polarizers to the front and back of the Paintable LCD stack (Fig. 4.11d). The LC material in the capsules (Fig. 4.11e) is switched by in-plane electrical fields originating from interdigitated electrodes that have been structured on the substrate (Fig 4.11f).

The outlook for Paintable Displays is anticipated to be as follows. A Paintable LCD owes its thin form factor, non-rectangular shape and flexibility to the fact that it is made by a single substrate technology. In addition, the array of LC-filled capsules provides the necessary robustness to the Paintable LCD. These properties open the doorway for LCDs to new applications like wearable displays, rollable displays and electronic wallpaper.

4.7 Polymers with a Mechanical Response

This section will give a concise overview of a class of smart materials that are key enablers for the realization of responsive hardware in Ambient Intelligence applications, namely materials that are able to shape our surroundings by mechanically deforming as a response to an external stimulus. Apart from the effect of shape change, the deformation may also be used to apply a mechanical load. The stimulus may be of varying nature: electric field, temperature, humidity, magnetic field, light, etc.

Traditional materials that show this mechanical response, and that are being applied as actuators in many applications (e.g. inkjet printing heads), are electroactive piezoelectric ceramics such as barium titanate, quartz and lead zirconate titanate (PZT). These materials respond to an applied electric field by expanding. The typical magnitude of the shape change can be expressed as the actuation strain, which is the relative change in size, and which is about 0.1–0.3% for these materials. When significant effects are required, this may not be sufficient. Another drawback of electroactive ceramics is that they are brittle, that is, they fracture quite easily. Finally, an important disadvantage is that the processing technologies for electroactive ceramics are rather expensive and cannot be scaled up to large surface areas.

A more recently explored class of responsive materials is that of SMAs (www.cs.ualberta.ca/~database/MEMS/sma_mems/sma.html). These are metals that demonstrate the ability to return to some memorized shape or size when they are heated above a certain temperature. The stimulus here is thus change in temperature. Generally those metals can be deformed at low temperature and will return to their original shape upon exposure to a high temperature, by virtue of a phase transformation (martensite to austenite) that happens at a critical temperature. A relatively wide variety of alloys are known to exhibit the shape memory effect. The most effective are NiTi (Nickel–Titanium), and some copper–aluminum-based alloys, for example CuZnAl and CuAl. The medical industry has developed a number of products, such as surgical tools, bone plates and vascular stents using NiTi alloys because of their excellent biocompatibility. The actuation strain that can be reached is significantly higher than for electroactive ceramics, namely up to 8%. However, SMAs have important drawbacks. The alloys are relatively expensive to manufacture and machine, and large surface area processing is not possible. Also, most SMAs have poor fatigue properties, which means that after a limited number of loading cycles, the material will fail.

To overcome the drawbacks of the ceramic and metallic smart materials mentioned (brittleness, low actuation strains, poor cyclic loading resistance, cost and especially lack of large-area processing perspective), much effort is currently being invested in exploring the possibility of making polymeric smart materials. Polymers are, generally, tough instead of brittle, relatively cheap, elastic up to large strains (up to 10%) and offer the perspective of being processable on large surface areas with simple processes. Polymer materials can often be applied in solution using spinning or various printing techniques, they can be structured down to on sub-micrometer scales by processes such as hot-, liquid-, and photo-embossing, nano-imprinting and lithography. There are infinite possibilities to vary their chemical/molecular structure to tailor their (mechanical, electrical) properties. In the remainder of this section we will briefly review the most important polymeric smart materials that are being developed, both in- and outside Philips Research laboratories.

4.7.1 Electrically Stimulated Responsive Polymers

Polymers that respond to an applied electric field or an electric current by changing in shape or size are generally called electroactive polymers (EAPs). Electroactive polymers can be subdivided into ionic EAPs and electronic EAPs. Generally speaking, ionic EAPs display an actuation due to the formation and/or displacement of charged species whereas in electronic EAPs due to the presence of an electric field, dipoles are created, phase transitions are induced or Coulomb forces are engendered.

Perhaps the best-known electronic EAP is the ferroelectric polymer polyvinylidene fluoride (PVDF). The molecular structure of PVDF, in the so-called all-*trans* configuration, is shown in Fig. 4.12. The chain has permanent dipole pointing from the (electronegative) fluorine atoms to the hydrogen atoms. If the processing of the material is done properly one obtains a semi-crystalline material with crystalline structures ordered such that a macroscopic permanent polarization is present. Just as for ferroelectric ceramics, an external electric field will couple to the electrical dipoles, thereby exerting a force onto the dipoles. The internally induced stresses will lead to a certain strain that depends on Young's modulus of the PVDF, which is on the order of 1 GPa. Figure 4.13 shows a bimorph actuator based on PVDF both at rest and in activated configurations, which demonstrates that a rather large effect can be obtained (Cheng et al. 2001). A main disadvantage of this polymer smart material is that large electric fields are required for significant actuation strains, namely on the order of $100\,\mathrm{V\,\mu m^{-1}}$. Additionally, at such high field strengths the risk of dielectric breakdown is very high.

Polymers with low elastic stiffness and high dielectric constant can be used to induce large actuation strain by subjecting them to an electrostatic field. The so-called dielectric elastomer is sandwiched between two compliant

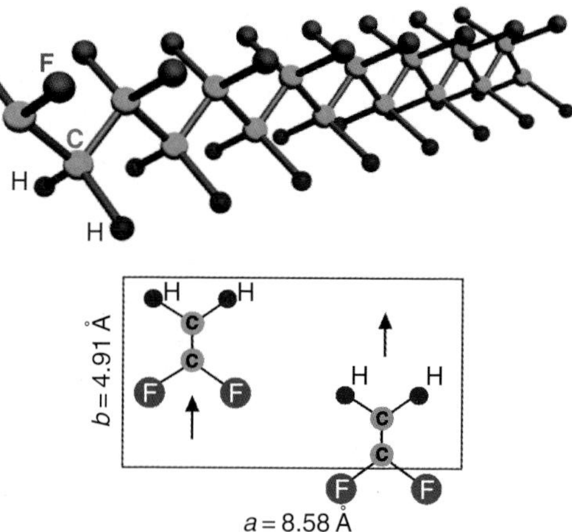

Fig. 4.12. Molecular structure of PVDF, a ferroelectric polymer

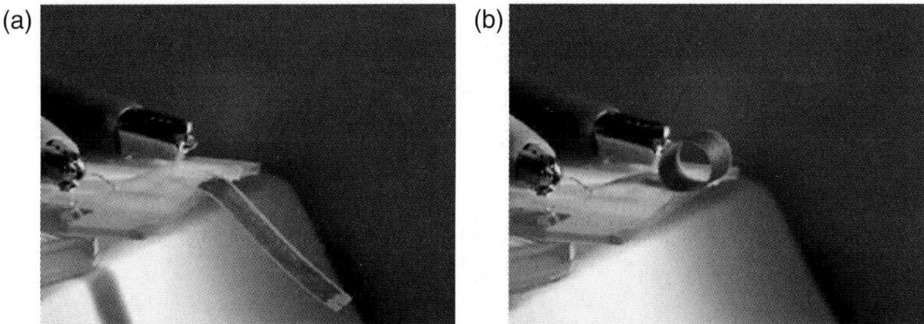

Fig. 4.13. PVDF-based copolymer layer (22 µm thick) bonded to an inactive polymer of the same thickness: (**a**) with no voltage applied and (**b**) after the voltage is turned on, $65\,\mathrm{V\,\mu m^{-1}}$ (Cheng et al. 2001)

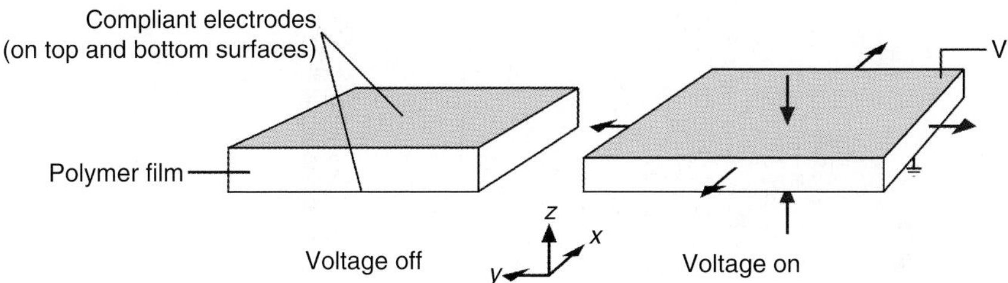

Fig. 4.14. Principle of operation of a dielectric elastomer

electrodes, as is sketched in Fig. 4.14. The principle of dielectric elastomers is simple: the structure forms a capacitor, and as a potential difference is put over the capacitor, the electrodes experience an attractive electrostatic Coulomb force, squeezing the elastic dielectric material, which is accompanied by a lateral expansion. The resulting stresses are often referred to as Maxwell stresses. Silicone elastomers and acrylic elastomers are being explored as dielectric elastomeric materials. By introducing polarizable moieties into the elastomeric materials, the effect may be enhanced. These materials are known as electrostrictive graft elastomers. They consist of two components, a flexible backbone polymer and grafted crystalline groups. The schematic in Fig. 4.15 shows the structure and molecular morphology, respectively, of the graft elastomer. The crystalline phase provides the polarizable moieties and also serves as a cross-linking site for the elastomer system (Su et al. 1991). In addition to the effect of the electrostatic Maxwell stress, the reorientation of the polar phases in response to an applied electric field then gives an additional electrostrictive straining effect.

The deformation of the polymer film may be used to produce various types of shape changes, as sketched in Fig. 4.16. For example, the film and electrodes can be formed into a tube, rolled into a scroll, stretched over a frame, or laminated to a flexible substrate to produce bending. As for ferroelectric polymers,

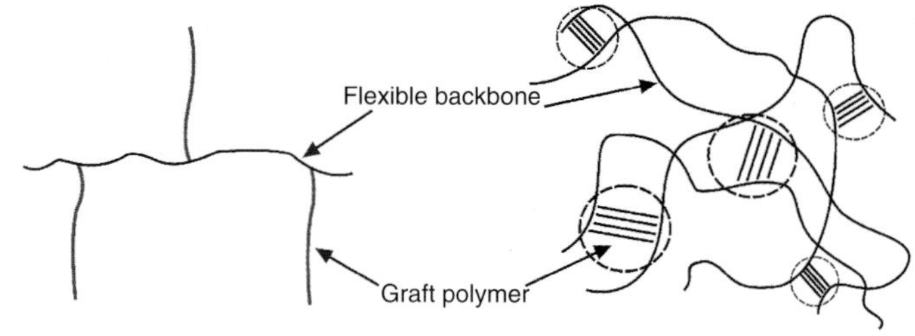

Fig. 4.15. Structure and morphology of a graft elastomer (Su et al. 1991)

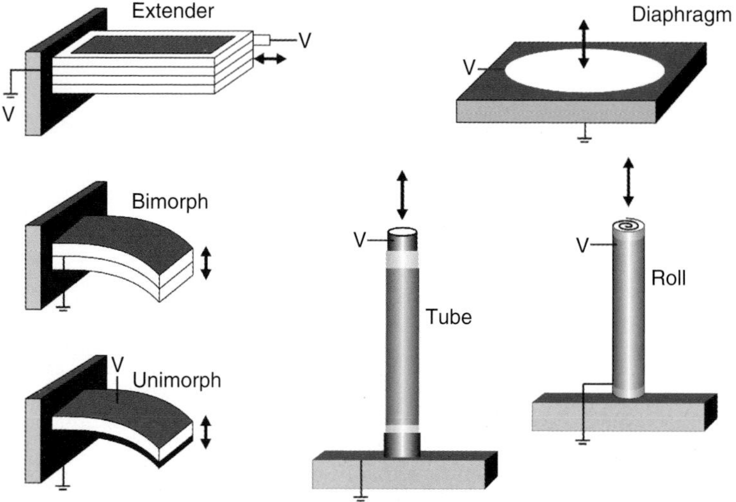

Fig. 4.16. Dielectric elastomers actuator configurations (Kornbluh et al. 2001)

dielectric elastomers as well as electrostrictive graft elastomers require large electric fields (about $100\,\mathrm{V\,\mu m^{-1}}$), which is a disadvantage, but this field can induce significant levels of strain (10–200%).

Recently, another class of electroactive polymers have been shown to work as actuators, namely ferroelectric liquid-crystalline elastomers (Lehmann et al. 2001). As shown in Fig. 4.17, this material consists of liquid-crystalline (LC) molecules (the green parts), that are attached, in a comb-like fashion, to a (elastomeric) siloxane backbone (the blue parts), with flexible alkyl spacers (the pink parts) as linking molecules. Hence, one has an LC-elastomeric (LCE) network in which the LC molecules are ordered as shown on the right in Fig. 4.17. The ordering structure results in a one-dimensional, long-range order that is preserved on a macroscopic length scale in the LCE. The specific LC molecules used are ferroelectric, i.e. they have a permanent polarization (Lehmann et al. 2001), so that they can be tilted by applying an electric field.

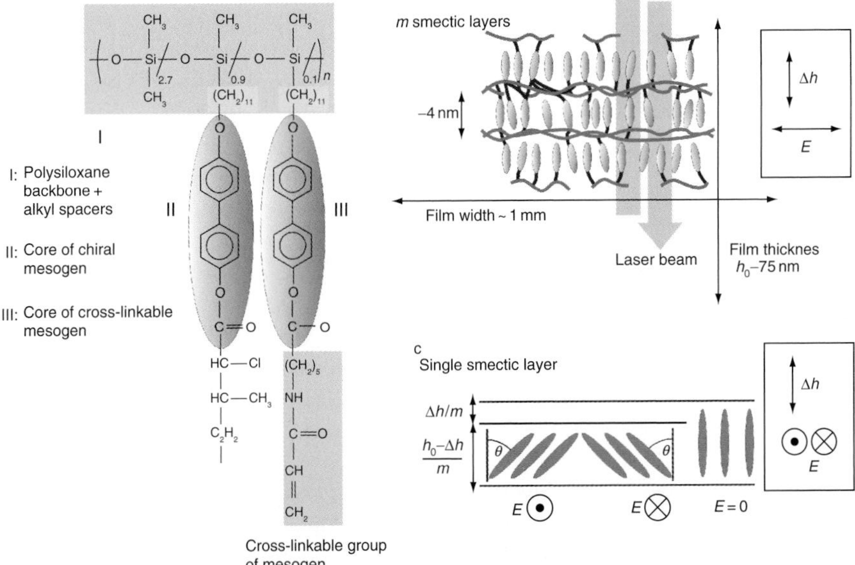

Fig. 4.17. Molecular structure and ordering of a ferroelectric LC elastomer (Lehmann et al. 2001; reproduced with permission of *Nature*)

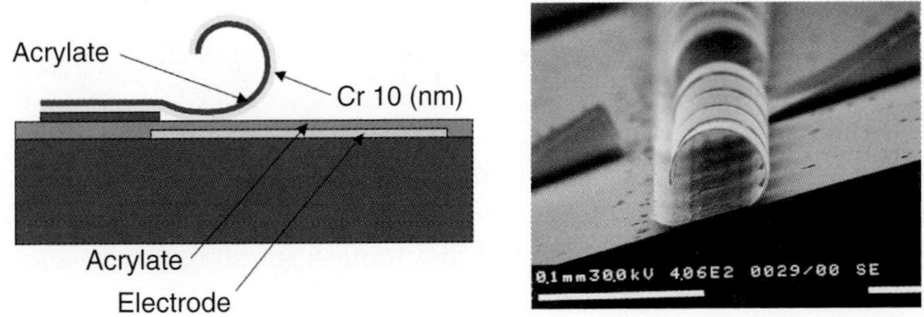

Fig. 4.18. Electrostatically actuated polymer composite structures: left: principle; right: actual structures

Since they are anchored in the ordered elastomeric network, this results in a macroscopic strain, as illustrated in Fig. 4.17. Lehman et al. have shown for a 100 nm film a thinning of 4 nm (i.e. 4% strain) at an applied electric field of 1.5 V μm^{-1}. This value of field is substantially lower than those necessary for other EAPs, as described earlier. A potential drawback of this material is that the response is sensitive to changes in temperature. Also, much work is still needed to scale up the effect to micron-scale or larger, instead of nanometer-scale.

One can also make polymer structures that are electrostatically actuated. Figure 4.18 shows a double layer composite structure consisting of a polymer film (in this case an acrylate) and a conductive film (in this case chromium) made in our laboratory. The processing is tuned such that the structure curls upward, being attached at one end. When a voltage difference is applied between the electrode underneath the actuator and the conductive film that is part of the actuating structure, an electrostatic force will pull the structure towards the substrate. Consequently, it will roll out and flatten out on the substrate. When the voltage is removed the slab will return to its original curled shape by elastic recovery.

These types of structures can be processed on large areas forming rows and columns of microactuators that can be addressed individually. They can be used for active surface texture modification, or optical property changing and/or patterning of large surface areas. The time constant of actuation is very fast, namely, substantially smaller than a millisecond. The voltages needed depend on the mechanical design of the actuators; these typically are tens of volts.

The other class of electroactive polymers is the ionic EAPs, and polymer gels form a subclass of that. A polymer gel consists of a cross-linked polymer network and a liquid solvent. One could loosely describe these systems as polymer networks filled with a liquid. The networks can change volume by expelling or taking up the solvent. Several stimuli are used for different material systems: temperature, pH and electric field. The change in volume can be dramatic: volume changes of over 350 times have been observed for temperature-induced transitions, whereas volume changes of certain gels of up to 100 times have been measured due to an applied electric field as low as $1\,\mathrm{V\,cm^{-1}}$ (Tanaka et al. 1982).

The basic structure of a typical electroactive polymer gel is sketched in Fig. 4.19. In the case of electrically stimulated polymer gels, the volume change is caused by the transport of mobile ions entering or leaving the gel structure from the outside liquid environment. This changes the so-called "osmotic pressure" inside the gel leading to shrinkage or swelling.

Figure 4.20 depicts a photograph of an actual EAP gel, in this case a cylindrical segment of a partially hydrolysed acrylamide with a length of 3 cm, that is placed between Pt electrodes and immersed in an acetone–water mixture. The volume of the gel segment decreases under an increasing electric field. The effects obtained with polymer gels are large, and the necessary electric fields are small. They have, however, at least two disadvantages. Obviously, they work only in liquid environments. And the effect is rather slow because the osmotic pressure driven diffusion takes its time.

Ionomeric polymer metal composites (IPMCs) form another example of a polymer smart structure based on ionic mobility. Such composites consist of a polymer film sandwiched between two electrodes. The general working principle of an IPMC actuator is shown in Fig. 4.21 (Shahinpoor and Kim 2001). The polymer material contains fixed (negative) anions, which are covalently bonded to the polymer molecules. In addition, the material contains free counter-cations, positively charged, such as sodium (Na^+) ions. Finally, it

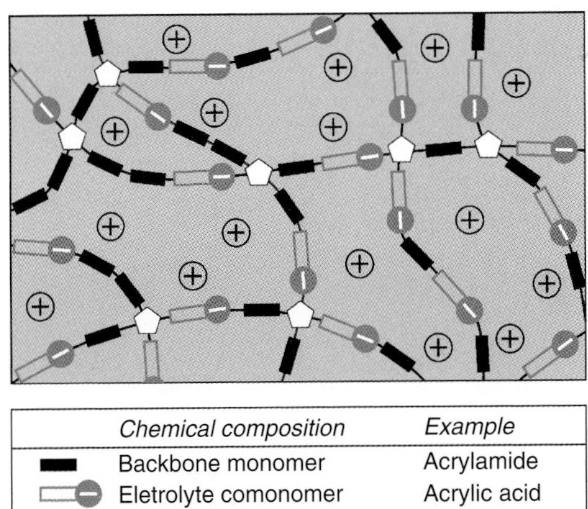

	Chemical composition	Example
▬	Backbone monomer	Acrylamide
▭⊖	Eletrolyte comonomer	Acrylic acid
⊕	Mobile counterion	H^+
⬠	Cross-linking comonomer	Bisacrylamide
▢	Solvent	H_2O

Fig. 4.19. Schematic drawing of a typical electroactive polymer gel

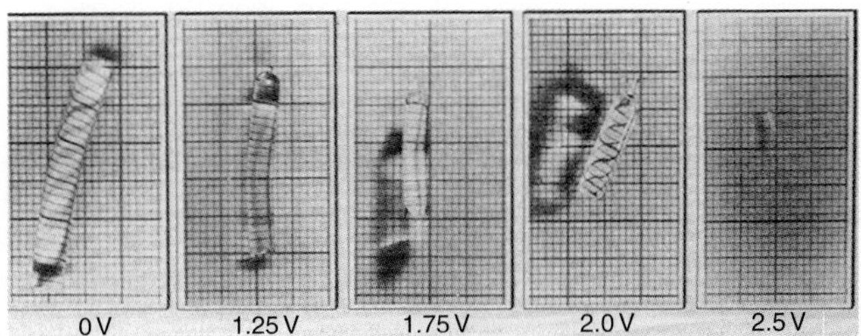

Fig. 4.20. A cylindrical polymer gel segment between two platinum electrodes, immersed in an acetone–water mixture. The gel contracts under the influence of an applied electric field (Tanaka et al. 1982; reproduced with permission of *Science*)

is essential that water-molecules are present in the material. The counter-cations are hydrated. When applying an electric potential to the electrons, two effects come into play. The cations are redistributed, which results in an internal stress acting on the backbone polymer. And the mobile cations move towards the negatively charged electrode, dragging along water molecules. The latter leads to the swelling of the film at the negative electrode, which results in bending.

Examples of polymer materials that are used for the IPMC actuators are perfluorsulphonate and perfluorcarbonate. Such materials are commercially available, for example from Dupont (Nafionâ) or from Dow Chemical. IPMC actuators give large effects at relatively low electric fields, as illustrated in Fig. 4.22 where a voltage potential of 2 V over a 200 μm thick structure gives

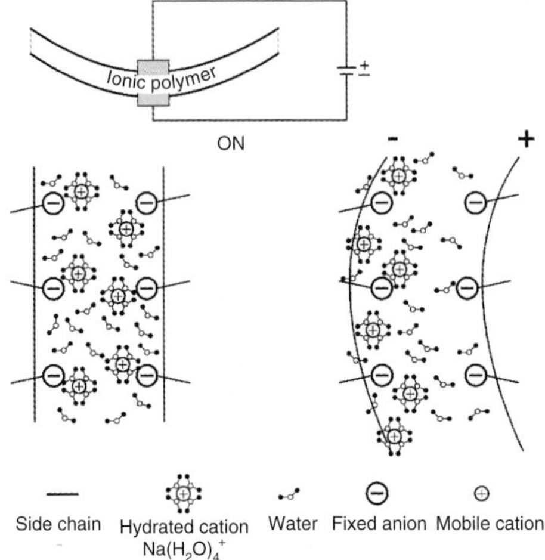

Fig. 4.21. Working principle and schematic structure of an IPMC actuator (Shahinpoor and Kim 2001)

Fig. 4.22. A 200 µm thick IPMC strip that shows significant bending due to a 2 V potential

significant bending. Obviously, it is a big disadvantage that the operation requires the presence of water, and IPMCs therefore do not work in dry environments.

Also conductive polymers such as polypyrrole, polyaniline and polythiophene, can be used to make electrically responsive actuators. The basic structure, again, is a sandwich structure, consisting of two films of the conductive polymer forming two electrodes, separated by an electrolyte, as sketched in Fig. 4.23. When a voltage is applied between the two polymer electrodes, by the nature of the material, oxidation occurs at the anode and reduction at the cathode. To balance the charge, ions (H^+) migrate between the electrolyte and the electrodes. Addition of the ions causes swelling of the polymer and their removal results in shrinkage. The result is that the structure bends.

Some interesting demonstrators have been made using actuating conductive polymer structures. For example, small boxes (300 µm × 300 µm size) that close when activated and paddles driven by a hinge actuator have been constructed (Fig. 4.24) (Smela et al. 1995).

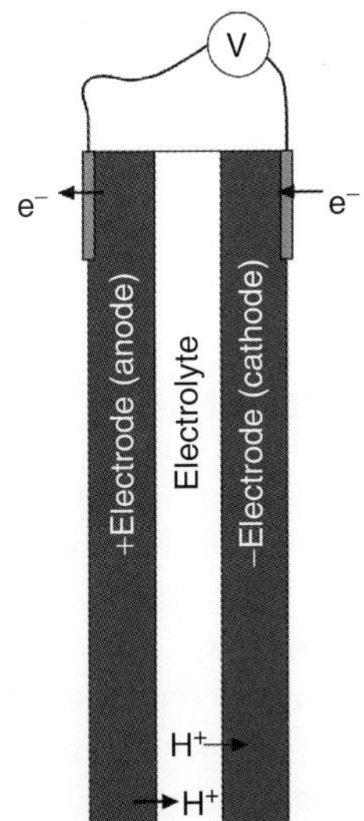

Fig. 4.23. The electromechanical mechanism of actuation in conductive polymers

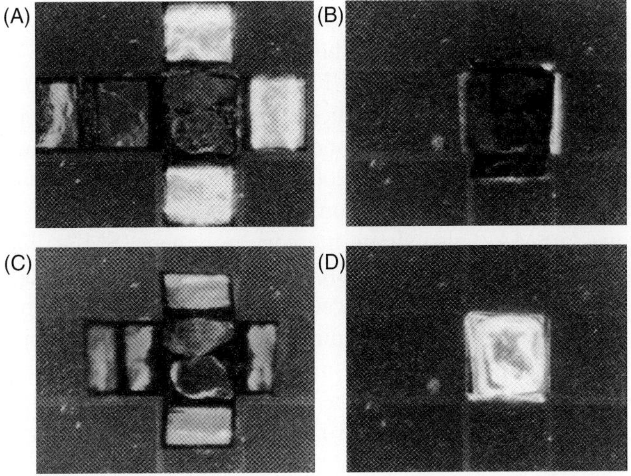

Fig. 4.24. "Closing box" based on conductive polymer materials. Size of the sides is 300 μm × 300 μm (Smela et al. 1995; reproduced with permission of *Science*

The force density that can be applied by conductive polymer actuators is relatively large, however the actuating strain is modest although this may be amplified by a proper geometrical design of the actuator. Conductive polymer actuators generally require voltages in the range of 1–5 V. A big disadvantage is the rather slow response; the limiting factor is the diffusion time of ions into the conductive polymer films and the ionic conductivity of the electrolyte. Furthermore, it is necessary to use a liquid electrolyte since the use of a solid polymer electrolyte would reduce the response time even further. Hence, conductive polymer actuators only work in a liquid medium, or they should be fully encapsulated but this solution causes the mechanical stiffness to increase impeding the actuator's movement.

4.7.2 Temperature-Driven Responsive Polymers

As already mentioned before, the volume of polymer gels may also be controlled by temperature changes. The mechanism behind this is that, if we have a polymer molecule in a solvent, the balance between, on one hand, attractive forces between the polymer segments – that tend to shrink a polymer coil – and, on the other hand, the attractive forces between the polymer segments and the solvent molecules, changes with temperature. When the polymer–solvent system is carefully chosen and controlled, this effect can be magnified and appears as a rather sudden phase transformation. If the polymers are "tied together" forming a polymer network, it is possible to produce a microactuator. Such thermally activated gels are mostly polyacrylamide derivatives and polypeptides. Figure 4.25 shows photographs of a specific polymer gel material that collapses as a function of temperature.

As can be seen from Fig. 4.25, the effect can be quite large (in this case the volume change is a factor of 30). The application of these types of gels has been considered for various fields, including drug delivery systems and agriculture, automatic valves that respond to cold and hot water, and actuators that simulate muscles (Bar-Cohen 2001). A main disadvantage of thermally activated polymer gels is that they only work in solution.

Analogous to SMAs, also shape memory polymers (SMPs) have been explored as smart polymer materials. In fact, the thermally responsive polymer gels just discussed can be viewed as being SMPs. Other material systems are polyurethane open cellular (foam) structures and poly(styrene-block-butadiene). In practice, these materials are deformed at temperatures above their softening (glass transition) temperature, T_g; cooled below T_g they maintain their deformed state when the force is removed. Heating above T_g

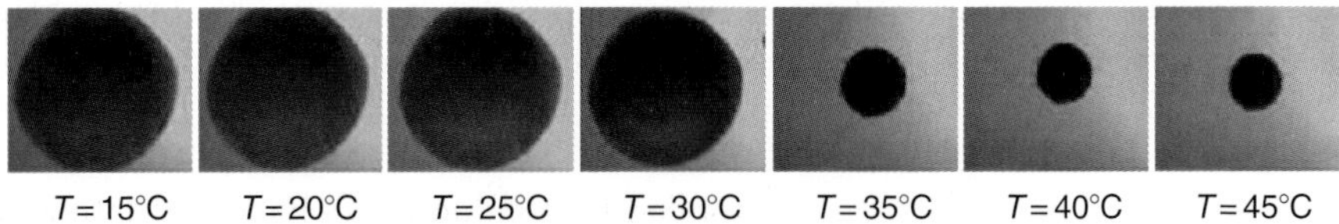

Fig. 4.25. Collapse transition of a bead of poly(N-isopropylacrylamide) hydrogel, abbreviated as NIPA gel. The diameter of the gel bead is 1.87 mm before the collapse transition (Zrínyi 1998; reproduced with permission of *IEEE*)

restores the original shape. A schematic of the mechanism is shown in Fig. 4.26. This figure also shows a photograph of an actual foam-like SMP, exhibiting quite a large shape and volume change.

SMPs exhibit a one-way memory effect. This means that these materials spontaneously change from their temporary shape to their memorized shape by heating, but upon cooling, the memorized shape is maintained. The temporary state can only be obtained by performing work onto the SMP, as is summarized in Fig. 4.26. For reversible actuation the SMP needs to be combined with some other actuator system that deforms the memorized shape. Therefore the one-way character of an SMP hampers to some extent its application as an actuator. Many SMPs are biodegradable, and they can for example be fashioned into threads, and applied in medical surgery as stitches that tighten and melt away as the body heals. SMP (polyurethane) is also applied in garments to obtain thermo-responsive apparel.

Thermally responsive polymer smart materials can also be made on the basis of liquid crystal molecules. The key is that the geometrical shape (often rod-like) of LCs causes them to pack in an ordered manner at low temperatures, for example in an oriented fashion (the so-called nematic phase) as shown in Fig. 4.27. The order is lost when the temperature is increased above a critical value, the nematic to isotropic temperature $T_{n,i}$. The material structure becomes isotropic above $T_{n,i}$ as sketched in Fig. 4.27.

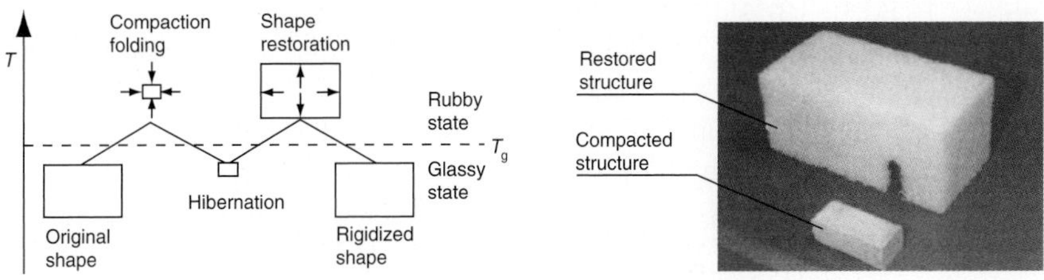

Fig. 4.26. Shape memory polymer process (left), and a polyurethane foam structure in its restored and deformed shape (Sokolowski et al. 1999)

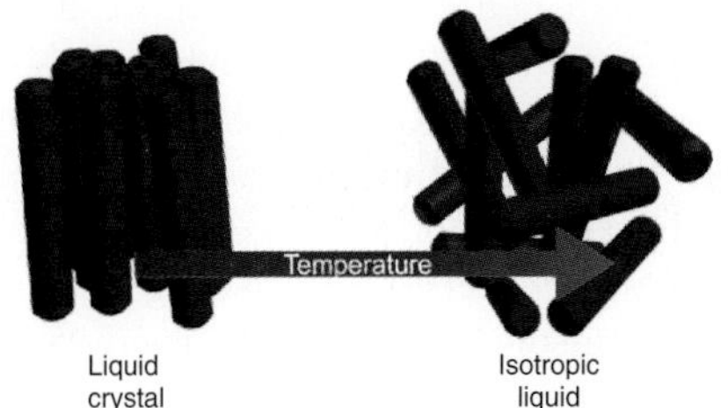

Fig. 4.27. Liquid crystal nematic (left) to isotropic (right) transition with increasing temperature

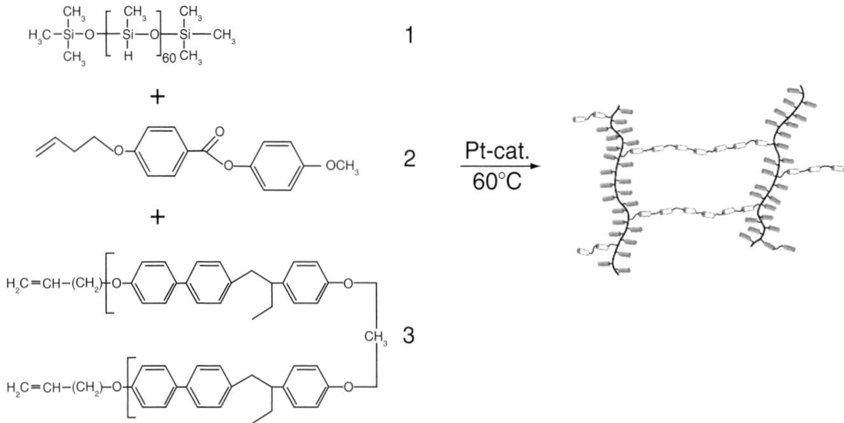

Fig. 4.28. Structure of a nematic liquid crystal elastomer (Wermter and Finkelmann 2001)

To get to a temperature-responsive actuating material, the LC is incorporated in an elastomeric network. The structure of the material corresponds to the ferroelectric LC elastomer shown in Fig. 4.17. In Fig. 4.28 the basic structure of a studied thermally responsive LCE is depicted. LCEs are composed of three main components: (1) an elastomer backbone (here silicone rubber, PDMS), (2) side-chain liquid crystal molecules and (3) side-chain cross-linkers which interconnect the elastomer backbone strands and may also exhibit liquid-crystalline properties.

An LCE has two characteristic temperatures, the glass transition temperature T_g above which the polymer network softens, and the nematic to isotropic temperature $T_{n,i}$. The LCE is designed such that $T_g < T_{n,i}$. Due to the physical connection of the LC molecules to the backbone polymers and the interconnection of the backbone polymers via the cross-link molecules, the nematic to isotropic phase transition will lead to a conformational change of the backbone molecules. As a response, the material will shorten along the direction of the nematic director when becoming isotropic. The obtained strains are rather spectacular, as it can amount to almost 400%. Because the LCE can be considered as incompressible the diameter of the sample will increase by $\sqrt{L_n/L_i}$ where L_n and L_i are the lengths of the nematic and isotropic samples, respectively. The larger the extent of cross-linking, the higher the strain but the slower the response of the LSCE as the mobility of the chains is suppressed by cross-linking. Stresses that can be generated by LCEs can be as high as 200–300 kPa (Thomson et al. 2001), outperforming skeletal muscles.

Liquid crystal polymer networks can be used in a similar manner to make bending structures. Increasing the molecular order of the LC molecules even further can enhance the effect. The way we have achieved this in our laboratory is shown schematically in Fig. 4.29. By a careful control of processing

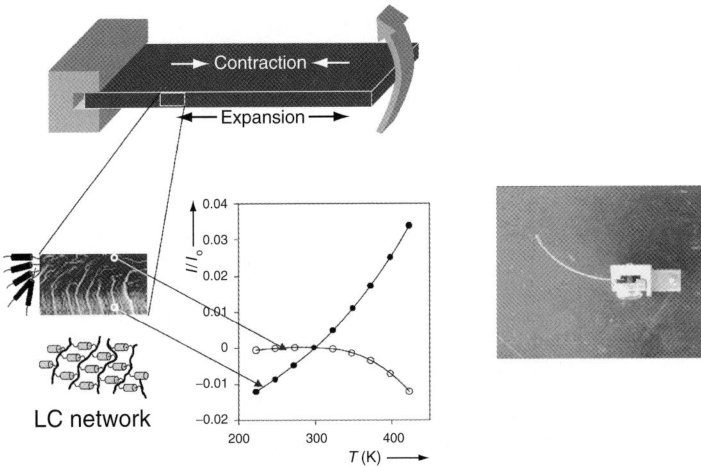

Fig. 4.29. A temperature-driven bending actuator formed by an organized LC network. Left: working principle; right: a bending actuator

conditions (Broer et al. 2005), it is possible to obtain a gradient in orientation of the LC molecules over the thickness of the bending actuator, for example, as in Fig. 4.29, the molecules are oriented parallel to the structure's surface at the top, and oriented perpendicularly to it at the bottom, and subsequently connected by a polymer network. When heating a free film with this molecular organization, the extent of the molecular order will decrease, as indicated in Fig. 4.29, and this results in a contraction parallel to the surface at the top of the structure, and, simultaneously, in an expansion at the bottom. The net result is that the polymer structure bends. The photographs in Fig. 4.30 illustrate that the bending can be quite large. An advantage of these types of structures is that they can be processed on large surface areas using established manufacturing processes and materials.

4.7.3 Light-Driven Polymer Smart Materials

There are some examples of polymer smart materials that can be actuated using light. One is a polymer–water gel based on N-isopropylacrylamide (NIPAM) that changes volume when illuminated with a focused laser beam (Becker and Glad 2000). The beam excites polymer molecules locally, causing functional groups on nearby molecules to temporarily attract each other. This causes the polymer gel to shrink at the point where the beam is focused. Relaxing the beam releases these attractive forces and the gel returns to its original size. The effect is quite large as shown in Fig. 4.31. A disadvantage, as for all polymer gel systems, is that this only works in a specific liquid environment.

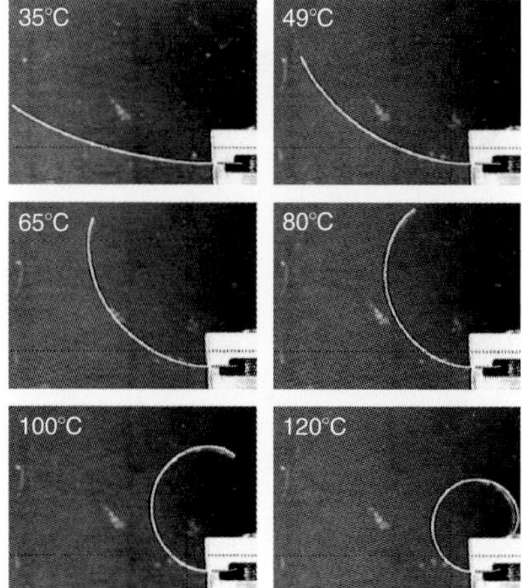

Fig. 4.30. Temperature-driven deformation of a monolithic film of a splayed liquid-crystalline network

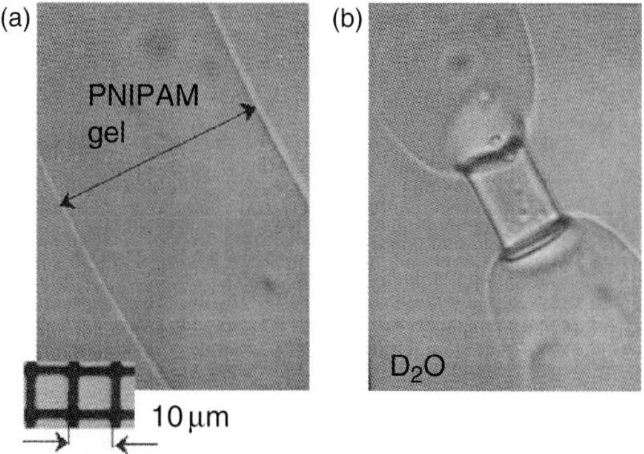

Fig. 4.31. A poly-NIPAM gel rod in D_2O before (**a**) and after (**b**) illumination by 0.75 W power laser at wavelength 1,064 nm (Becker and Glad 2000)

There are also liquid crystal elastomers (LCEs) that respond to light by deforming. When (part of) the LC molecules in LCEs are replaced by azo-containing molecules these can undergo a shape change under the influence of UV radiation. The radiation (wavelength 365 nm) provokes a so-called *trans*- to *cis*-photoisomerization of the azo-groups thereby inducing a large change of the shape of the molecules from rod-like to a kinked form as shown in Fig. 4.32, the latter lacking the ability to form an ordered liquid-crystalline phase (Hui et al. 2003) or affecting the nematic to isotropic transition temperature (see Fig. 4.27) and the order parameter of the nematic phase (Hogan et al. 2002). The contraction speed of a sample under UV radiation is typically on the

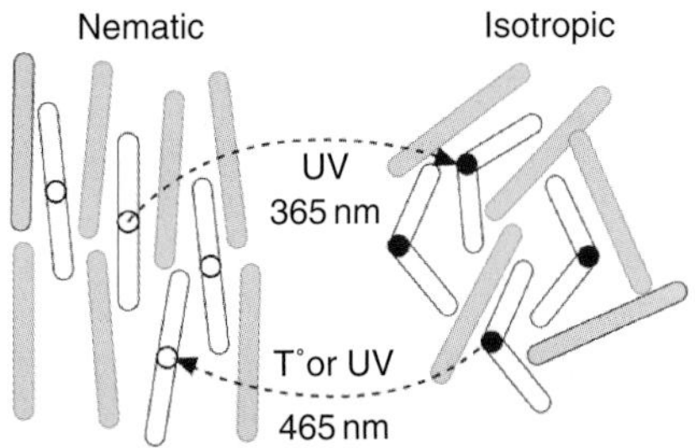

Fig. 4.32. Nematic–isotropic phase transition in a liquid crystal containing photo-isomerizable LC molecules, turning from rod-like *trans*- to a kinked *cis*-conformation upon irradiation of light with the appropriate wavelength

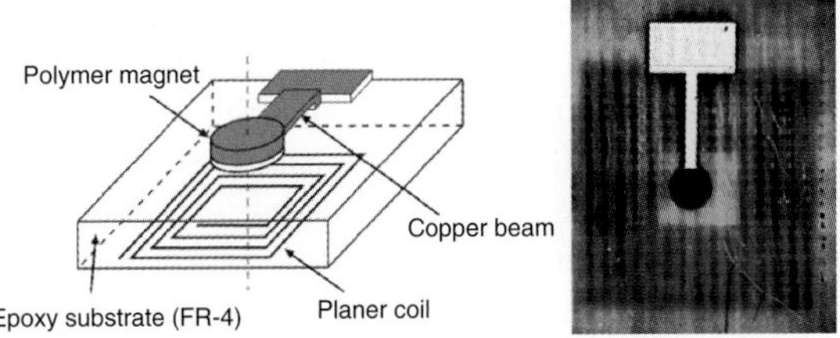

Fig. 4.33. Cantilever beam actuator based on a magnetic polymer composite. Working principle (left) and actual photograph (right) (Lagorce et al. 1999; reproduced with permission)

order of 1 min, decreasing when the intensity increases and also decreasing when the azo-content decreases. The process is reversed by irradiating the sample with light of another wavelength (450–700 nm). When the molecules are built into an elastomeric network to obtain an LCE, the molecular shape change can be transformed to the deformation on microscopic or larger scale.

As indicated already, a disadvantage may be the relatively slow response. The light-induced expansion is at least ten times slower compared to the contraction. Reported strains are up to 20%.

4.7.4 Other Stimuli

Another interesting way of actuating structures is by a magnetic field. Polymer magnetic materials have been made by incorporating magnetic beads into a polymer matrix. For example, hard magnetic ferrite particles were imbedded in an epoxy resin (to a volume loading of 80%) to obtain a hard magnetic polymer composite (Lagorce et al. 1999). On the basis of this material, a microactuator was made as shown in Fig. 4.33. This consists of a copper beam with the polymer magnet disc, which is actuated by running a current through a coil that is integrated in an epoxy substrate on which the microactuator is attached. The polymer discs were magnetized in the thickness direction.

They can be applied for example by (screen) printing. Some groups are working on synthesizing polymer materials that have intrinsic magnetic properties, but it turns out to be very difficult to reach significant effects.

In conclusion, one can say that much research is being done on developing mechanically responsive polymer materials. Many of these overcome disadvantages of conventional mechanically responsive materials, which are ceramics and metals. Polymers are, generally, tough instead of brittle, relatively cheap, elastic up to large strains (up to 10%) and offer the perspective of being processable on large surface areas with simple processes. There are infinite possibilities to vary their chemical/molecular structure to tailor their (mechanical, electrical) properties.

Polymer materials can be made responsive to various stimuli. Electrically responsive polymers (EAPs) can be subdivided in two subclasses: both have their pros and cons. The electronic EAPs generally require large electric fields, which is a big disadvantage, but show reasonably large effects. An exception is the ferroelectric liquid crystal elastomers that require fields that are 20–50 times smaller. Although the induced strain depends on the temperature, this class of material is a very interesting candidate material that is still in an explorative stage. The ionic EAPs show large deformations at low electric fields but the fact that they only work in liquid environments is a huge drawback for many applications. Another problem is their relatively slow response.

When temperature is as desired or suitable to stimulate the mechanical response, liquid crystal polymer networks are very interesting materials that can be processed on large areas. Liquid crystal elastomers can also be tailored to respond to light by changing shape, and this may be a very interesting feature for ambient intelligent applications: it would enable surfaces that change their texture or appearance depending on the ambient light conditions. The same holds for humidity (water) sensitive materials, responding to changes in the ambient humidity.

Most of the work on smart polymer materials is in the exploring phase, and application issues such as large-area processing are hardly touched, except for the LC-based materials we are developing at our laboratory and exemplified for instance in Figs. 4.18 and 4.29. These examples, however, do indicate the large potential for large-area processing of mechanically responsive polymer microstructures. Together with the intrinsic advantages of many polymeric smart materials mentioned earlier, this will make them very attractive for application in the area of Ambient Intelligence in the future.

5

Electronic Dust and e-Grains

H. Reichl and M. J. Wolf

"Small is beautiful"

5.1 Introduction

With microelectronics as the technology and Ambient Intelligence as the application driver, the changes to all areas of our professional and private lives will be increasingly far-reaching. Today, visions of the future are all founded on a common theme: seamless communication based on the Internet and mobile multifunctional terminal devices. As is apparent with mobile phones, the trend is moving towards smaller, more complex, autonomous systems. These systems are emerging as universal information and communication terminals, characterized by extreme miniaturization and the implementation of a wide range of services. This trend will inevitably be carried forward into the future. In the process, present-day system integration requires the production of ever-more complex systems, comprising an increasing number of active and passive components while reducing production costs.

In order to reduce the size of a system, it is necessary to develop new integration technologies, which also use the third dimension for system integration, and, secondly, to develop an alternative to replace the conventional method of system integration through rigid connections. This represents the framework for a new concept of system integration, based on so-called "electronic grains" or "e-Grains" or "Smart Dust." e-Grains are highly miniaturized sensor nodes. They are tiny, autonomous, functional units, and are distinctive, not only through their ability to communicate with each other, but also because they are freely programmable and to a certain degree modular. At the same time, these units are universal and partly specialized, for example, through the integration of selective sensors.

In terms of hardware, technology, and software design, system integration based on the innovative e-Grain concept represents an entirely new challenge and requires a synergy of individual technologies at an exceptionally early stage. And naturally, through the differing nature of the approaches used to find a system solution, it also holds a particular techno-scientific appeal, while simultaneously yielding extremely high innovative potential.

This technology poses particular challenges with regard to the desirable sizes (a few cubic millimeters), the need to achieve continuous operation through an integrated or external wireless power supply, and the necessity of allowing multiple e-Grains to communicate. The system is characterized by a large number of individual interconnected e-Grains. The e-Grain vision therefore represents a new approach to system integration that will help to develop complex, flexible, and cost-efficient integrated systems with black-start capability, based on ultra-small sub-components. Due to the high density required for integration, the realization of e-Grains necessarily involves stacking thin functional films, which are then bonded together. To do this, a new 3D bonding technique based on polymer films is under consideration. The development of highly integrated 3D wiring calls for a vertical-integration technology that allows interconnections to be fabricated and precisely aligned on ultra-thin substrates below 50 µm. The production of ultra-thin functional films as well as their integration to form a complete wafer-level (WL) system is therefore a key technology. In addition to reducing the size, the miniaturization of passive components is also fundamentally important for the complete system.

The e-Grain system requires that the tiny electronic units be capable of autonomously meeting their own energy requirements over a certain period of time or during their entire operating life. This necessitates energy storage devices with high power density compatible with the system in terms of dimensions and production technology.

Another important issue of the e-Grain system is wireless linking of individual e-Grain cells. Microwave front end technology will therefore assume a key position as a connecting link between data-processing electronics and the transmission channel. In view of the low range of 1–10 m and the already oversubscribed utilization of frequencies in the range below 10 GHz, which restricts the flexibility required by the e-Grain system, transmission will lie in the millimeter wavelength range (10–60 GHz).

Communication between a large number of tiny, highly integrated e-Grains presents a particular challenge in the design and implementation of this communication system. One of the important parameters for communication between a number of e-Grains is the available energy capacity which is greatly limited by the small volume.

Special attention is also given to the way in which two individual e-Grains communicate, address assignment within a self-organized network, a flexible semantics-based grouping of messages within an e-Grain network as well as interfaces to applications and management. The creation of a self-organized ad-hoc network that can comprise a large arbitrary number of e-Grains requires a suitable operating-system environment that provides the necessary administration service. In its entirety, this form of ad-hoc network can be seen as the configurable system, whereby the networked components representing the configurable resources. The actual operating system provides the necessary administration services to monitor and control this type of system. This also includes the programming of individual e-Grains.

In addition, the operating system offers a μ-network interface through which the applications themselves can access the network components. For linking the e-Grain μ-network to other networks (LAN, Internet), the operating system provides gateways with the corresponding address and protocol conversion functions.

5.2 Basic Construction of Self-Sufficient Wireless Sensor Nodes – e-Grains

Figure 5.1 shows the principle architecture of a wireless sensor node such as an e-Grain. It is divided into an analog part (sensors, amplifiers, and AD-converter) in digital signal processing, a wireless interface, and a power supply. The functions of the different elements are described in Fig. 5.1.

- The *sensors* are the basic components of the e-Grains. Depending on the sensor requirements the preferred technology for realization is MEMS, with a mandatory prerequisite of ultra-low power.
- The *low power analog interface* is in charge of interfacing the analog physical world seen by the sensors to the digital world of the microcontroller. The main constraint on this function is the minimization of the internal components to reduce the power consumption, while being compatible with noise levels adapted to the resolution of the sensors, the necessary bandwidth, and the dynamic range.
- The *microcontroller* is in charge of interfacing with almost all the functions of the e-Grains, particularly for sequencing all the operations, while controlling the power dissipation to reduce it to the lowest possible level. In some applications its functionality is extended to data compression

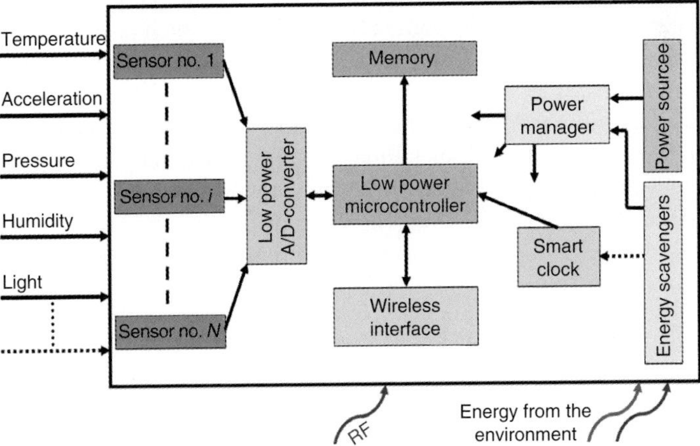

Fig. 5.1. Principle architecture of self-sufficient wireless sensor nodes, e.g., e-Grains

operations, self-organizing procedure with other e-Grains for data transmissions and relay, calibration operations, or other specific operations (such as localizing other e-Grains).

- The *wireless interface* is in charge of allowing the best possible link budget for communication, using different techniques, from specific antenna schemes to dynamic impedance matching. This interface is bi-directional: uploading programs and getting data back from the sensors. The main challenge here is the optimal compromise among the highest transmission efficiency, the lowest power dissipation, and the reduced dimensions acceptable in a given domain.
- The *events storage memory* is a generic memory with specific properties of very low power consumption for a given memory space, data/state remanence, possibly dynamic programmability, and dynamic sharing between data and programs. In the simplest case, it can be a standard flash memory.
- The *smart clock* is a specific function useful for different cases. It can be used as an "awakener" of the microcontroller in the case of ultra-low power specific applications such as transient event monitoring where, most of the time, there is nothing (no event) to be detected, but from time to time specific short time events occur, for which a high resolution is necessary to track the data to be monitored. Alternatively, it can be used as a self rate-switchable clock (for power saving objectives) or a smart clock for localization purposes.
- The *micropower source* is the basic local energy source of any e-Grain. In the simplest case, it can be a micro-battery or a supercapacitor. In more elaborate versions, it can also be a micro-fuel cell. Its main characteristic is that its lifetime is necessarily limited in time, contrary to the energy scavenger described below:
- The *energy scavenger* is a function designed to get energy from the environment: either from the real natural physical flows of energy (optical flux, EM, thermal flux, etc.) or from artificial sources, such as in the situation of remote powering, where a source of power is directed towards the e-Grains from an external location.
- The *power management* can be realized by a smart circuit able to handle the power management issues linked to the multiplicity of energy sources. This will be achieved by analyzing the available energy densities at different points of time, deciding on the procedure for maximizing the collected energy, switching in time between the different energy sources, while storing this energy, e.g., in micro-batteries.

The future e-Grain is a 3D arrangement of functional sub-modules (layers), each of which is a heterogeneous functional layer with embedded components as illustrated in Fig. 5.2.

The minimal physical size of an e-Grain is determined by three main factors: the size of the constituent integrated circuits, the power storage and

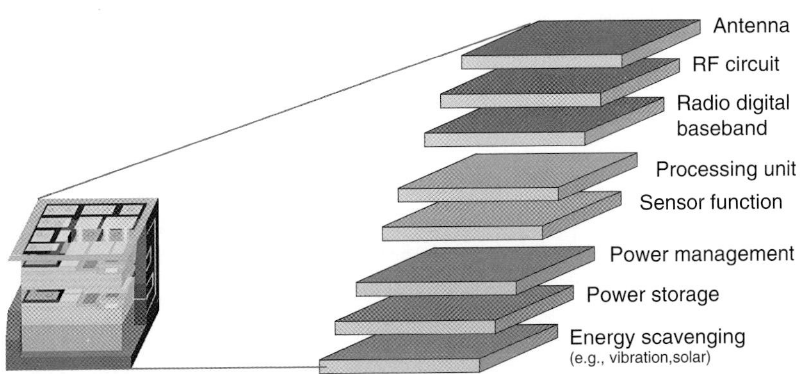

Fig. 5.2. Schematic of an e-Grain structure based on heterogeneous functional layers

harvesting subsystem, and the size of the antenna required for power-efficient communication. The scaling roadmap (ITRS) of integrated circuit technologies enables the realization of complex circuits in rather small sizes. The main size limitations for e-Grains are therefore the power system and the antenna size (e-Grain, online).

5.3 System Design

After consideration of the functional units, it is required to specify the architecture by designing the circuitry and looking for bottlenecks of the complete system. Behavior models of the whole system regarding costs, energy, and size help to choose materials, components, and technological processes (Niedermayer et al. 2004). The selection of materials and technologies is also influenced by environmental conditions. Ambient parameters such as force magnitudes, temperature maxima, and humidity result in a very different focus of the system design. Additional design restrictions follow from the process flow. Thus, the position of several components is not arbitrarily due to effects such as field coupling and shielding, if the functional layers are folded or stacked afterwards. According to the design rules, the layout is generated by defining the component positions and routing their interconnections; see Fig. 5.3.

The interdependencies between different functional units have to be considered in the case of higher component densities. For instance, the transmission frequency has to be defined for designing the communication hardware. A sole consideration of the communication circuitry would only include parameters such as antenna efficiency, antenna directivity, RF-losses, polarization, and mismatch. Resulting in a changed structure of transmission data or a modification of the data rate, other modules influence the hardware optimization of the communication unit. For instance, technological improvements of an applied sensor could allow an accelerated data acquisition. In this case a faster

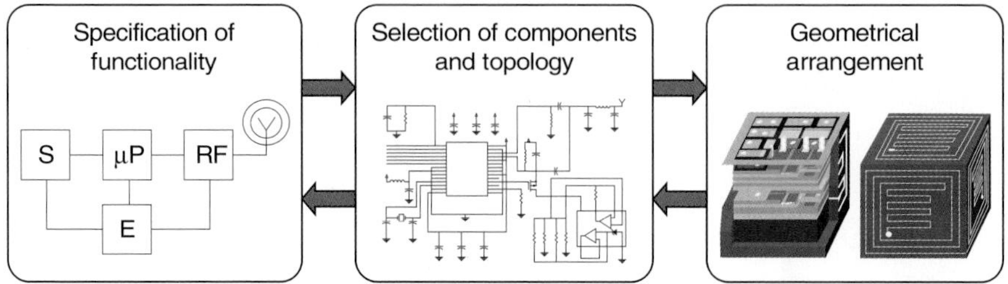

Fig. 5.3. Abstraction levels of system design

architecture with shorter duty cycles would be recommended. Thus, the optimal transmission frequency regarding power consumption will move to higher values. The suitable solution for the whole system, however, also has to reflect environmental conditions such as absorptions, reflections, noise, and RF coverage.

Depending on the reliability requirements, monitoring of some internal and external conditions guarantees safer functioning in security sensitive cases. Therefore, appropriate algorithms calculate the probability of a malfunction. As a result, an observation and evaluation of operating voltage gradients, temperature variations, and vibrations can significantly improve failure predictions. A very compact layout with dense interconnections results from the system design. The debugging of highly integrated prototypes can become very challenging and requires system verification at different levels of the design process.

An integrated treatment of physical couplings is especially important for small systems with tightly positioned components. Apart from the realization of prototypes and their characterization by measurements, the whole system or adequate subsystems have to be verified in a computerized model (Mentor Graphics 2001). Distributed microsystems are often heterogeneous systems and possess both spatially distributed and concentrated elements. Properties resulting from different physical domains have to be simulated to describe the behavior of the whole system. As a consequence, the modeling and simulation of these devices is more complex. Due to the limited computation time for simulations, details which are of less significance for the system behavior often have to be neglected during model generation.

Besides the verification of the analog, digital, and mixed-signal circuitry, the parasitics have to be considered for the components, package elements, and interconnections. This includes a quantification of reflections, cross talk, and radiation. Furthermore parasitic charging of insulation layers can substantially influence the system behavior.

Small distances between several components are often critical regarding heat spreading, conductance, convection, and radiation. Thus, the heating of a microprocessor can cause a thermal drift in an adjacent sensor. These thermal effects can be reduced by solutions such as thermal vias, an efficient energy management, or a sensor data processing with drift awareness. To achieve the latter, macromodels of the sensors are helpful, which describe the coupled thermo-mechanical and electrical behavior. Often, for the electrical behavior white (or glass) box models are used, which are transparent and represent the "physics," while for the thermal behavior black box models are preferred, which describe the I/O behavior "merely" mathematically.

Diverse computations from other domains such as acoustics or fluidics may also be required depending on the application. Mechanical simulations addressing shock and stress are often relevant for reliability estimates of packaging elements like solder bumps. Some verification examinations can result in coupled simulations of different domains. This is required, for instance, for a characterization of forces generated by electrostatic actuators.

Highly integrated systems certainly need to be tested by measurement but this meets with two main problems. The first arises from the high degree of complexity, the second one stems from the extreme miniaturization, which makes interfacing to standard measurement equipment difficult. For example, even the capacitance of an RF probe may cause erroneous results.

It would thus be desirable to perform the verification of the complete system by virtual prototyping, in which the entire design is represented in computer memory and verified in the virtual world before being implemented in hardware. Software testing is not, however, immune to problems arising from complexity either, as has already been pointed out.

The more precisely the effects can be characterized, the more correctly the counter measures can be dimensioned without high design overheads. Therefore, efficient system verification allows increased component densities and smaller energy supplies due to optimized power consumption. Those energy savings can be achieved by means such as reducing timing slack or lowering voltage levels. Additionally, the increased effort for system verification can result in cheaper products by permitting higher component tolerances.

The errors, detected during the verification process, can necessitate different efforts. At best, only small changes have to be made in the layout. More serious errors require a replacement or addition of components. In the worst case, the design must be restarted at the concept level, demanding a completely different architecture. Therefore, prototyping can require several iterations to discover and fix problems. Furthermore, a problem solution can introduce new errors. If the verified prototypes meet the application needs, it is often necessary to additionally adapt to the requirements of high volume manufacturing.

5.4 System Integration Technologies

Several companies and research institutes all over the world are currently working on the development of 3D system integration. Besides the approaches which are based on silicon technologies (SoC) like fabrication of multiple device layers using recrystallization or epitaxial growth, the large spectrum of technological concepts can be classified in the following categories: stacking of organic substrates, integration of flexible functional layers, embedded components in organic substrates, and WL integration by stacking of chips or wafers.

5.4.1 Stacking of Organic Substrates

Based on organic substrates (e.g., FR4) and existing assembly and interconnection technologies (e.g., chip on board), COB Fraunhofer IZM has developed a 3D stacked configuration (Grosser 2004; Wolf et al. 2005). Single components (SMDs, quartz), integrated circuits as bare die, and micropackages available on the market were used for this approach.

The basic construction for the sensor module is shown in Fig. 5.4. All required functions and components are integrated in the package within a size of 1 cm^3.

The advantage of this modular concept is that each sub-module can be tested separately. By exchanging sub-modules different applications can be reached, new components can be implemented, and the production yield can be increased as well. The configuration of the individual sub-modules is shown in Fig. 5.5.

Figure 5.6 illustrates a 3D visualization of the printed circuit board (PCB) sensor node (1 cm^3). An FR4 frame acts as a vertical interconnection between the individual sub-modules, e.g., sensors and microprocessors.

Figure 5.7 shows the single sub-modules (microcontroller, RF unit, sensor unit, clock). Solder joints interconnect the sub-modules via a frame. A cross section of the sensor node and the vertical solder interconnects is shown in Fig. 5.8.

Figure 5.9 shows a sensor node (PCB technology), which is able to measure the temperature of the environment. A number of these systems can be

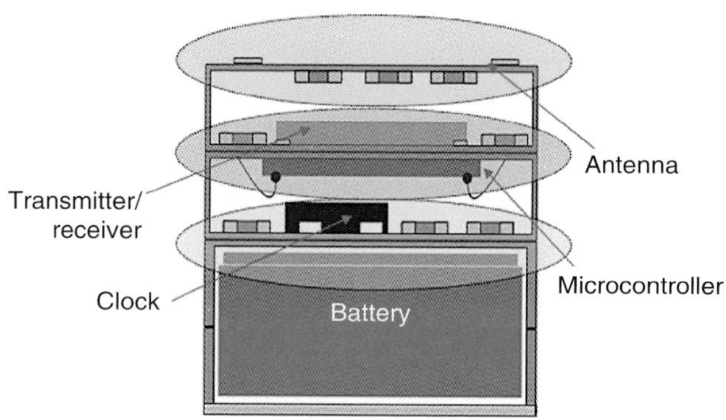

Fig. 5.4. Layered structure of a wireless sensor node in stacked PCB technology

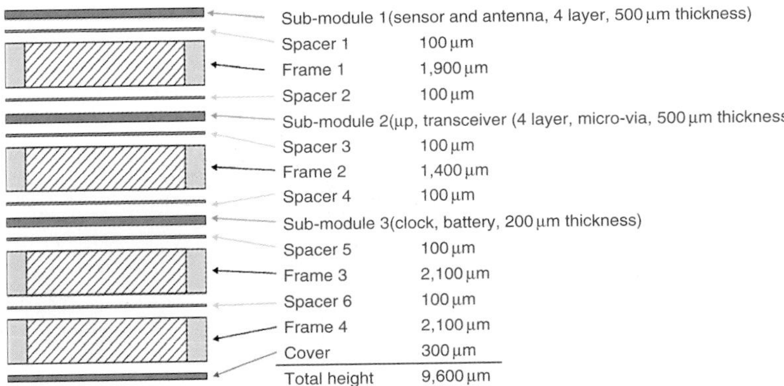

Fig. 5.5. Sub-module specification and arrangement for the PCB-sensor node (Kallmayer et al. 2005)

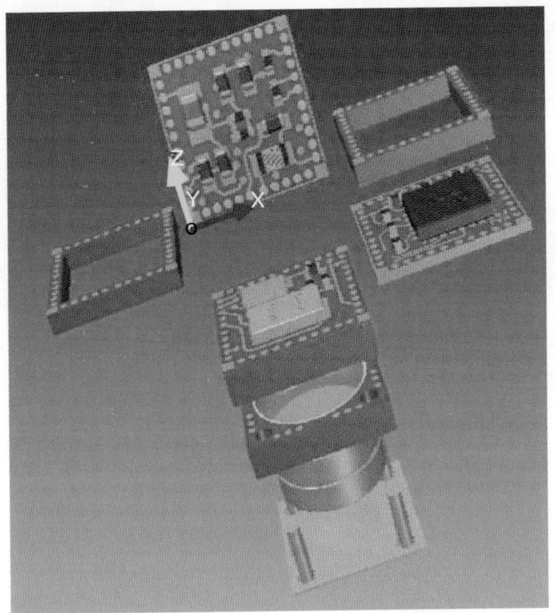

Fig. 5.6. 3D visualization of the functional sub-modules of the sensor module

combined and wireless linked for specific applications (e.g., logistic,environment, process control, etc.). The technical parameters of the PCB-sensor node are:

- Operating range: > 1 m
- Operating frequency: 2.4 GHz
- Repetition rate: 1/s
- Power supply: two batteries (2 × 1.5 V)
- Operating time: > 500 h

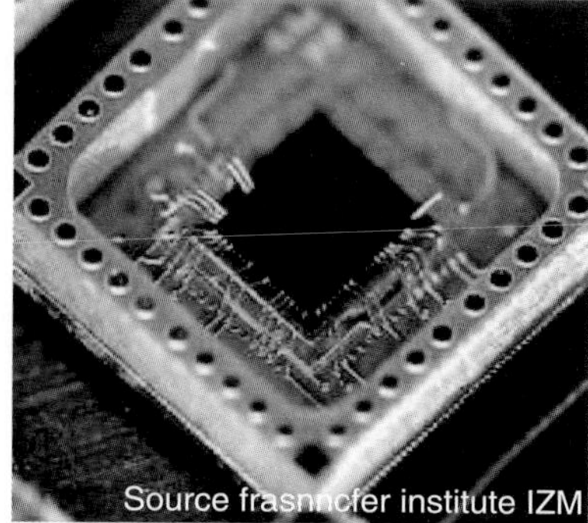

Fig. 5.7. Assembled functional sub-modules and details of microprocessor unit

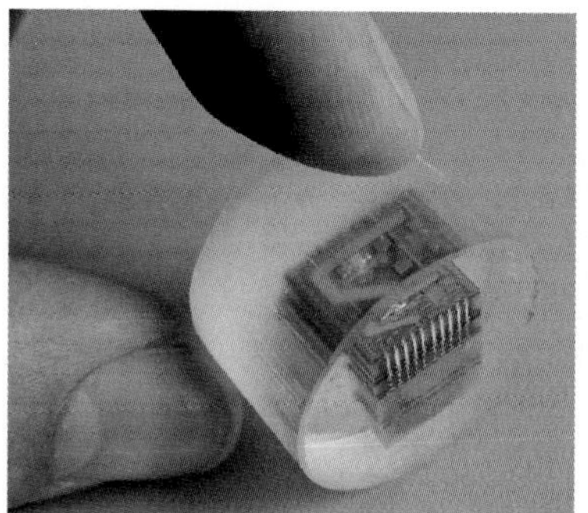

Fig. 5.8. Solder interconnection between the individual functional unit and the frame for 3D assembly

Fig. 5.9. Wireless temperature sensor node in stacked printed circuit board technology (size: $1\,\mathrm{cm}^3$)

5.4.2 Integration of Flexible Functional Layers

The integration of chip components on flexible substrates offers the possibility to create high dense sensor nodes by folding of the flexible substrate. Figure 5.10 shows a schematic of a folded sensor node. An essential advantage of this technology is that the sensor node can be tested before folding. To reduce the system height and the footprint of the node flip-chip interconnects are preferred.

As long as ICs with standard thicknesses between 250 and 600 µm are used, conventional flip-chip technologies with contact heights of approximately 50–100 µm may be applied, forming reliable interconnects on various organic or inorganic, rigid, or flexible substrates.

A reduction of the contact height towards ultra-thin flip-chip interconnects gains in importance in connection with thin chips and substrates. Figure 5.11 shows the disproportion between the thicknesses of the structural elements of an assembly consisting of a thin IC (50 µm), bonded on a thin substrate (50 µm) with conventional flip-chip technology. In consequence, the reduction of the contact heights as well as the thicknesses of chip and substrate to 10 µm holds an impressive potential for a further reduction of the total module thicknesses to approximately 100 or 30 µm, respectively. This makes ultra-thin flip-chip interconnects a substantial contribution to meeting the requirements of future highly integrated and miniaturized modules that may consist of several, stacked or folded, ultra-thin active and passive layers.

By using solder technology the thicknesses of the UBM, the solder bumps and the substrate metallization can be reduced. The standard electroless nickel UBM for high reliability has a thickness between 5 and 10 µm but only a minimum thickness of 1 µm is necessary to have a closed and void free nickel layer (Nieland et al. 2000). Technologically, it is possible to reduce the Ni layer to almost the thickness of the last chip passivation layer.

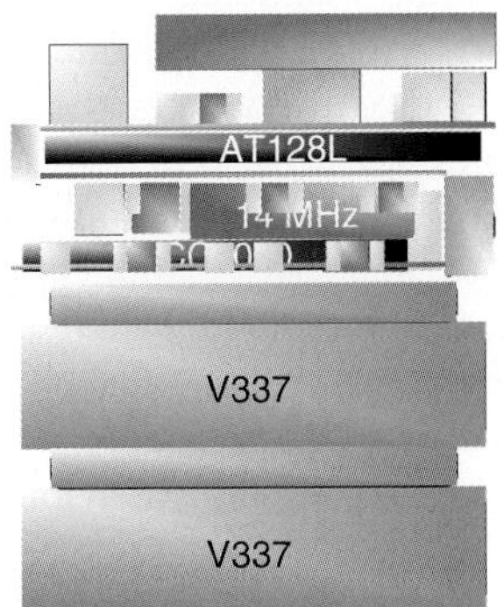

Fig. 5.10. Schematic of a sensor node with a folded flexible substrate, antenna, and battery (Kallmayer et al. 2005)

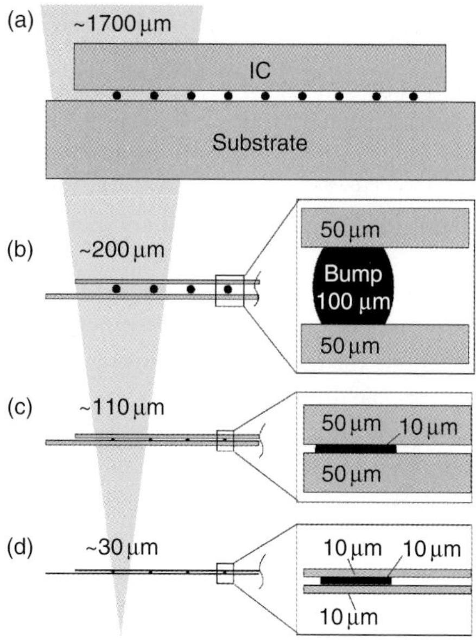

Fig. 5.11. Miniaturization potential of ultra-thin flip-chip interconnects (Kallmayer et al. 2005)

For the application of very small solder volumes and bumping of ICs with pitches $< 150\,\mu m$ conventional printing methods are not suited. Electroplating is one choice, but the just lately developed "immersion solder bumping" (ISB) is a step ahead under low-cost aspects. This maskless process uses the wettability of the UBM by a liquid solder alloy through which the entire wafer is moved. The solder cap height is typically in the range of a few micrometers; see Fig. 5.12.

Regarding adhesive bonding technologies, NCA and ACA will be within closer consideration, as they are the most promising technologies. UBM and substrate metallization can be reduced even more than in the case of soldering as adhesive bonding does not lead to dissolution of the metal layers. For both, NCA and ACA, leaving out bumps completely is an option for further reduction of contact height. Some ACA applications today are making use of nickel bumps. In this case the contact height is determined by the nickel bump, the substrate metallization, and the size of the trapped particles, commonly ranging from 3 to 15 μm. With these possible ACA and NCA contact configurations, investigation of the ageing behavior for both technologies gains much importance.

An impressive step for further miniaturization of any 3D-chip- or multichip-assembly can be taken by the use of thin chips. Today, wafers can be thinned from 600 μm standard thickness to approximately 10 μm. In these thickness ranges silicon shows great mechanical flexibility comparable with that of polymeric foils.

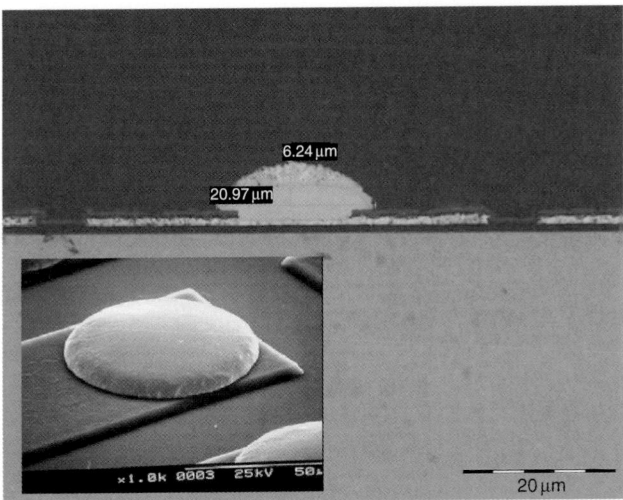

Fig. 5.12. SEM picture of a cross section of an immersion solder bump

Figure 5.13 shows SEM pictures of cross sections of thin soldered and ultra-thin ACA bonded flip-chip test assemblies (Pahl et al. 2002). The realization of extremely space saving system–level integration requires technologies allowing 3D packaging while taking into account the space required for passive components. The minimization of the area needed for passive components such as coils, resistors, capacitors, and filters is an important factor for volume reduction of such wireless sensor systems as well.

The application of thin flexible circuit interposers stacked by solder bump connections is a possibility to combine high density routing with passive device integration and to realize a 3D approach by bending, folding, or stacking (Wolf et al. 2005).

The thin flexible circuit interposer units are based on a temporary release layer, a polyimide dielectric with a built-up high density thin film copper wiring as well as passive element integration at wafer level; see Fig. 5.14. After removing the units from the carrier wafer by applying a temporary release layer a stacking process can be realized by using standard assembly equipments like a flip-chip bonder; see Fig. 5.15.

Besides conductor lines and passive elements such as resistors, capacitors, coils and filters, active chips thinned down to 20 µm can also be integrated into the polymer layers.

The advantages of WL processing of polyimide-based flexible interposers can be formulated as follows:

- Realization of ultra-thin ($\approx$ 50 µm) high density stackable flexible substrates with via chains.
- Application of the conventional thin film technology, e.g., spin-on of photo-definable dielectrics (polymer) and photoresist, sputtering of metal seed

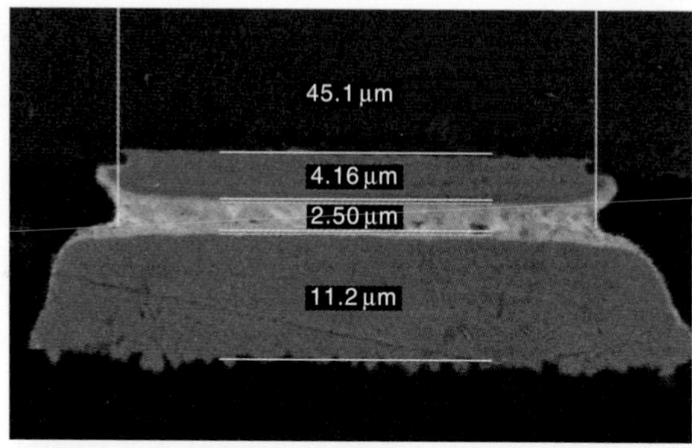

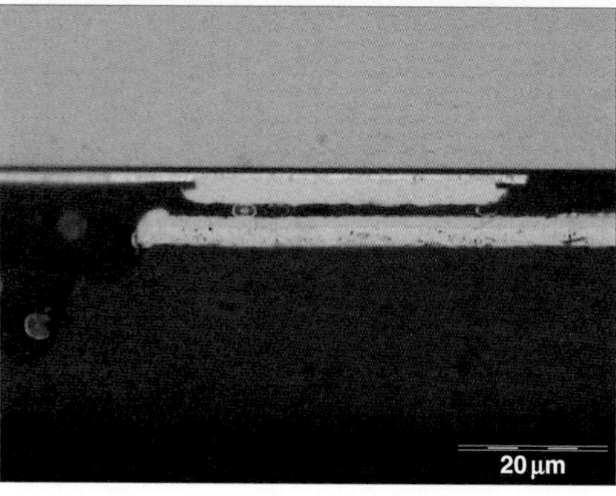

Fig. 5.13. SEM pictures of thin soldered contact (upper) and ultra-thin ACA contact, $\sim 3\,\mu m$ Ni, $\sim 3\,\mu m$ deformed particle, $\sim 3\,\mu m$ Cu/Au (lower)

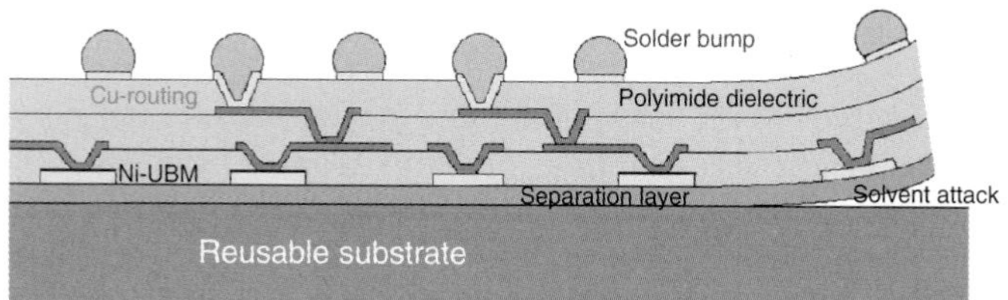

Fig. 5.14. Principle of thin film flexible interposer

layers and resistor layers, electroplating of routing metal (e.g., Cu, Ni, Au) and solder alloy (e.g., AuSn, SnAg, CuSn).

- Small dimensions of conductor lines and vias (high density).
- Arbitrary shape, foldable, stackable for 3D integration.
- Integration of passive elements (R, C, L) and active devices (ICs).
- Metal routing completely encapsulated.
- Excellent chemical and high temperature stability.

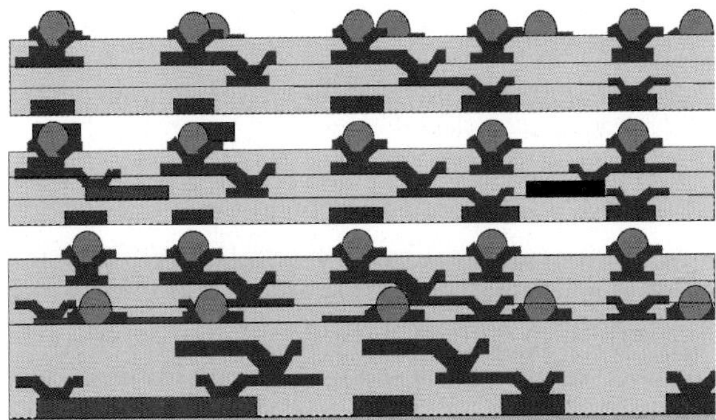

Fig. 5.15. Stacking of flexible interposer with embedded devices

The realization of integrated passive elements into the flexible interposer can be achieved by adapting the thin film process.

Resistors

Most resistor requirements can be met with thin films having a sheet resistance R_s of 100 ohms sq^{-1}, which can be achieved by an approximately 20 nm thin NiCr layer. Low resistor values can be realized by straight lines, while higher values have to be laid out as meanders.

Inductors

The only way to realize inductors in a thin film build-up of only two routing layers is the planar coil approach. A good magnetic field coupling is reached by a conductor, which is arranged into a spiral form in one metal layer. A second metal layer is necessary to route the inner coil contact to the outside of the coil. This layer can also be used for additional windings so that a larger inductance value can be realized within the same area. The characteristic values, e.g., inductance, quality factor, resonance frequency, are influenced by the material and design parameters like number of turns, diameter of coil window, line pitch/space, etc.

Capacitors

In a thin film build-up with two routing layers integrated capacitors can be realized by simple parallel plate structures. For the parallel plate type the interlayer polymer serves as the dielectric material. The characteristic values like capacitance, quality factor as well as their resonant frequency depend on the electrode area A, the interlayer polymer thickness as well as the electrode form and the dielectric constant ε_r which is around 3.3 for polyamide.

Filters

Based on the single components (R, L, C) more complex passive structures like matching or resistor networks, oscillators or filters can be implemented (Zoschke et al. 2005). The necessary interconnections between the elements can be realized as high density conductor lines as well as transmission lines for high frequency application. As one example Fig. 5.16 shows an LC low pass filter using Cu and polyimide ($\varepsilon_r = 3.3$). The filter was designed in Tschebyscheff style of third order to have cut off frequencies of 2.45 and 5 GHz. Microstrip transmission lines were used for elements interconnection. The coils have an inductance of 5.8 nH, the capacitor has a capacitance of 1 pF.

Figure 5.17 shows the measured attenuation behavior of this filter structure up to a frequency of 8 GHz. At 2.45 GHz a clear bend in insertion loss can be observed. The structure has a total size of 3.3 mm × 1.95 mm (Zoschke et al. 2004).

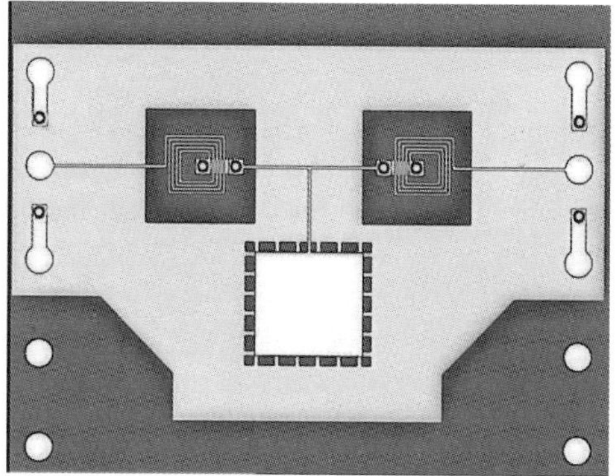

Fig. 5.16. Attenuation of thin film low pass filter

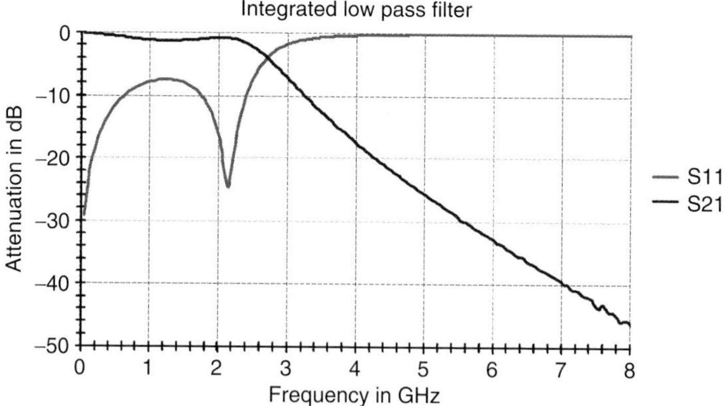

Fig. 5.17. Integrated low pass filter with two inductors and one capacitor

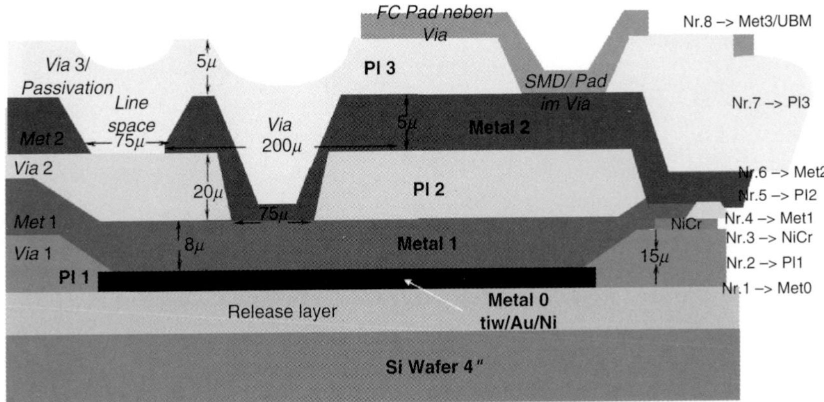

Fig. 5.18. Schematic cross section of the flex interposer

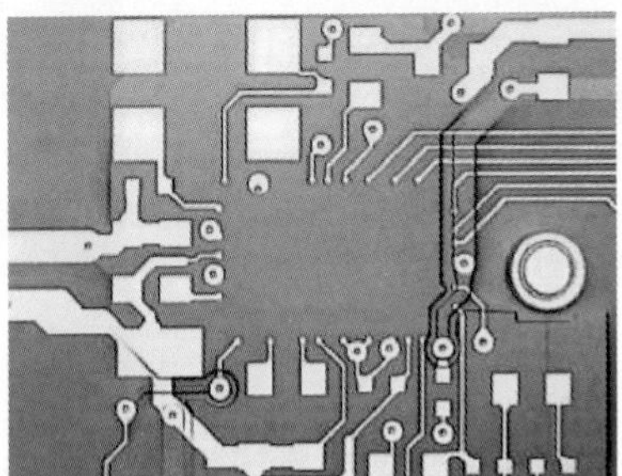

Fig. 5.19. Detail of RF circuit pads

Based on the release process for a functional layer a thin film interposer for a wireless sensor node was realized at Fraunhofer IZM. A schematic cross section is shown in Fig. 5.18.

Figure 5.19 shows a detail of the IO pads of the receiver component, which will later be flip-chip assembled.

The interposer consists of four metal layers (Cu) and three polymer layers (PI) with an overall thickness of 35 µm and a size of 9.8 mm × 31.2 mm; see Fig. 5.20 (Wolf et al. 2005).

The substrate with the assembled components as shown in Fig. 5.20 will be folded, and connected to the antenna and batteries. The overall size of the module including antenna is around 1 cm^3 from which the electronic components take up approximately 20% of the volume.

Modified configurations such as stacked devices instead of folding the flex interposer are implemented. Figures 5.21 and 5.22 show examples of assembled stacks of flex substrates. The stacks were aligned and soldered using a flip-chip bonding tool. The stack units were designed for the assembly of thin flip-chips on top of them.

Fig. 5.20. Flex polymer interposer assembled with all electronic components for a wireless sensor node (868 MHz) (Fraunhofer IZM) (Wolf et al. 2005)

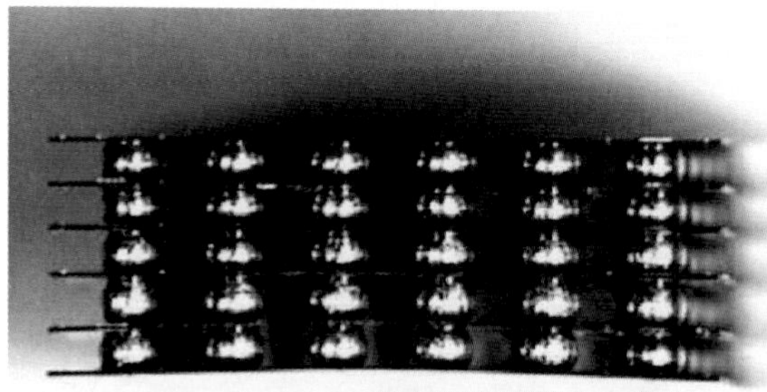

Fig. 5.21. Stack of six flexible substrates with solder ball interconnects, total thickness approximately 1.3 mm

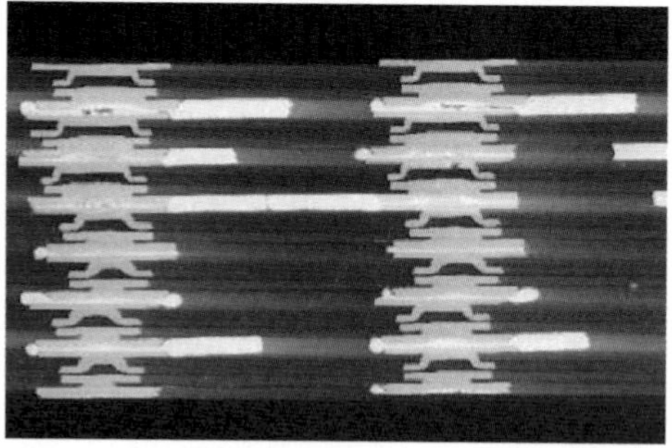

Fig. 5.22. Stack of eight flexible substrates with thin solder interconnects, total thickness approximately 0.53 mm

5.5 Embedded Components in Organic Substrates

A number of approaches concerning the integration of components have lately been presented. One of the first was the integration of power devices, shown by General Electric (Filion et al. 1994). The Technical University of Berlin later presented an embedding of chips into ceramic substrates for a signal processing application. Recently Intel introduced its bumpless build-up layer (BBUL) as packaging technology for their future microprocessors (Waris et al. 2001). At Helsinki University of Technology the so-called integrated module board (IMB) technology has been developed, which embeds chips into holes in an organic substrate core (Braunisch et al. 2002; Tuominen and Palm 2002).

At Fraunhofer IZM and the Technical University of Berlin the chip in polymer (CIP) technology is under development (Ostmann et al. 2002a, b). Its main feature is the embedding of very thin chips (50 µm thickness or less) into built-up layers of PCBs, without sacrificing any space in the core substrate. The embedded chips can be combined with integrated passive components; see Fig. 5.23.

A substantial advantage of the CIP approach is the embedding of components, using mainly processes and equipment from advanced PCB manufacturing. The realization of the process is illustrated by the cross section in Fig. 5.24.

Before introducing such a technology into production, a number of challenges have to be mastered. An acceptable yield has to be obtained. Since no subsequent repair of an embedded component is possible the use of known good dice is essential. Furthermore, via formation and metallization process must be highly reliable. The reliability of the CIP package concept still has to be proven. CIP is basically a compound of materials with quite different Young's moduli and thermal expansion coefficients, which might give reason for cracking and delimitation. This requires a detailed understanding by modeling and possibly an adaptation of technological parameters. The RF characteristics have to fulfill requirements of future applications, i.e., several GHz of bandwidth. This requires the use of polymers with low dielectric losses and

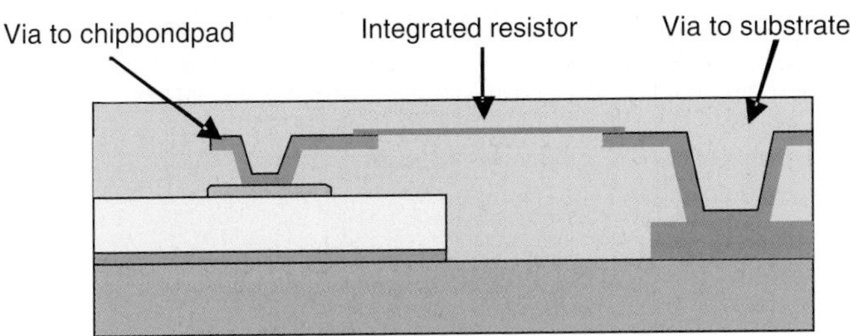

Fig. 5.23. Principle of the chip in polymer structure with embedded chip and integrated resistor

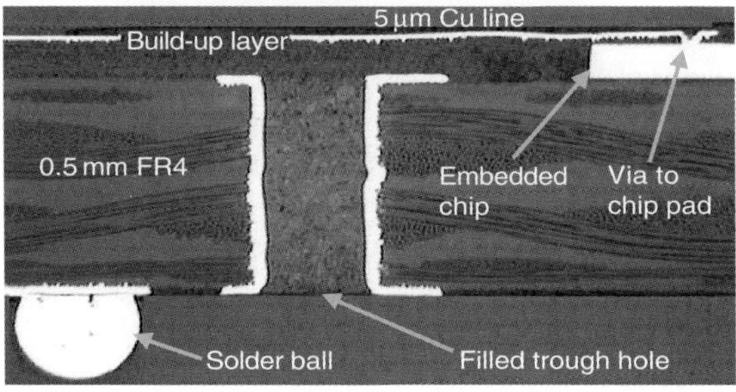

Fig. 5.24. Chip stack and cross section of a stackable package in CIP technology

low dielectric constant as well as very low tolerances for conductor geometries. A power management concept for components with high power dissipation has to be developed. Since no convection cooling of embedded active chips is possible, materials with low thermal resistance have to be used. Integrated resistors have to achieve sufficiently low tolerances, especially in order to be suited for RF circuits. This requires a processing technology with high geometrical resolution and tight control of deposition parameters. Finally, a new production flow has to be established in which chip assembly and testing will be part of substrate manufacturing.

However, CIP technology is a promising approach to obtain 3D assemblies using board integration technologies to meet the target in terms of low cost. Only technologies with the potential to be applied for high volume production are suitable for microelectronic modules integrated in everyday objects and products, and thereby the prerequisite for ubiquitous computing and Ambient Intelligence.

5.6 Wafer-Level Integration by Chip or Wafer Stacking

Most of the currently available industrial 3D stacking approaches of chips or packages only interconnect the layers on the perimeter of the devices. Due to the required high 3D interconnectivity in e-Grains, real 3D interconnections have to be realized within the bulk of the e-Grains.

The 3D inter-chip via (ICV) technology allows a very high density vertical inter-chip wiring. Especially vertical system integration technologies with freely selectable ICVs have a strong demand for 3D system design methods enabling the high performance of extremely miniaturized 3D systems. Manufacturing technologies that largely rely on wafer fabrication processes show a comparatively favorable cost structure. On the other hand wafer yield and chip area issues may be an argument against wafer stacking concepts. In consequence, the so-called chip-to-wafer technologies mainly based on WL processes and utilizing only known good dice will be of advantage. In Fig. 5.25 wafer-to-wafer and chip-to-wafer concepts for vertical system integration are shown in principle.

Vertical system integration in general is based on thinning, adjusted bonding, and vertical metallization of completely processed device substrates by ICVs placed at arbitrary locations. For wafer stacking approaches, the step raster on the device wafers must be identical. This is easily fulfilled for 3D integration of devices of the same kind (e.g., memories) but in the general case of different device areas, the handicap of processing with identical step raster would result in active silicon loss and in consequence increase the fabrication cost per die. In most cases this restriction is even more serious than the yield loss by stacking a non-functional die to a good die. For chip-to-wafer stacking approaches the starting materials are completely processed wafers

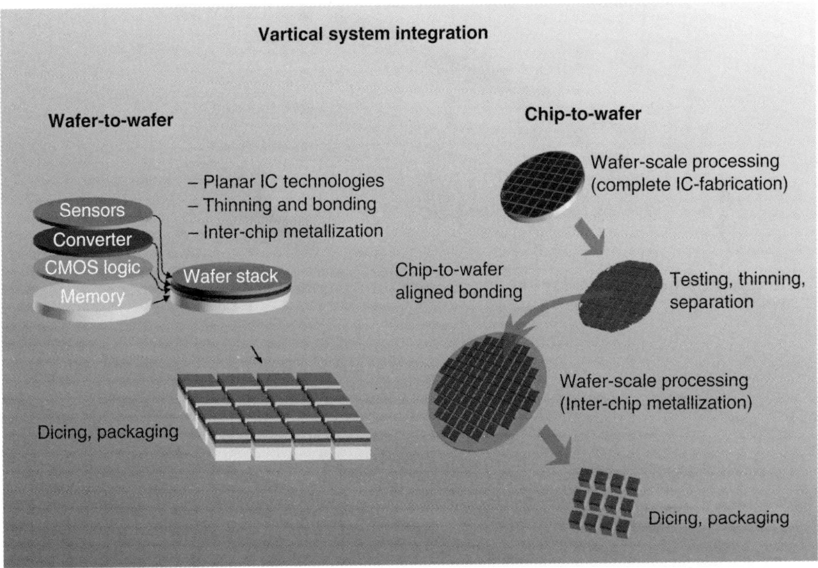

Fig. 5.25. Vertical system integration; wafer-to-wafer versus chip-to-wafer stacking concept for fabrication of vertical integrated systems

too. After WL testing, thinning, and separation, known good dice of the top wafer are aligned bonded to the known good dice of a bottom wafer. This process step represents the only one at chip-level within the total vertical system integration sequence. The subsequent processing for vertical metallization is on wafer-scale again. In consequence the development of a new vertical system integration technology with no need for additional process steps on stack level is in progress. The so-called ICV-SLID concept (Ramm et al. 2003, 2004) is based on the bonding of top chips to a bottom wafer by very thin Cu/Sn pads, which provide both the electrical and the mechanical interconnect. The new approach combines the advantages of the well-established ICV process and the solid–liquid-inter-diffusion technique, which is already successfully applied for face-to-face die stacking. The ICV-SLID concept is a non-flip concept. The top surface of the chip to be added is the top surface after stacking it to the substrate. The ICVs are fully processed – etched and metallized – prior to the thinning sequence, with the advantage that the later stacking of the separated known good dice to the bottom device wafer is the final step of the 3D integration process flow. As a fully modular concept, it allows the formation of multiple device stacks. Figure 5.26 shows the schematic cross section of a vertically integrated circuit fabricated in accordance with the ICV-SLID concept, also indicating the stacking of a next level chip.

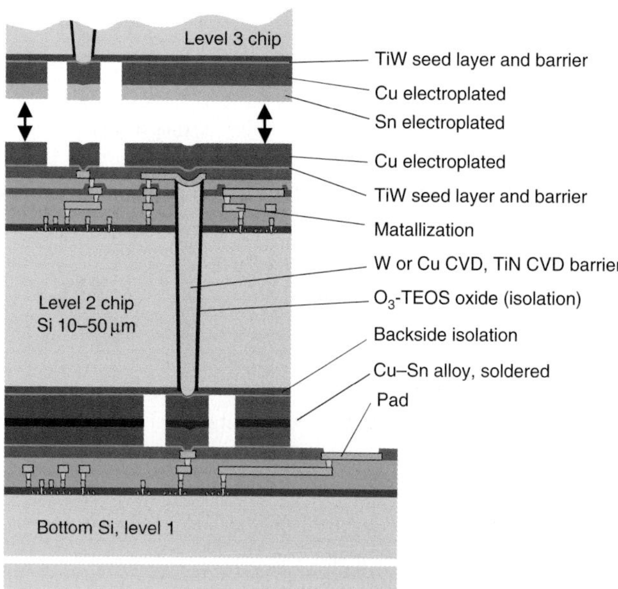

Fig. 5.26. ICV-SLID technology: schematic for the formation of multiple device stacks

For RF-application, low capacitance RF-vias have to be realized. Vias with diameters from 50 to 100 μm are generally required for this application. Metallization can be partially realized in the via hole, filling the hole with dielectrics in a subsequent step.

5.7 Autonomous Energy Supply for e-Grains

Effective power supply of autonomous electronic systems and sensor nodes is a major challenge and a limiting factor for the performance. In general two power supply methods exist: storing the needed amount of energy on the node or scavenging available ambient power at the node.

As the system size decreases, designing a sufficient energy supply is becoming increasingly difficult. Therefore, the power supply is usually the largest and most expensive component of the emerging wireless sensor nodes currently proposed and designed. Furthermore, the power supply (e.g., a battery) is also the limiting factor on the lifetime of the system. If wireless sensor networks are to become ubiquitous, replacing batteries in every device is simply too expensive. Using replaceable batteries in a system smaller than $1\,\mathrm{cm}^3$ may also be difficult.

The most important metrics for power supply technologies are power and energy density as well as lifetime for energy storage devices and power density for harvesting devices. The volumetric energy density of coin type primary batteries like alkaline (0.2–$0.3\,\mathrm{W\,h\,cm}^{-3}$), silver oxide ($0.3$–$0.4\,\mathrm{W\,h\,cm}^{-3}$), and lithium ($0.4\ \mathrm{W\ h\ cm}^{-3}$) batteries does not differ much from the energy density values of the active materials since packaging contributes much to the volume of small batteries. Zinc–air batteries have the highest energy density (1–$1.2\,\mathrm{W\,h\,cm}^{-3}$) of commercially available primary batteries, but the lifetime is limited to few months, which makes them unsuitable for most autonomous systems.

Rechargeable batteries like NiMH- or Li-batteries are used to store energy, which is supplied for example from solar cells. Power density of most coin type secondary batteries is small, but recently some systems have been developed which show good high current behavior. In comparison to Li-button cells Li-polymer batteries have a much higher power density because of the layered thin structure of electrodes and electrolyte, but they are currently not available at sizes below approximately $1\,\mathrm{cm}^3$.

Power harvesting of the environment to continuously fill the electrical storage device of the system can be achieved in different ways and strongly depends on the environmental conditions at the location of the device. Some examples of energy conversion methods are summarized in Table 5.1.

Practical applications are at present more or less limited to solar cells. Standard lighting condition, that means the incident light at midday on a sunny day, yields $100\,\mathrm{mW\,cm}^{-2}$. But in indoor conditions this value is easily reduced by a factor of 20. Single crystal silicon solar cells exhibit efficiencies of 15–20%.

Table 5.1. Examples of realized micropower energy conversion modules for the use of ambient energy to power autonomous systems

energy	conversion module	power density
light radiation	photovoltaic modules	$0.01 - 20\,\mathrm{mW\,cm^{-2}}$ (Hahn 2002)
thermal energy	thermoelectric devices	$10 - 100\,\mathrm{\mu W\,cm^{-3}}$ (Böttner 2002)
mechanical energy: vibrations, air flow, pressure variations, movements of humans, etc.	piezoelectric generators, rotary electromagnetic engines	$100–1,000\,\mathrm{\mu W\,cm^{-3}}$ (Roundy et al. 2003)
chemical energy	bio-fuel cells	$175\,\mathrm{\mu W\,cm^{-2}}$ (O'Neill and Woodward 2000, Palmore pal-more2000)

5.8 Wafer-Level Integration of Lithium-Polymer Batteries

For systems which receive their power from ambient energy as shown in Table 5.1 or from fuel cells a temporary energy storage is needed in most cases, since time and quantity of the supplied power do not correspond with the instantaneous power demand of the device. The electrical energy can be stored in a secondary battery or in a capacitor. Ultra-capacitors represent a compromise between batteries and standard capacitors. Their energy density is about one order of magnitude higher than standard capacitors and about one to two orders of magnitude lower than rechargeable batteries.

Lithium-polymer batteries have the highest energy density of all commercially available secondary batteries. They are in widespread use in portable electronics with capacities between 100 and approximately 1,000 Ah. Due to the layered structure of thin foils, their power density is quite high. They can easily be adapted to the dimension of the device and are flexible in size. Miniaturization of these batteries down to sizes below approximately $1\,\mathrm{cm}^3$ leads to a reduction of energy density, since the fraction of the battery package increases at the expense of active material.

Figure 5.27 shows the energy density of small Li-polymer batteries as function of battery thickness. There is a significant reduction of energy density at thicknesses below 1 mm because the thickness of the packaging foil becomes dominant.

High energy density of small (mm) size batteries can only be maintained, if the volume fraction of the battery package is reduced. Neither the coin type metallic casing nor the Al-polymer foil package is suited for e-Grain batteries. Small geometrical dimensions, minor discharge as well as a long lifetime, are

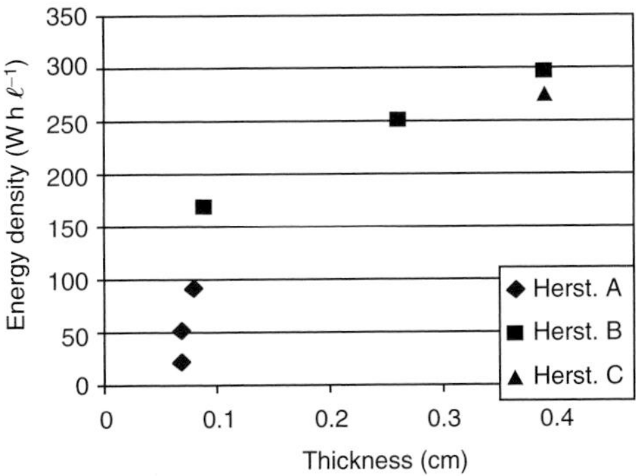

Fig. 5.27. Energy density of small lithium-polymer batteries as function of battery thickness

additional specifications for secondary batteries used in miniaturized electronic systems. Furthermore, a fast charge rate and cost saving production technology are desirable.

The packaging foil and the sealed seam used for battery encapsulation make it difficult to reduce the battery size of conventional Li-batteries. To achieve high energy density for small size batteries new encapsulation techniques are necessary. A new encapsulation technology based on parylene was investigated at Fraunhofer IZM. Parylene deposition followed by a final metallization allows the production of secondary Li-batteries with a maximum relationship of active material to the encapsulation material. A cross-sectional view and a picture of the WL Li-battery are shown in Figs. 5.28 and 5.29. The moisture vapor transmission of parylene is better than most polymers while the gas permeation of O_2 and N_2 is comparable to epoxies. Since most of the battery is covered with thin film metal, only a small area of some square micrometer of parylene at the current feed through are effective for gas permeation and a nearly hermetic package can be realized.

A high number of batteries can be produced and encapsulated at the same time using WL technology. The arbitrary footprint of the battery allows the best utilization of the available space in small electronic systems like e-Grains. The thickness of a single battery assembly is reduced because the encapsulation layer is very thin. A satisfactory way to boost the deliverable capacity while avoiding a raise of the internal resistance is by creating a battery stack.

5.9 Micro-fuel Cell Integration

As improvements in battery technology have so far been limited to energy density increases of only a few percent per annum, over the past few years many R&D activities have concentrated on alternative forms of miniature power supply. One of the most promising candidates is micro-fuel cell (FC) based on polymer electrolyte membranes (PEM). Compared to batteries the

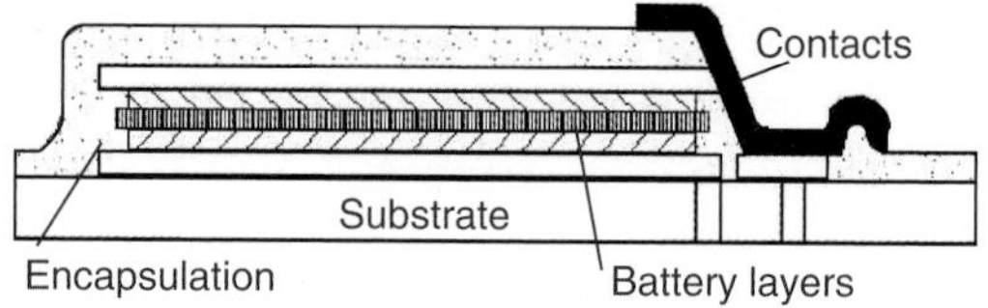

Fig. 5.28. Cross-sectional view of a WL Li-battery

Fig. 5.29. Prototype of a WL Li-battery

environmental impact of fuel cells is much lower (Hahn and Müller 2000). Table 5.2 gives an energy density comparison of fuel cells and batteries, showing the potential of this technology.

Fuel cells operate on the same principle as batteries, converting energy electrochemically, but are "open" systems where the reactor size and configuration determine the energy and power output. Since a single fuel cell has an operation voltage of approximately 0.5 V, a multitude of cells are needed which are typically assembled in stack configuration.

A fuel cell stack can be subdivided into three constituent component groups: the membrane electrode assemblies (MEAs) which fulfill the electrochemical function of the fuel cell, the bipolar plates (commonly consisting of graphite and carbon-filled polymers) which supply the MEAs with hydrogen and oxygen, provide cooling and conductive electronic paths, and the gas diffusion layers (GDL) which are inserted between MEAs and bipolar plates to distribute the reactants uniformly. Fuel cells of the air-breathing type use ambient air as an oxidant. PEM fuel cells operate with hydrogen. At the moment there is no hydrogen storage available which is suitable for miniature applications. For direct methanol fuel cells (DMFC) a better storage opportunity exists in the form of methanol cartridges.

At Fraunhofer IZM technologies for WL fabrication of planar PEM fuel cells between 1 mm^2 and approximately 1 cm^2 have been developed (Schmitz et al. 2003; Wagner et al. 2003). The investigations focused on patterning technologies for the fabrication of microflow fields, design studies for integrated

Table 5.2. Comparison of PEM micro-fuel cell with $NaBH_4$ hydrogen generator and Li-polymer secondary battery in 1 cm^3 size; 0.5 mm housing thickness was assumed

	micro-fuel cell with $NaBH_4$ hydrogen generation	wafer-level Li-polymer battery
1 cm × 1 cm × 200 μm		
voltage	(0.5 V) 1.5 V (at 40 mW)	4.1–3 V
power	40 mW	5–20 mW
energy	–	3.5 mW h
1 cm × 1 cm × 1 cm		
voltage	(0.5 V) 1.5–7.5 V	4.1–3 V
power	40 – 200 mW[a]	150–600 mW
energy	500 mW h	150 mW h

[a] Micro-fuel cells on five sides of the cube.

flow fields, material compatibility for fuel cells, patterning of membrane electrodes, serial interconnection of single cells in a planar arrangement, laminating, and assembling processes. Although wafer technologies were applied, foil materials were used which allow low-cost fabrication in future production.

Prototypes of self-breathing PEM fuel cells with a size of 1 × 1 cm^2 and 200 μm thickness were fabricated. V/I curves were measured at a variety of ambient conditions. Fuel cells with 40 mA output current at 1.5 V (= 120 mW cm^{-2}, 25°C, 50% RH) have been successfully demonstrated. Cell performance was validated under varying ambient conditions. Stable long-term operation at 80 mW cm^{-2} was achieved. The total performance of the micro-fuel cells is in the same range of current and power density as the best conventional planar PEM fuel cells (Wagner et al. 2005). At the same time this technology offers a high degree of miniaturization and the capability for mass production which is a clear success of the micropatterning approach. Figure 5.30 shows a micro-fuel cell based on foil technology.

Since storage of gaseous hydrogen is not practical in small size applications activities are focused on developing a micro-chemical reactor that produces hydrogen on demand from $NaBH_4$ and water, and yields an energy density of 800 W hl^{-1}. A possible integration of fuel cells and storage container into a 1 cm^3 e-Grain is schematically illustrated in Fig. 5.31.

Power supply for wireless sensor nodes like e-Grains is a real challenge. Storing the energy needed for long operation periods of the e-Grain is conflicting with the aim of miniaturization. Size reductions may be achieved, when micro-fuel cells are available with higher energy density compared to primary batteries, which are the well-established sources at the moment. For widespread use and miniaturization of the systems the power harvesting of ambient energy or use of alternative power sources is essential. In this context a small size secondary energy storage is needed which can be easily integrated. A combination of standard Li-polymer battery processing and wafer technologies is considered to reduce battery dimensions to chip size and keep

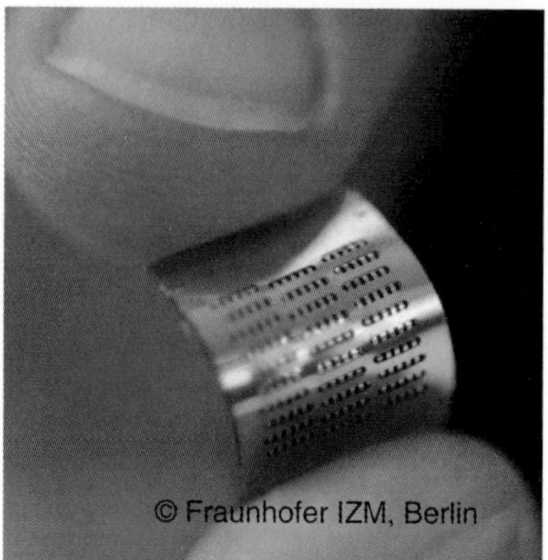

Fig. 5.30. Flexible micro-fuel cell demonstrator: size $1\,cm^2$, thickness $200\,\mu m$, $80\,mW\,cm^{-2}$

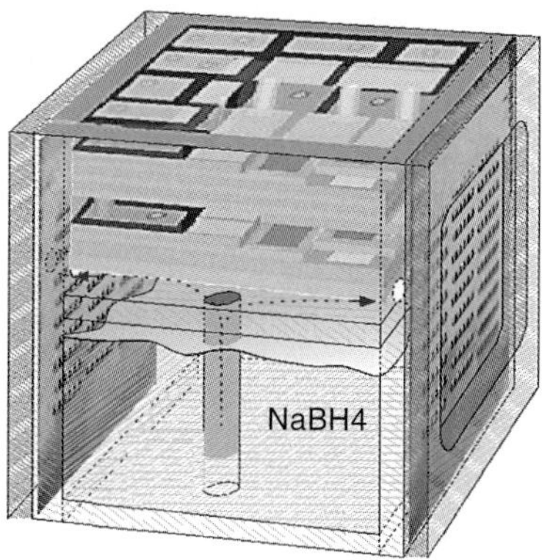

Fig. 5.31. Schematic micro-fuel cell cube, which comprises the electronic part on top, micro-fuel cells at the sides, and a chemical hydride hydrogen generation system

fabrication cost low. A great effort is still needed to realize the WL battery. Table 5.2 gives an overview of energy and power densities obtainable in $1\,cm^3$ size systems.

5.10 Summary and Outlook

The e-Grain concept attempts to link all types of system integration techniques and represents a full value chain from microelectronics and microsystems. The realization of these future ultra-miniaturized self-sufficient wireless sensor nodes requires heterogeneous technologies on wafer and substrate level, including advanced assembly and interconnection technologies, integration of passive devices, thin circuit device and sensor integration, flexible functional substrate and 3D vertical integration at WL, etc.; see Figs. 5.32 and 5.33.

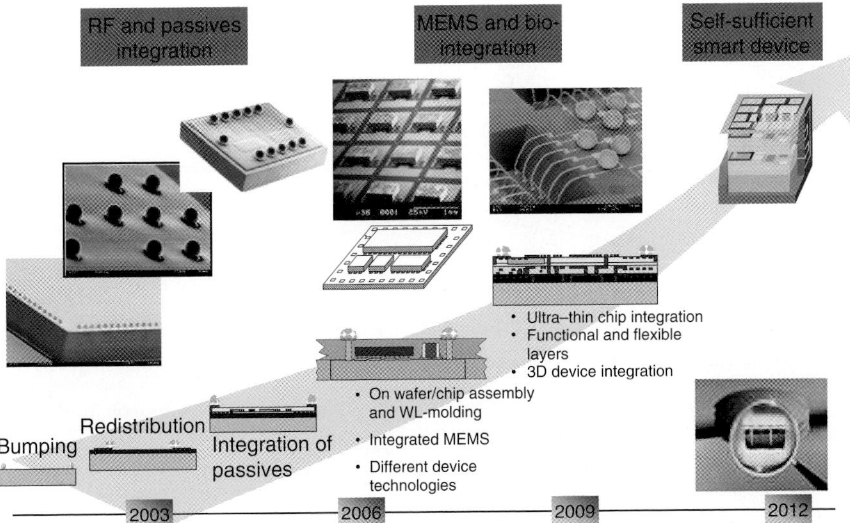

Fig. 5.32. Technology roadmap for WL integration

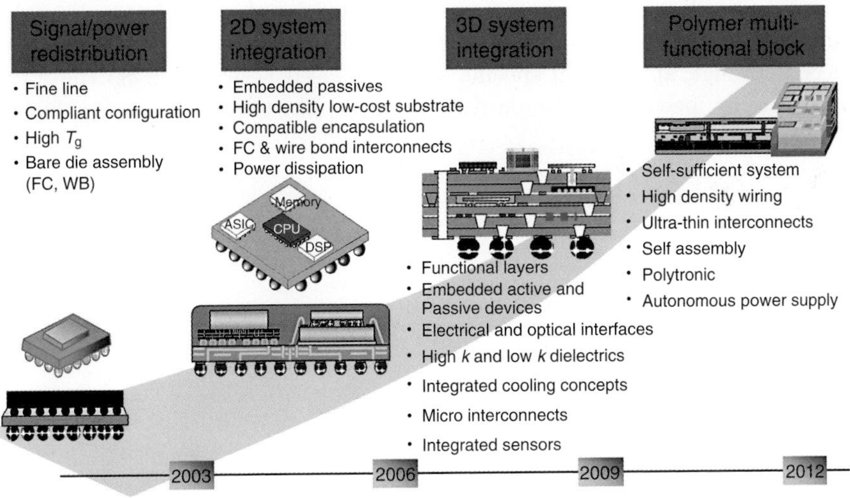

Fig. 5.33. Technology roadmap for substrate level integration

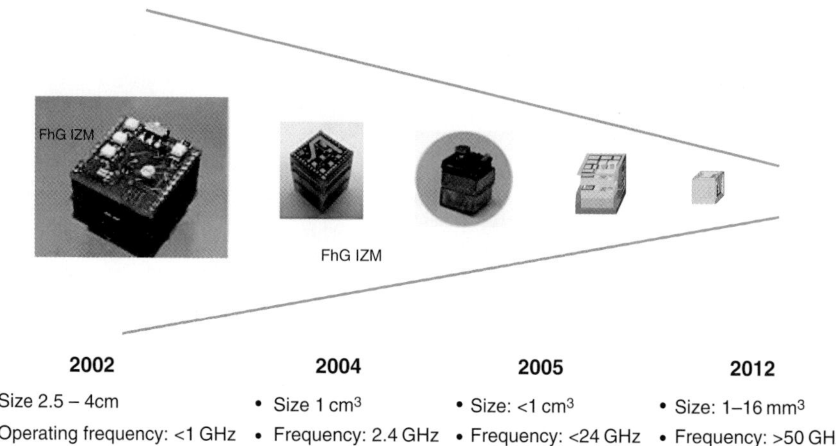

Fig. 5.34. Miniaturization roadmap for self-sufficient wireless sensor nodes (e-Grains)

Contemporary technologies at wafer and substrate level can be used to realize fully functional wireless sensor nodes in a volume of less than 1 cm^3. The progress in silicon and integration technologies will result in e-Grains with a size of few cubic millimeters in the next decade; see Fig. 5.34.

Perhaps in just a few years time, a system for an individual specific application will no longer be developed in the currently known way of today. Instead, it will use few grams of universal e-Grains that can be integrated in any human environment and will meet specific information or communication needs through application specific software-based networking. So the future e-Grain may become a "standard component" like microprocessors today.

6

Electronic Textiles

H. Reichl, C. Kallmayer, and T. Linz

"Electronic means that textiles are capable of exchanging information."

(Wearable Computing Lab, ETH Zürich)

6.1 Introduction

In the context of Ambient Intelligence the integration of electronics into textiles has gained much interest recently. With the next generation of mobile devices it can be expected that there will be no more separation between local/mobile data processing and data exchange. Along with the improved infrastructure, the mobile devices, e.g., mobile phone, PDA, and MP3 player, have to change to meet the future requirements of mobile communication and computing. This will also lead to a significant change in the construction of these devices.

One logical consequence is the integration of the mobile devices in clothing, using the fabric as the circuit board providing bus structures and interconnection between the modules. All devices can share input/output (I/O) modules, such as displays, keyboards, and loudspeakers, thus leading to comfortable ergonomic handling and further miniaturization. The complexity of microelectronic systems in textiles can vary in a broad range. From textile passive transponder systems up to sensor shirts, e.g., ECG, and wearable computers, there are few limits in developing new ideas for integrating electronic function in textiles and clothing. An overview on possible applications and the functionalities that might be included in the future is given in Table 6.1. But a lot of other applications are also possible even if they cannot be realized with the technologies currently available.

The highest dynamics in development are currently seen in health care, mobile energy supply, and transponders. Another important driver seems to be automotive industry, which needs electronic textiles for the car interior. Due to the relatively small market, protective and alerting clothes do not appear as very important areas for early application although the products would be very useful. The applications closer to fashion, leisure, and entertainment will probably only face increased interest after the technologies are mature enough for mass production (Strese 2005).

Table 6.1. Applications for electronics in textiles

application	functionality
identification, access control, originality protection	transponder
medical	sensors, data processing and storage, wireless interfaces, etc.
sport/wellness	sensors, data processing and storage, actuators, etc.
protection/safety	sensors, visual and acoustic output, wireless interface, actuators
professional	textile keyboards, acoustic and visual output, wireless interfaces, data processing and storage
entertainment	visual and acoustic output, MP3 players, video players, etc.
air condition	sensors, actuators, heating pads
technical textiles	sensors, wireless interfaces
textiles for decoration	luminescent textiles, sensors, actuators, wireless interfaces, etc.

The wearable devices that have been developed in research laboratories are often highly sophisticated, very complex, and large. On the other hand it is often claimed that electronic assistants integrated into our clothes are unobtrusive and easily accepted. However, so far the electronic gadgets have not really been hidden. This is crucial for a broad acceptance and consequentially for the market success of wearable electronics. Currently technologies have been developed to overcome this bottleneck of wearable computing and realize seamless integration of electronics.

The technologies involved aim at lightness, wearability, manufacturability, cost-effectiveness, and reliability. The technologies comprise weaving, embroidering, and snap fasteners as well as soldering adhesive bonding and encapsulation open a completely new chapter in electronic packaging.

The washability of electronic modules in textiles is an important issue. For a simple transponder system, the combination of suitable interconnection and encapsulation can provide washability even under industrial conditions. For more complex modules this task is not yet solved. The concepts for miniaturization and material combination together with a suitable location on the body still have to be developed (Tröster et al. 2003).

A general problem is still the lack of standardized suitable reliability tests for wearable electronics, which correspond to the real working conditions and provide accelerated aging tests. The mixture of tests from the textile and the electronics industry, which are still performed, will not be sufficient when the smart textiles finally enter the market.

6.2 Conductors

The integration of electronics into textiles is an interdisciplinary field, and attracts textile engineers and electrical engineers alike. Therefore, it comes as no surprise that often two mainly different philosophies and approaches concerning developments can be found, e.g., for textile conductors. Some have more wire character while others behave more like threads.

6.2.1 Textile Wires

Textile wires consist of a very thin wire, which is spun loosely around a textile thread that serves as a core; see Fig. 6.1. Being shorter than the wire the thread takes all the mechanical stress along the construction. Typically the textile thread is made of polyester or polyamide and is more than ten times bigger in diameter than the wire which is between 20 and 80 μm. Depending on the application the wire can be isolated or not. Silver-coated copper wires with isolating Kevlar coating are common.

While tension along the thread can be handled well, bending or sheer forces tend to break the wire or strip it off the core. Therefore, it is suitable for a weaving process both in warp and in weft. However, it cannot withstand the mechanical demands of sewing or embroidery.

6.2.2 Metallized Threads

The textile engineers approach the subject by metallizing filaments. The filament could for example be polyamide or polyester. These materials are already commercially available in different strengths of thread and were initially meant for electrical shielding applications as well as for antibacterial purposes. Such metallized threads can be woven as well as sewn or embroidered; see Fig. 6.2. Furthermore they can be washed.

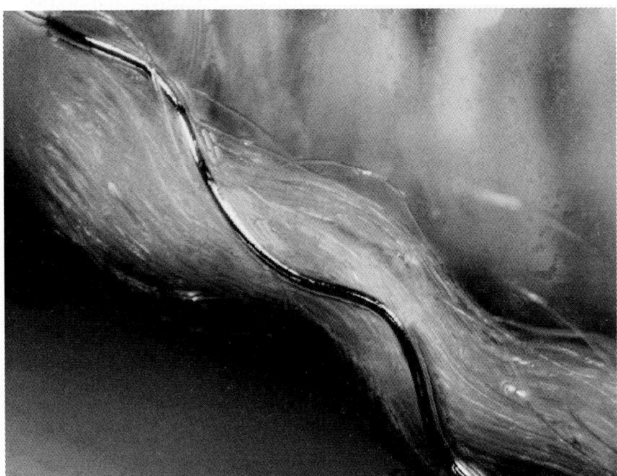

Fig. 6.1. Textile wire (Snowtex)

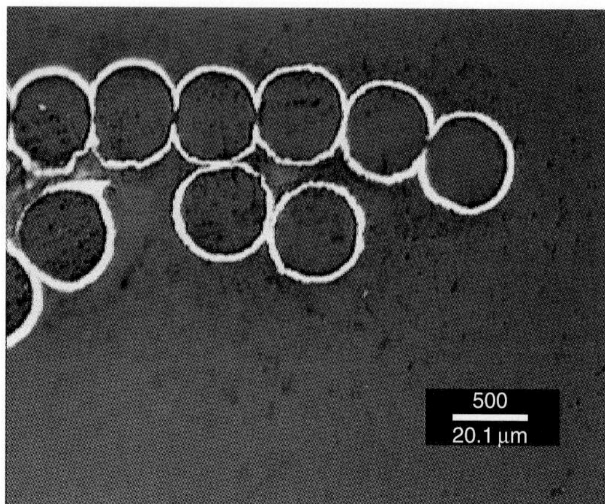

Fig. 6.2. Metallized polyamide fibers in yarn

Fig. 6.3. Steel thread (27000 dtex)

6.2.3 Intrinsically Conductive Yarn

There are also materials with intrinsic conductivity that are suitable to produce fibers and yarn. For instance, steel and copper with silver coating can be used; see Fig. 6.3. On the other hand there are polymers filled with conductive particles. The properties of these materials regarding the textile processes and their washability are very promising but the conductivity is poor in the case of filled polymers and the interconnection is difficult for steel yarn.

6.2.4 Resistance

Unfortunately most types of conductive yarn come with a very high resistance as the cross section contains mainly non-conducting material. The silver-coated polyamide fibers are around 20 µm in diameter with an extremely non-uniform silver thickness < 0.3 µm. It was not intended for use as a conductor after all. It is possible to deposit about 1 µm of gold or silver on top of the silver and thereby achieve an increase in conductance of around a factor 10 without decreasing the mechanical properties (Neudeck et al. 2004). Depositing even more gold will result in substantial brittleness (Kallmayer et al. 2003). A more uniform coating could improve conductivity without compromising mechanical properties.

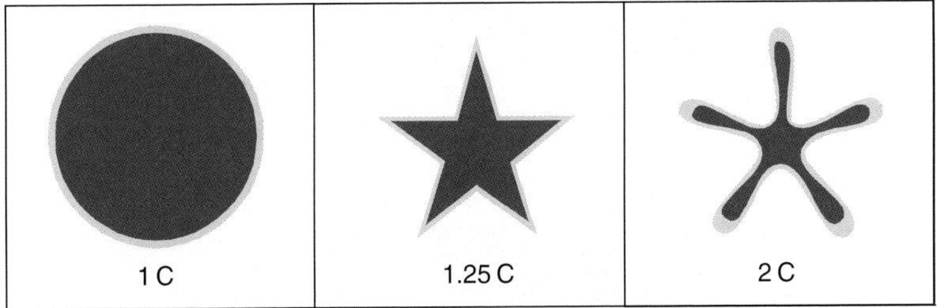

Fig. 6.4. Alternative fiber cross sections for improved conductivity

Improvements may also be achieved by changing the geometry of the fiber. Figure 6.4 shows the given circular core form and more complex structures with bigger circumferences and thus a larger percentage of metallization in the cross section. Feasibility and reliability (brittleness) have not yet been investigated. Possibly nano-fibers may foster conductivity.

6.3 Textile Processing with Conductive Threads

Applications for textile "wiring" can be roughly classified as follows:– Straight parallel conductors like in a ribbon cable

- Straight conductor lines crossing (may be interconnecting)
- Arbitrarily formed lines not crossing (may be interconnecting)
- Arbitrarily formed lines crossing (may be interconnecting).

More or less all types can be produced with weaving as well as with embroidery. However, weaving generally favors straight lines and its capability of growing arbitrary forms is limited to rectangular structures. Embroidery can create any form in 2D very much like printed circuit boards (PCBs). Interconnecting several layers is a bit more difficult with embroidery. It is also much more demanding on the thread as mentioned earlier.

6.3.1 Weaving

Similar to a ribbon cable, Infineon developed a woven ribbon "cable." It is used to interconnect different modules of an MP3 player that can be integrated into a jacket (Weber et al. 2003). Besides normal (non-conductive) threads it also consists of thin copper wires in the warp. The weft threads are all non-conductive.

ETH Zurich introduced textile wires also in the weft. The copper, which is spun around a thread, has been isolated beforehand to prevent short-circuiting in this single layer construction. By removing the isolation locally at a crossing point of two wires and soldering these together, wiring can be achieved (Cottet et al. 2003).

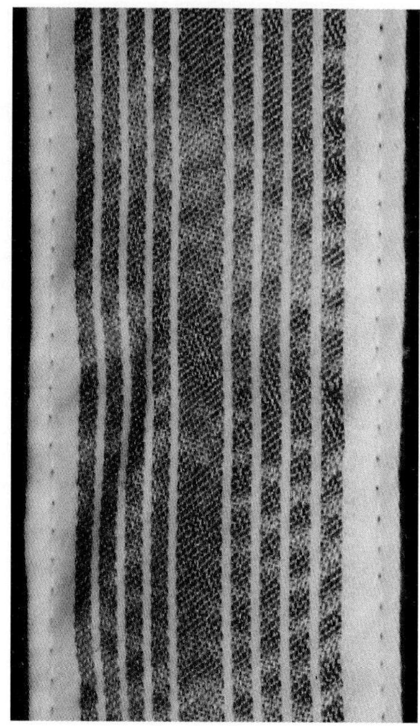

Fig. 6.5. Ribbon with woven conductor lines (TITV)

TITV developed a weaving process that allows interconnecting conductive threads of different layers; see Fig. 6.5. Essentially the sandwich construction consists of three layers, which can be enhanced by more layers if a higher level of integration is desired. One layer has conducting lines in weft direction, another layer has conducting lines in warp direction, and an intermediate layer isolates these. Weaving "errors" are introduced at those points where an electrical contact between the two outer layers is desired. It causes the weft threads to go to the other side of the sandwich and makes contact with the warp threads. The contact may be supported by electrically conductive adhesive or solder (Kallmayer et al. 2003). If it is necessary to use the same wire for two or more connections, it can be cut into sections. This applies equally for the ETH approach.

6.3.2 Embroidery

Often the primary task is to make electrical connections between different modules. Such modules could be flexible substrates, snap contacts, textile keyboards, etc. Therefore, 2D structures are sufficient if the modules can be designed to fit to each other without crossing cables. Such structures can be laid out with a standard embroidery process. An example of such an embroidered structure for a textile mobile phone is shown in Fig. 6.6.

Fig. 6.6. Embroidered circuit for textile mobile phone

Special cases like modules that cannot be adapted or textile RFID antennas compel a 3D design. This may be accomplished with two (or more) layers of embroidered circuits. At specific points the layers can be interconnected with conductive threads. In this case both the sewing thread (top side) and the bobbin thread (bottom side) should be conductive.

6.4 Interconnection and Packaging

The first approach for the electrical engineer is to modify well-known interconnection technologies from microelectronics to connect electronic modules with conductive textiles. It has been shown that the soldering with low melting solders, adhesive bonding with conductive adhesives, and even wire bonding are suitable for the assembly of electronics on textiles. Welding can also be applied with suitable material combinations. But there are components in smart textiles that require new developments closer to textile processes.

6.4.1 Embroidered Contacts

Textile keyboards, for instance, are made of woven or knitted conductive ribbons. A certain area of these ribbons may be contacted by sewing over it with a conductive thread. So, the thread is both the conductor as well as the contact material.

Flexible substrates can be connected similarly. Metallized contact areas with drill holes in the middle can be applied. The needle stitches through the drill hole, and the thread fixes the module and at the same time makes an electrical contact. Detailed research concerning the design rules and the reliability for this new type of interconnection is on the way; see Fig. 6.7.

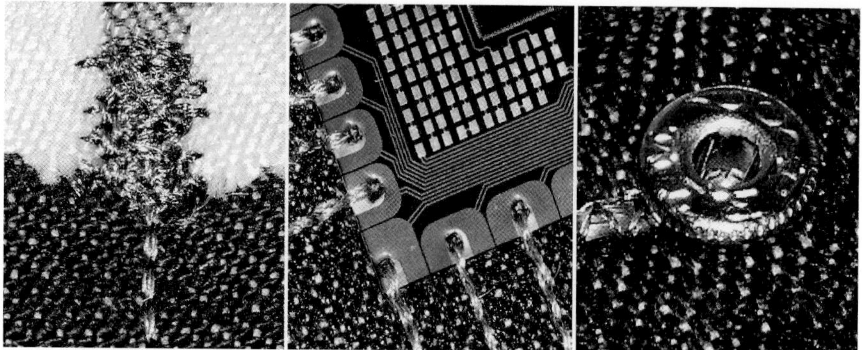

Fig. 6.7. Embroidered contacts for textile, flexible, and rigid components

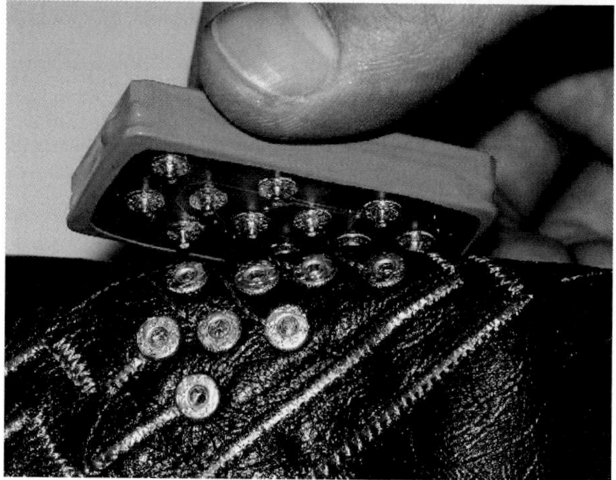

Fig. 6.8. Detachable interconnection by snaps for a display module

6.4.2 Detachable Interconnections

Sometimes it is desirable to interconnect electronics and textile substrate in a separate process. This may be the case when the textile manufacturer does not have the facilities to integrate the electronics or when the electronic modules need to be detached by the end-user. A display may not be washable or could be used in different pieces of clothing.

For such applications the use of metallic crimpable standard snaps has been investigated. The smallest snaps on the market are around 6 mm in diameter and 2 mm in height (total). The ball part can be soldered onto metallized pads on the back of an FR4 substrate. The female counterpart is crimped through an embroidered conductive pad; see Fig. 6.8.

The electrical contact is very reliable. It varies from 0.35 to 0.45 ohm. Even after months of use there are no degradation effects. As for the other interconnection technologies on textiles, the reliability investigations have yet to be carried out.

6.4.3 Dimensions and Precision

The dimensions in the textile world are some orders of magnitude larger than those in the electronics world. Chips with typical contact dimensions of 100 µm and smaller cannot be interconnected directly with a textile substrate with its larger structures and its dimensional instability under mechanical stress. This leads to the need for interposers or modules that carry the components and distributes the contacts to uncritical dimensions (Kallmayer et al. 2003); see Fig. 6.9.

While embroidery machines have a relatively high precision (below 0.5 mm) the silver-coated thread typically tends to fuzz which leads to shorts. Improved material can reduce this problem but it limits the distance between adjacent contacts to $> 1\,\mathrm{mm}$.

6.4.4 Encapsulation

Special care has to be taken to protect the electronic modules and to isolate the conductors. Small electronic modules that are permanently attached to the fabric may need to be washable. Therefore, it is necessary to encapsulate these modules. This can be achieved with a textile pocket around the module, which can be filled with low viscous encapsulation material. On top of the textile, modules can be encapsulated with screen printable or dispensable materials as described in Kallmayer et al. (2003); see also Fig. 6.10.

A different but very promising approach for larger modules and high reliability requirements is transfer molding. This is acceptable for clothing or other textiles as long as the module is reasonably small and thin (shirt buttons are also not flexible).

A generic problem with encapsulating fabric is to hinder the threads from sucking water (or other liquids) into the package by capillary effects. To avoid

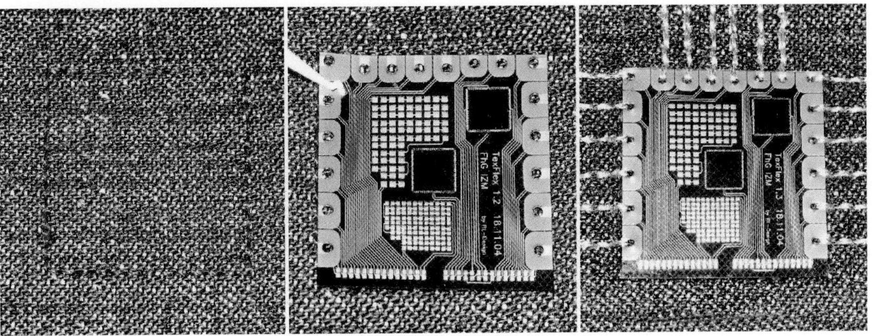

Fig. 6.9. Embroidered interconnection for flexible circuits

Fig. 6.10. Encapsulated module on textile

this effect to some extent the mold compound must interpenetrate the threads themselves. Nonetheless with polymer encapsulation the package will never be entirely waterproof but for many applications sufficient.

The metallized thread itself is quite robust and washable. Whether it has been woven or embroidered it needs no protection in general. However, precautions must be taken to prevent short-circuiting when the fabric is draped or when it gets wet. Covering the conductors with another fabric is one solution. The electronics and the embroidery were constructed on the lining. Electronics and embroidery hide in-between the lining and the shell.

A stronger isolation for the conductive threads can be achieved with polymer coatings. Such coatings can be applied in a bath. This is a common process in the textile industry to make fabric water repellent. Unfortunately this may also change the feel of the fabric. Thin layers can also be applied by electroplating.

An even stronger mechanical protection, e.g., against abrasion, can be reached with a lamination process also common treatment for textiles. Sometimes this is already part of the product anyway.

6.5 Applications

The possible applications for electronics in textiles cover a wide range from textile transponders to sensor clothing or technical textiles with functionalities as sensing, lighting, etc. Therefore, three examples have been chosen to illustrate the opportunities of the technology.

6.5.1 ECG Shirt

The continuous remote monitoring of vital signals from patients is part of a new health care concept, which has become necessary by the increasing number of elderly and chronically ill people. It is only possible due to the recent rapid developments in sensor and micro-systems technology. Only the integration of such devices in clothes will allow unobtrusive monitoring without changing people's usual way of behavior. One possible application is a T-shirt that measures bio-signals. The shirt is an ideal place for such electronics because it measures the signals in an unobtrusive way right where they appear. Several approaches to realized ECG shirts are currently reported. Other sensors for SpO2, EMG, temperature, acceleration, breast movement, etc. will follow later.

Three problems are essential for developing such an ECG shirt: making the ECG pads textile-like (i.e., drapable, long-lasting, and washable), miniaturizing the electronics to a reasonably small size, and integrating the electronics into the shirt. Using flip-chip technology and thin silicon on a flexible substrate these modules can be miniaturized to less than 2 cm side length. The modules have to be embroidered to the shirt to be connected with the battery and the ECG pads. To provide enough energy for one day an ultra flat, flexible accumulator with an area as big as three business cards will be required.

Achieving a good skin contact is as challenging as it is crucial for a good signal. The most common ECG pads are made of an AgCl electrode and a wet gel electrolyte (mostly containing a 0.9% NaCl solution). A flat plastic cup with an adhesive ring holds the gel in place and in contact with the body (Grimnes and Martinsen 2000). These pads are mostly disposable. Reusable pads require the application of new gel for each use. Neither is appropriate for the application in a T-shirt.

Using silver-coated polyamide threads, ECG pads can simply be embroidered. However, without any contact electrolyte, the signal will be too noisy. This is a result of many factors: movement artifacts, bad contact between ionic conductor (skin) and electronic conductor (silver), dry skin cannot be wetted, small contact area (fractal surface of the skin is not entirely used), and high junction potential (Grimnes and Martinsen, 2000). An embroidered pad made of very fussy conductive thread (or at maximum nano-tubes) could tackle some of these problems. Movement artifacts can be significantly lowered with a tight-fitting shirt. Wetting may only be achieved with wet gels or with pre-treating the skin.

Rather than trying to make the silver threads fussy it is possible to integrate a hydrogel into the embroidered pads. A hydrogel is a solid electrolyte and beside many other applications can be found in commercially available body electrodes mostly combined with an adhesive. However, the ion mobility in hydrogel and thus also conductivity of hydrogel is lower than that of wet gel. Therefore, they are mostly used for high frequency applications rather than for ECG (0.1–100 Hz) measurements. Nonetheless promising results have already been achieved with hydrogel. The material properties of the hydrogel have to be modified for the permanent use in a T-shirt, i.e., washable, slightly adhesive to reduce movement artifacts; see Fig. 6.11.

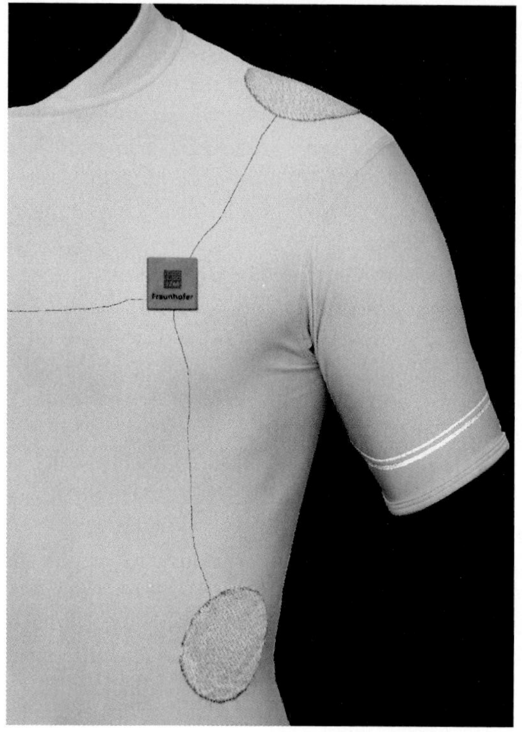

Fig. 6.11. Prototype of ECG-T-shirt with embroidered pads and flexible module

6.5.2 Communication Jacket

In the recent years we have seen a tremendous increase in portable devices. A PDA and a cell phone are usually the minimum set of devices a businessperson carries around. Music players, digital cameras, and laptops add to this. Each of these devices has its proper I/O devices. Often these I/O devices limit further miniaturization as humans cannot handle smaller screens and keypads.

To overcome this restriction the integration of such devices in textiles allows to separate the part with the application logic and control from the I/O devices, and to share a set of I/O devices among different devices. The trade-off between size and usability vanishes as the functional devices may be further miniaturized, and the I/O devices adapted to the users' need. If done well this step may even lead electronics to a more socially acceptable level. For instance, elderly people may want to use extremely large keypads.

Figure 6.12 shows a simple demonstrator of a set of I/O devices, display, keypad, microphone, and headset that have been realized and integrated into a jacket. At this point it is possible to control a mobile phone with this jacket. The phone, which is the functional part, is placed in a pocket. Controlling other devices, e.g., a music player, is just a question of implementation.

Fig. 6.12. Communication jacket with embroidered conductor lines

The communication jacket runs with any cell phone that uses the AT commands defined in the GSM standard, which is based on the Hayes AT commands for modems. The phone sits in the side pocket and is connected to a headset and the external display module. The display module is connected to a textile keypad. Both sit at the left sleeve. The display can be detached and moved to another piece of clothing or for washing the jacket; all the rest can be washed. An AT command transmission via serial RS-232, which is a cable-based interface, was given preference to BlueTooth or IrDA because a physical connection to the phone was desired anyway to transfer energy to the modules at the sleeve; see Fig. 6.13.

The jacket consists of shell and lining. The entire system has been set up on the lining. The shell has openings where the display, the keyboard, and the headset come out. Lining and shell were produced separately (see picture below) and later assembled together. Embroidered conductive lines interconnect the modules. The data transmission lines between cell phone and display as well as the lines between the keyboard and the display were made with a double run step stitch. Only for the energy supply of the display module a steel thread with higher conductivity had to be applied with a different embroidery process.

6.5.3 Textile Transponder

Passive transponder labels are available in a variety of types mainly for logistic purposes, which require higher functionality and/or higher reliability than the conventional bar codes allow.

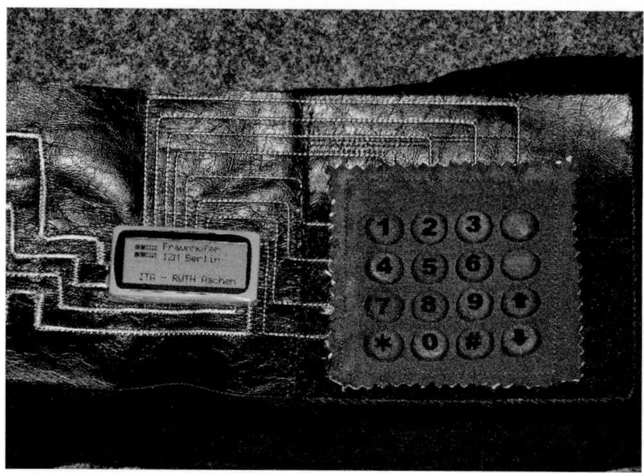

Fig. 6.13. Detail of embroidered circuits in sleeve

Smart labels for textiles either for production, logistics, or professional laundry have to meet different requirements from these applications. They have to be ultra thin and very reliable under conditions unusual for electronics, e.g., washing and high pressure. Therefore, the realization of textile transponders has become possible only recently with the development of thin silicon chips. Such chips are flexible and can overcome high mechanical load. They can be easily encapsulated and can resist the extreme conditions of a washing process.

In order to achieve high readout distances, e.g., 0.5–2 m, the antennas have to enclose an area of $>6\,\text{cm}^2$. The reliability of flexible substrates in this size during washing procedures is not sufficient. Therefore, textile-based antennas have to be produced. The transponder ICs have to be assembled to the conductive yarn. The module also needs encapsulation to protect the IC from humidity, chemicals, etc. This requires new technologies especially as the dimensions and tolerances between microelectronic components and textiles differ by orders of magnitude. The development of these technologies will be the basis for the realization of textile-based wearable electronic assistants, which will be a key component of future ubiquitous computing.

In order to be able to identify hospital textiles automatically for cleaning, such a textile transponder (13.56 MHz) was developed jointly by TITV and IZM. The textile carrier material that was realized by weaving of conductive and non-conductive yarn by TITV was not only the carrier for the microelectronic module but also provided the antenna and contact structures like a PCB. In order to achieve higher conductivity the metallization on the conductive yarn has been increased by electroplating.

The transponder chips were assembled in flip-chip technology on thin flexible polyamide interposers. These interposers are needed to overcome the large

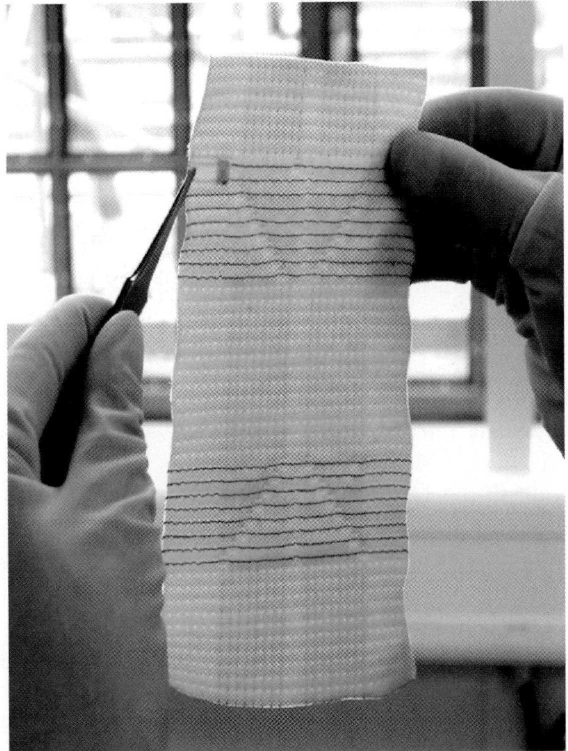

Fig. 6.14. Textile transponder without encapsulation

tolerances of the fabric. The small modules ($2 \times 4\,\mathrm{mm}^2$) are connected to the antenna by adhesive or solder (Kallmayer et al. 2003). A thin encapsulation was realized using commercial glob top material, which is often used in chip on board technology. This provided good reliability during washing; see Fig. 6.14. With these transponders a good readout distance of 80 cm was already achieved but can be improved with increasing conductivity of the conductive yarn.

For future applications transponders with higher frequencies (868 MHz and 2.4 GHz) will be needed – especially for purposes in logistics and for proof of originality. The antennas will be shaped differently so that they can be realized easily by weaving and embroidery.

6.6 Future Challenges

Although, there has been progress in technological developments lately, there are still a lot of problems to be solved and developments to be made. The components which can currently be integrated in clothes are mostly simple and yet do not lead to systems needed to provide the electronic assistance we expect.

In order to achieve this in the near future special focus has to be on the following areas of development:

- Sensors suitable for integration in textiles
- Actuators
- Energy saving robust displays, e.g., bi-stable LCD displays, e-ink, and electroluminescence
- Energy supplies, e.g., micro-fuel cells, energy harvesting, and solar cells

Smart electronic textiles should be able to sense, react, and adapt to the environment (Tröster et al. 2003). An important step toward more useful systems – especially in the medical area – is the integration of sensors. These can be micro-systems but also textiles with sensing capability, e.g., elastic threads, optical fibers, and piezoelectric material.

In future applications also actuators will play an important role. These can also be micro-systems and textile structures. Piezoelectric and shape memory material could be applied for such purposes. Luminescent textiles provide another interesting functionality. Woven conductive double-comb structures are the basis for electroluminescent light sources. The distance of the integrated thread electrodes is less than 300 mm, which leads to a thread density of about 60–70 per cm. This is under the critical distance, where electroluminescence can be created with commercial EL converters (Gimpel et al. 2005).

With the selected production of conductive structures and their coating with special pastes by screen-printing, plain light sources can be created; see Fig. 6.15. Although, the desired and demanded properties of textiles like softness and sleaziness, and a high break durability are still given.

Fig. 6.15. Woven textile light sources

Fig. 6.16. Contacted LED on a textile structure

Another textile light source by using woven conductive yarn was developed at TITV with light emitting diodes (LEDs). These diodes are contacted with the fabric on several spots as shown in Fig. 6.16. Supply lines and interconnections will be realized by newly developed textile circuit and connection technologies.

Besides these developments there is also the necessity for new technologies in the area of energy supply. Wearable electronics – as all mobile technologies – require suitable energy supplies. Besides secondary batteries and solar cells there are no commercial solutions available yet. Energy harvesting is often mentioned but is still far from commercialization (Paradiso et al. 2005). Another option might be micro-fuel cells but these are also still under development.

6.7 Conclusion and Outlook

A variety of methods to seamlessly integrate electronics into textile environment are currently developed. Depending on the type of textile and the desired interconnection of electronic modules one can choose between embroidery and weaving. In future research knitting must also be analyzed for this purpose. At the moment poor thread quality is an issue for all of these. Threads need to be stronger, less fussy and have a higher conductivity. Besides permanent contacts even detachable contacts to textiles can be realized using simple technologies, e.g., snaps. Further research concerning the reliability and contact quality of such interconnections through flexible substrates is required.

Encapsulation technologies from the electronic industry have been transferred to electronics on textiles. Tests and modifications for the improved washability are on the way.

The idea of wearable electronics often leads to the misinterpretation that one has to choose the clothes by their functionality. This is not the case when functionality is separated from the interface. The functional part may be moved among the jacket, the shirt, the car, the home, or office environment, etc. The I/O devices are tailored in style and behavior according to their proper environment and become part of it.

Although, electronics in textiles can already be realized to some extend, research and development is required in several interdisciplinary fields in order to achieve the unobtrusive wearable electronic assistance everybody would want to use in everyday life.

7

Computing Platforms

H. De Man and R. Lauwereins

"Designing hardware platforms is about bridging the hell of physics and the heaven of Ambient Intelligence."

7.1 Reconsidering the Vision

In the past, mass production of PCs and communication infrastructure fueled the growth of the semiconductor industry. Today, we are entering the "embedded-everywhere" world in which all objects around us become intelligent micro-systems interacting with each other, and with people, through wireless sensors and actuators. This culminates in the vision of Ambient Intelligence (Aarts and Rovers 2003), a vision of a world in which people will be surrounded by networked devices that are sensitive to, and adaptive to, their needs. Figure 7.1 provides a glimpse of the AmI world.

This world can be partitioned into three connected classes of devices characterized by their energy sources, cost, and computational requirements.

- *Stationary devices.* Applications in the home, office or car, require computing platforms that take care of major information processing needs, and execute computationally intensive tasks, such as providing interactive audio-visual infotainment in the home, or advanced safety, navigation, and engine-control roles in cars. Computational power for these applications can reach 1 Tops in the future, but packaging and cooling costs limit power for such consumer products to less than 5 W. Therefore such devices are called "Watt nodes" (Aarts and Rovers 2003). Clearly Watt nodes require a power efficiency (PE) of 100–200 Gops W^{-1}, which is three orders-of-magnitude higher than today's PC microprocessors. However, AmI components are embedded systems. They need not be general-purposeprogrammable, but must be "just flexible enough" within the intended set of applications. This selective limitation must be exploited to reduce power.
- *Nomadic devices.* Every person will carry a universal personal assistant (UPA), powered by battery or fuel cell. It will have natural human interfaces. It will provide adaptive multi-mode wireless and broadband connectivity to the Web, to the personal space, and to a body area network (BAN), for health monitoring, security, and biometric interaction. Natural interfacing requires full multimedia capabilities. Notice that a UPA has to

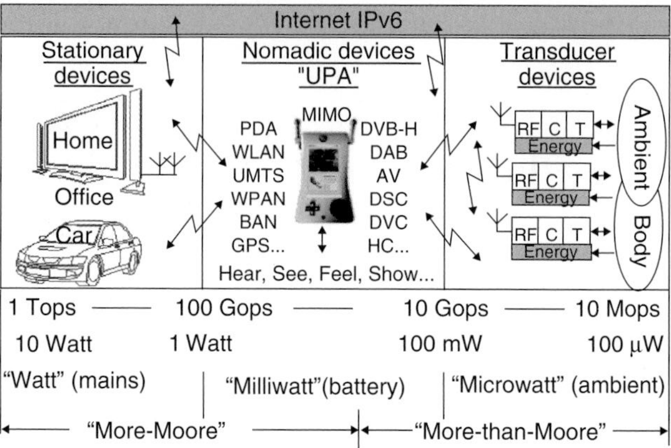

Fig. 7.1. Ambient Intelligence system view

perform a set of concurrent dynamic real-time tasks that adapt to the user's wishes, and to the services provided by the environment. Therefore, such a system must be flexible (programmable and reconfigurable). Peak computational power of a UPA is between 10 and 100 Gops, but peak power for the silicon parts should be less than 1 W. Therefore nomadic devices are called the "Milli-Watt nodes." They require a PE of the order of 10–100 Gops W^{-1}.

- *Autonomous wireless transducers.* Such devices will be empowered by energy scavenging or by a lifetime battery. They will observe and control our surroundings, and form ad-hoc networks communicating with the two other device classes. Today, energy scavenging is limited to 100 μ W cm^{-2}. So, in spite of the low duty cycle ($< 1\%$) and low bit rates (1 bps to 10 kbps), these devices must provide wireless sensing for less than 100 μ W on average. Therefore such devices are called "Micro-Watt nodes."

The digital parts of Watt and Milli-Watt nodes demand "more-Moore", that is a continuation of scaling, if we can manage the huge NRE design cost, and as long as it leads to further reduction of power and cost. Smart autonomous transducers and Milli-Watt wireless systems, on the other hand, demand "more-than-Moore." The challenge is in finding novel combinations of technologies above and around CMOS, for the design of ultra-low-power, ultra-simple, ultra-low-cost sensor motes for Ambient Intelligence.

7.2 "More-Moore": Managing Giga-complexity

Watt and Milli-Watt devices are consumer products. Unlike general-purpose processors, they are not designed for raw performance, but for two-orders-of-magnitude lower power for a given task set, at one twentieth of the cost. The computational bottleneck is usually in the memory-intensive digital-signal-processing parts for communication and processing of multi-media streams.

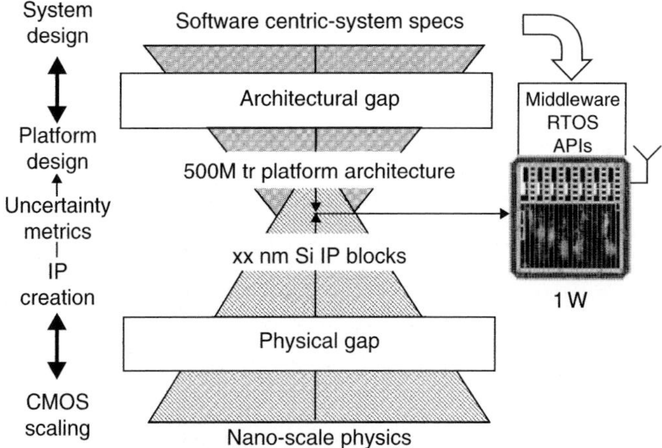

Fig. 7.2. Growing gaps between dreams and nano-scale realities

Up to 70% of chip area consists of embedded memory, which is also responsible for most of the power dissipation (Gonzales et al. 1996). Energy-efficient platforms are needed that can be adapted to new standards and applications, preferably by loading new embedded system software, or by fast incremental modifications to obtain derived products.

Figure 7.2 shows that, for the digital Watt and Milli-Watt nodes, two major gaps are popping up between AmI dreams and further nano-scaling.

First, there is an architectural gap between AmI dreams and platform architectures: System houses deliver reference specs in C++ or MATLAB to semiconductor or fabless companies. But, such reference specs show functionality without much concern for implementation. The task of platform designers is to map these into energy, and cost-efficient platforms of hundreds of processing devices and megabytes of embedded memory. Therefore, semiconductor houses have to migrate from pure component manufacturers to domain specialists, able to link system knowledge to PA nano-scale architectures.

Second, there is a growing physical gap between process technology and platform design, caused by nano-scale phenomena, such as increased leakage, intra-die variability, signal-integrity degradation, interconnect delay, and lithographic constraints on layout style. These effects jeopardize the digital abstractions now used for complexity management.

7.2.1 Managing the Architectural Gap

Improving PE of AmI systems by two orders-of-magnitude requires a rethinking of domain-specific computer architectures. Figure 7.3 shows the PE in Gops W^{-1} versus feature size, for 32-bit signal processing (Claasen 1999).

Mono-processors executing temporal computing are the easiest to program, but have the lowest PE. The way to increase their computational power is to increase clock speed to the GHz level. But, this is a poor use of scaling,

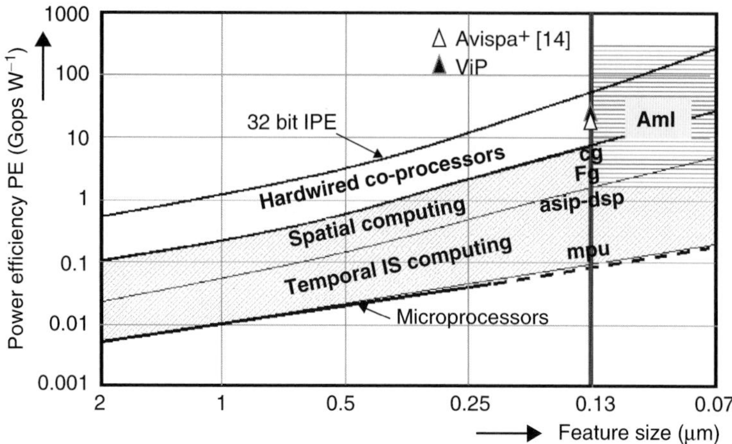

Fig. 7.3. The power-flexibility conflict

since more or less all additional hardware goes into more levels of cache and in extending instruction-level parallelism (ILP), which falls far short of the inherent task-level parallelism (TLP) and data-level parallelism (DLP) present in multi-media workloads.

Figure 7.3 shows that general-purpose microprocessors can dissipate up to 500 times more power for the same task than pure hardware accelerators in which all register-transfer operations are executed in parallel using distributed local storage. The latter is called the intrinsic power efficiency (IPE) of silicon. The challenge is to approach the IPE, but with just enough flexibility for the application domain. To reach this goal, spatial computing is indispensable, i.e., parallel architectures of communicating (temporal) processing elements that are configured at run-time under software control. Best PE is obtained if data and instructions are local to the processing elements. AmI computing platforms require a careful mixture of spatial computer architectures and accelerators at sub-90 nm with just enough embedded programmability to meet the power budget.

Next, we discuss three application-to-platform mapping techniques to combine flexibility with PE for data-intensive computing devices.

The Need for Data Locality and Code Transformations

The lack of temporal and spatial locality of data in system-level reference code for multi-media and communication leads to substantial power dissipation. This is especially so for data-dominated systems, since most of the power dissipation results from data-transfers to memory. Transfer of data from main memory, cache, and local registers, costs, respectively, at least 10, 5, 2 times as much energy as the actual operation on them by an ALU.

Catthoor et al. (2001) propose methods and tools to transform sequential C or C++ reference code into power optimized C or C++ code, helping the designer to optimize temporal and spatial locality of data production and consumption. In addition, tools have been developed to design optimal memory architectures for low power (De Man et al. 2002), and to minimize cache misses for software-controlled caches (Catthoor et al. 2002).

These data-transfer and storage exploration (DTSE) techniques have led to power gains between 2 and 10 on real-life designs. As an example, Fig. 7.4 shows the impact of DTSE on the implementation of VGA quality MPEG-4 on a Pentium-M running at 1.6 GHz. A first set of transformations of the initial code considerably reduces power-hungry memory accesses to external memory. The second set of transformations improves data locality in the buffer arrays stored in on-chip cache. The last set optimizes the storage order to reduce address computation. In total, the number of memory accesses is reduced by a factor of 18, which leads to considerable power savings. DTSE tools are now entering the EDA market (Powerscape, online).

The Need to Exploit Parallelism at Lowest Possible Clock Speed

As will be motivated when discussing the physical gap, decreasing static and dynamic power consumption requires reducing power supply voltage and clock frequency. Consequently, concurrency must be increased to meet the computational requirements. Concurrency should be detected and exploited at all levels: task level (TLP), instruction level (ILP), and data level (DLP). Since system-level reference code is mostly described in single threaded sequential C code, methods and tools are necessary to extract concurrency and to map it onto the parallel structures of the computing platform, see Gupta et al. (1999) for TLP, Schlansker et al. (1997) for ILP, and Banerjee et al. (1993) for DLP.

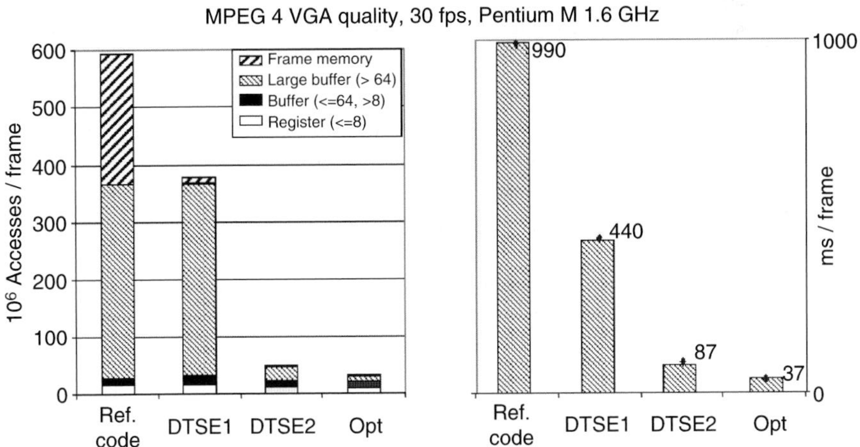

Fig. 7.4. Software for low power

The Need to Exploit Task-Level Dynamism

AmI applications are dynamic. Computational power depends strongly on image, music, or speech content as well as on the number of concurrently running applications and user quality requirements. Dynamic voltage/clock-frequency scaling (DVFS) can exploit this (Sakurai 2003). Conventional DVFS is based on a run-time scheduler employing worst-case task profiling in order to meet real-time constraints (Jha 2001). Recently, Yang et al. (2001) presented a task-concurrency method (TCM), employing a design-time energy-time trade-off Pareto analysis. Based on the restricted set of Pareto points, a simple run-time scheduler performs optimal DVFS scheduling on a multi-voltage multi-processor platform while still guaranteeing real-time behavior. It shows a factor of 2 power reduction with respect to traditional DVFS for an MPEG21 graphics application running on two StrongARM processors on different supply voltages (De Man 2002). A similar technique can be applied to cope after fabrication with process variation effects or even at run-time with dynamic (e.g., temperature induced) access time variation for embedded SRAM (Papanikolaou et al. 2005).

7.2.2 Managing the Physical Gap: Nano-scale Realities Hit Platform Architects

Scaling below 90 nm disturbs the digital abstractions, affecting both IP design and platform architecting, at a time when NRE cost is exceeding \$50 M. The main challenges are formulated below.

Gate and Sub-threshold Leakage Power Challenges

Leakage power starts to exceed dynamic power. Hi-k gate dielectrics combined with metal- or fully silicided-gates (FUSI) will be mandatory to solve the gate-leakage problem for high-performance applications, but not until 45 nm will it be used in mass production. For AmI applications, it may be cheaper to keep a thicker gate oxide and maintain performance by using strained silicon and architectural innovation.

On the other hand, sub-threshold leakage is hard to solve by semiconductor technology innovation. Scaling for performance (low V_t and V_{DD}) leads to a tenfold increase in sub-threshold leakage current per technology node. This is unacceptable for AmI applications. Scaling for low leakage (high V_t and V_{DD}) increases gate delays. Hence, larger computational power must come from more transistors rather than faster ones. Architecture and technology must be tuned to find an optimum trade-off between clock frequency, degree of parallelism, total power in active mode, and leakage power in idle mode.

Coping with Uncertainty

Nano-level scaling increases the intra-die variability of threshold voltage, drive- and leakage-current as they become dependent on the statistical distribution of doping atoms. These effects can be modeled by the V_t variability $\sigma_{\Delta Vt}$. Pelgrom's law states that $\sigma_{\Delta Vt} = A/\sqrt{(W \cdot L)}$ which shows its deterioration with scaling. Figure 7.5 shows the 3-sigma V_t spread versus technology node for minimum-size low-standby-power transistors, for both bulk CMOS and metal-gate dual-gate-fullydepleted SOI (DGFDSOI) (Yamaoka et al., 2004). Clearly, voltage headroom $V_{DD} - V_t$ (and thus I_{on} and t_d) becomes very unpredictable even for neighboring identical transistors. Figure 7.6 shows that gate delay becomes a stochastic variable. This jeopardizes timing-closure techniques and requires statistical timing-analysis methods, instead. So, in the coming years, CMOS technology will go through numerous changes that will strongly affect circuit, IP, and architecture design.

Platform architects will have to come up with new methods to design reliable electronics with uncertain components, and worst-case design must be avoided. One way to do this, especially suited for real-time applications, is by providing a run-time controller that minimizes the impact of the variability of the individual system components, as shown in Fig. 7.7 (Papanikolaou et al. 2005). This technique is based on first, at design-time, computing Pareto optimal schedules for a multi-task multi-processor architecture. This delivers a discrete set of operating points for a simple run-time task scheduler that guarantees the required cycle budget as the system-level for minimal total system energy, and selects optimal V_{DD}, clock frequency, and back-gate bias, for IP blocks (Yang et al. 2001).

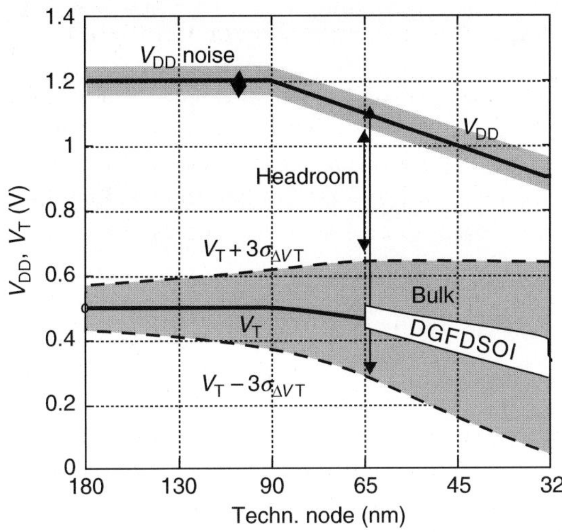

Fig. 7.5. Threshold voltage variability challenge

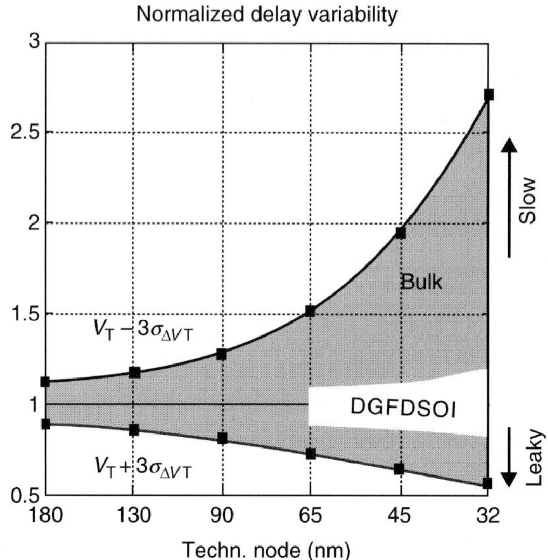

Fig. 7.6. Gate delay variability challenge

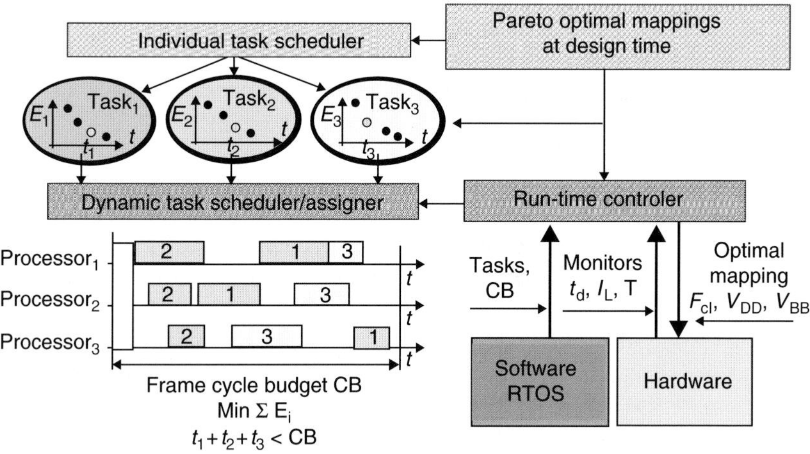

Fig. 7.7. Run-time control

Colwell et al. (2004) present another way to control V_{DD} at the clock cycle level in microprocessor contexts. Variability transforms the nominal Pareto points into point clouds in which the correct point needs to be sampled by on-chip monitors for IP block timing and energy consumption. If timing is not satisfied, a move is made to the next higher Pareto point for one of the IP blocks, satisfying the global timing constraints. This methodology impacts all stages of design, and "more-Moore" will critically depend on the availability of platform architects skilled in these new design methods.

Interconnect Challenges

Scaling causes faster logic but slower global interconnect. Chemical–mechanical polishing techniques cause thickness variability up to 40%, depending on the wiring context. And strong capacitive interline coupling leads to poor signal integrity. This will impact the way to design and lay out on-chip communication. Global bus structures do not scale well to higher complexity, and global synchronism will have to be abandoned in favor of *globally asynchronous locally synchronous* (GALS) architectures.

7.2.3 Future Computing Platforms: Towards Highly Parallel Tile-based Network-on-a-Chip (NoC)

Summarizing the above, future computing platforms will need to avoid communicating over long distances, will be as parallel and as slow as possible, and will dynamically manage whatever mother Nature brings them instead of being worst case designed. This allows us to make the following conjectures about their architecture.

Avoid Long Distance Communication

At the 45 nm node, even at a modest 500 MHz clock rate, 2 mm local copper wires represent a delay of 30% of the clock cycle. Hence, iso-synchronous zones must be restricted to compute tiles of less than 4 mm^2, or about 30 M transistors. Each tile can easily contain a synchronous bus-based multi-processor (BBPE) with local clocking, memory hierarchy, local power control, and a BIST processor.

Synchronous tiles communicate globally in an asynchronous way over low swing interconnects to guarantee both low power and ease of design. Since global buses do not scale, Dally (2002) suggested that new designs should "route packets, not wires." This leads to the NoC concept (Jantsch and Tenhunen 2003). Compute tiles route data-packets over a structured network fabric of shared wires, while a standardized network interface processor (NIP) per tile decouples communication and computation. The NIP implements the communication protocol programmed upfront from a software service layer. It hides the network details from the computing tile and thus allows for a plug-and-play design strategy necessary for fast-turn-around design of derivative platforms.

Rijpkema et al. (2003) present the first silicon implementation of synthesizable NoC IP. A router IP block can be programmed for guaranteed and best-effort services. In 120 nm, an aggregate bandwidth of 80 Gb s^{-1} in 0.26 mm^2 is reported for a 5 by 5 router. Low swing global interconnect is advantageous for low power, but is more sensitive to supply and cross-talk noise. So, we must learn to live with errors by applying error-correction (ErCorr) techniques that guarantee a bit error rate consistent with the required S/N ratio of the digital signals (Bertozzi et al. 2002). Power-hungry communication to external DRAM memory can be reduced by 3D stacking of DRAM on top of the platform chip (Beyne 2004).

The tree-like memory hierarchy within a single tile needs to aggressively exploit data and instruction locality; see Fig. 7.8. For the data hierarchy, the DRAM 3D stack feeds into a single level-2 on-chip data memory per tile, which in its turn feeds many small level-1 memories (Lambrechts et al. 2005). If memory has to be shared between tiles, a level-3 on-chip memory should also be provided but this is unlikely. For the instruction hierarchy, the DRAM 3D stack feeds into a single or potentially a few level-1 on-chip L1I caches per tile. Each of these serves several L0 buffers including local controllers to support local nested loops and conditions. The L0 instruction clusters do not necessarily correspond to the FUs combined in the data register file clusters (Jayapala et al. 2005).

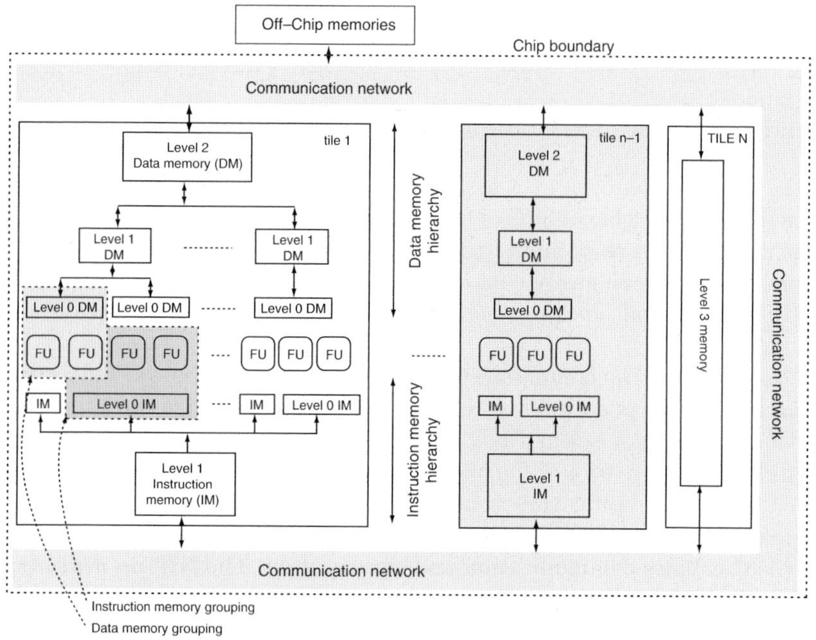

Fig. 7.8. Tree-like memory hierarchy

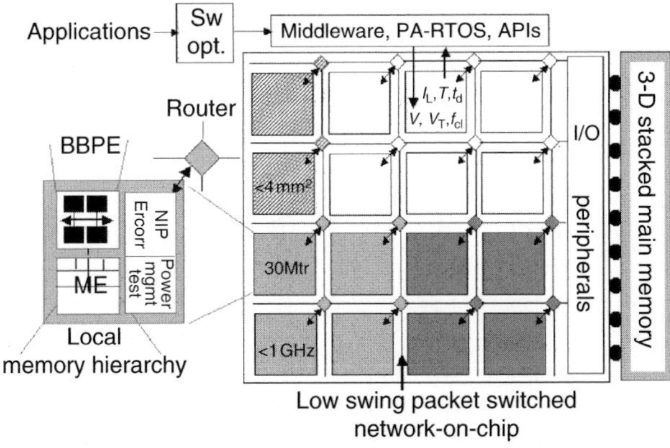

Fig. 7.9. Interconnected islands of heterogeneous multi-processors

Make It Parallel

Concurrency should be exploited at all levels. Future computing platforms will consist of interconnected islands of heterogeneous multi-processors to implement TLP, where each processor is specialized for a class of tasks; see Fig. 7.9. Within a task, various forms of parallelism exist. In data dominated applications, the majority of a task's code exhibits only moderate ILP, which can be handled efficiently by a VLIW processor; the number of instructions per cycle (IPC) is limited to less than 10. Specializing the instruction set for the targeted application domain further increases its PE. Sub-word parallelism boosts parallelism in case process samples have different word lengths. However, a small part of the code, the data intensive loops, heavily dominates execution time and possesses much more parallelism than can be handled by a VLIW processor: IPC is between 10 and 100.

Loop Level Parallelism (LLP)

LLP is a combination of the DLP available in the arrays and the ILP apparent in the processing of scalars within the loop. It can be exploited by mapping the loop kernels on a coarse grain reconfigurable array that is loosely coupled to an RISC processor (Singh et al. 2000). Such loosely coupled structures are hard to program since they need explicit copying of data and synchronization between the RISC and the array; procedural languages such as C do not offer the necessary synchronization and copying constructs. Excessive copying of the status information between the RISC and the array also leads to too high-power consumption. The RISC processor itself does not utilize the available IPC in the non-kernel code. For these reasons, a VLIW architecture template with embedded coarse grain extensions has been proposed;

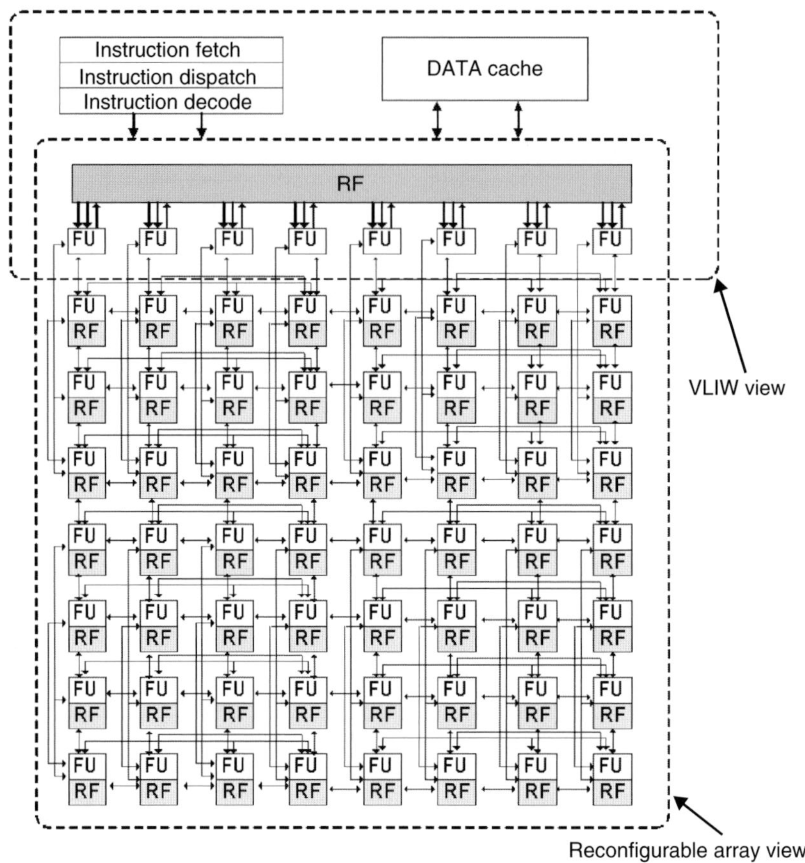

Fig. 7.10. VLIW architecture template with embedded coarse grain extensions

see Fig. 7.10 (Mei et al. 2003). The non-kernel code is executed on the VLIW processor, thereby fully using the available ILP. When entering a data intensive loop, the processor switches to 2D coarse grain array mode and offers an order-of-magnitude higher concurrency. The array mode shares with the VLIW mode the register file and the functional units, which act as the first row of processing elements of the array. This facilitates compilation from plain C since no explicit copying or synchronization needs to be done.

For each application domain, an optimized instance of the architecture template is designed. The architecture instance is described in XML and parsed by the compiler framework, which is capable of generating efficient code for the VLIW as well as the array mode directly from plain ANSI C; see Fig. 7.11. The compiler framework also generates an instruction set simulator, which gives timing and power consumption feedback. Experiments on the minimum mean squared error (MMSE) function of a multiple antenna wireless receiver indicate that the coarse grain extensions to a VLIW allow to lower the clock frequency from 2.1 GHz to 140 MHz, boosting up the average IPC to 35.

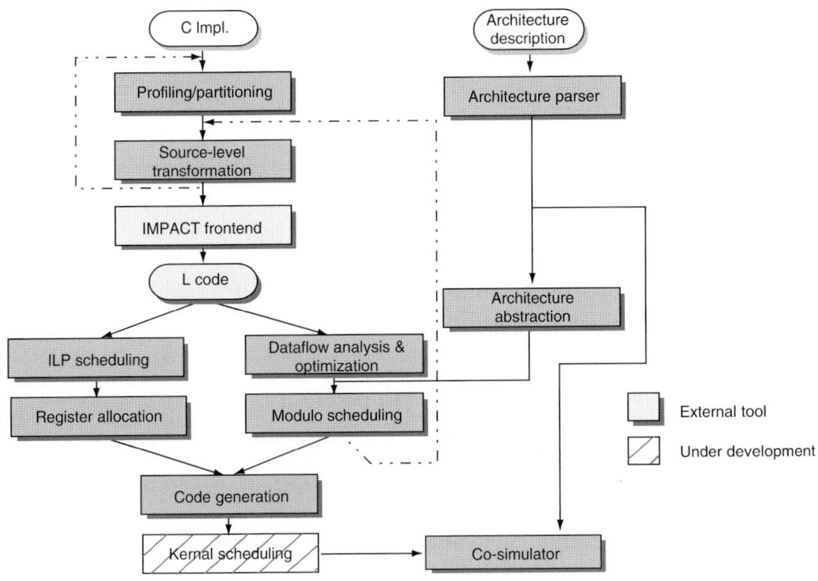

Fig. 7.11. Compiler framework for the coarse grain architectural template of Fig.7.10

Enable Dynamic Management

Temporal aspects, whether caused by leakage variation, dynamic application load variations or changes in the ambient temperature, require run-time calibration of energy and time to determine the instantaneous position in the Pareto clouds and enabling the run-time Pareto controllers to select the optimum combination of Pareto points that lead to the lowest energy consumption but still meeting the real-time deadline (Yang et al. 2001, Papanikolaou et al. 2005). Innovative circuits have to be developed to perform this in-circuit run-time calibration, for instance, a calibration circuit for on-chip SRAM (Geens and Dehaene 2005).

7.2.4 The Devil Is in the Software

AmI systems must be adaptive to new software services, standards, communication protocols, etc. As a result, a number of software layers must be co-developed. First, similar to the Java virtual-machine concept, a middleware layer is required to ensure interoperability. Second, a platform-specific, power-aware (PA) RTOS is needed to schedule tasks dynamically on the computing tiles, and to minimize power consumption at run-time. Finally, low-level APIs are needed for the processors in the tiles. The development cost of this Hardware-dependent Software (HdS) easily exceeds that of hardware development itself, and adds considerably to the NRE cost. Therefore, the development of such complex systems will be restricted to a few grand industry alliances that can organize disciplined armies of engineers to design the AmI products of the future.

7.3 "More-Than-Moore": Ultra-creativity for Ultra-low Power and Cost

Design of stationary and nomadic devices is all about managing giga-scale system complexity implemented in nano-scale CMOS. In contrast, design of wireless-transducer-network devices requires creative engineering to get to the ultimate limits of miniaturization, cost reduction, and energy consumption. This leads to the need for "more-than-Moore", which is a cost-effective integration of CMOS with MEMS, optical- and passive-components, new materials, bio-silicon interfaces, lifelong autonomous energy sources, and grain-size 3D packaging. The complexity is not in the number of transistors, but in combining technologies, circuit- and global-networking architectures to obtain utmost simplicity for the sensor nodes themselves.

Microwatt devices are low-duty-cycle ($< 1\%$) low-throughput (1 bps to 10 kbps) micro-systems that unify nearly all design art in one package: sensor, signal conditioning, A/D conversion, signal processing, PA MAC layer, pico-radio, antennae, energy management, and energy scavenging.

Figure 7.12 shows an IMEC SiP realization of a 1.4 cm^3, 2.4 GHz EEC, ECG sensor mote using laminate packaging of bare dies, a solar-cell battery charger, and integrated antenna. This system consumes 500 μW at 400 bps and 1% duty cycle. However, further integration to e-grain size and lower power will be necessary. Indeed, for true energy scavenging, only solar cells, piezo-electric MEMS (for detecting vibrations and shocks) and thermal generators have proven to be successful, but their average power capacity is limited to 100 μW cm^{-3} (Roundy et al. 2004). This means less than 10 mW peak power during active periods for a 1% duty cycle. So, peak power per subfunction should be below 2 mW. For 1% duty cycle, and in 90 nm technology this allows for about 5 M 8 bit ASIP operations per second for all data and signal processing, and less than 2 nJ bit^{-1} transmission energy for 10 kbps. This requires an order-of-magnitude power reduction with respect to Zigbee and Bluetooth, especially in the RF part. This can only be obtained by covering very short distances (< 10 m) and/or using multi-hop networking and

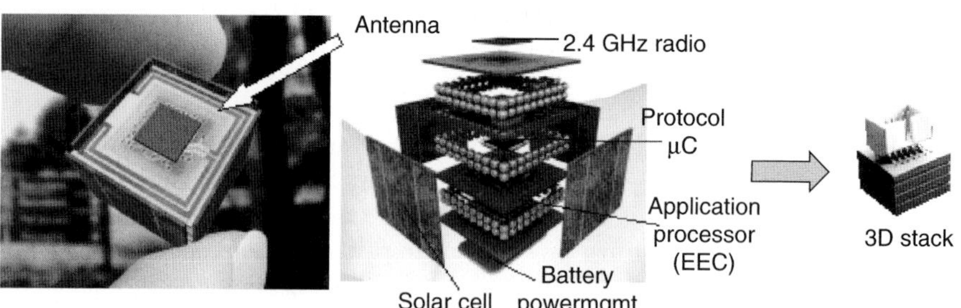

Fig. 7.12. SiP intelligent sensor node for epilepsy diagnosis

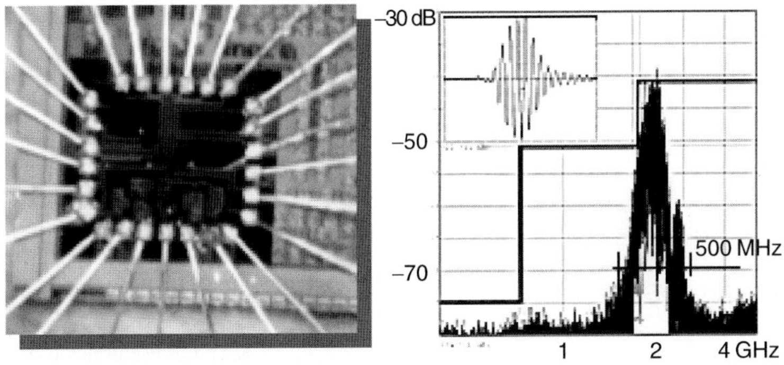

Fig. 7.13. Flexible ultra-low-power UWB transmitter

RF architectures of utmost simplicity. Pioneering work in this direction can be found in Rabaey et al. (2002) and Roundy et al. (2003).

For BANs and accurate positioning, UWB radio could be a solution. Figure 7.13 shows an IMEC design of a 180 nm, 0.6 by 0.6 mm^2 UWB transmitter in the 3–5 GHz band. All circuits are digital except for the power-output stage and a triangular wavelet shaper. This extreme simplicity leads to 0.5 nJ bit^{-1} transmit energy at 10 kbps and 10 pulses bit^{-1}. Active power is 2 mW. CSEM and Delft University approach UWB by using ultra-wide band FM (UWBFM), which leads to very simple Tx/Rx architectures (Gerrits et al. 2004).

Clearly, there are great engineering challenges in this domain, which is so crucial for Ambient Intelligence. Not the least of the challenges is security of sensor networks, which together with network protocol and data storage and lightweight operating system requires most of the computational power (Tinyos, online). New ideas in efficient, safe, and cheap encryption for sensors are urgently needed, and will require ultra-low-voltage (< 500 mV), low-leakage computation such as proposed by Wang et al. (2004) and Calhoun et al. (2005). Ishibashi et al. (2003) present a 32-bit adder in 130 nm that reaches 0.3 pJ/add at 300 mV with forward body bias. With such a circuit technique and using massive parallelism, one would be able to reach about 1 Gops mW^{-1}, which opens great perspectives if we can overcome the architectural and nano-scale challenges described above.

Massive deployment of sensors (or RFID tags) requires cost reduction to the single-dollar or even single-cent range, but this conflicts with very cheap integration of standard CMOS with non-CMOS devices, such as passive components and MEMS, and even with bio-tissues. SiP techniques are too expensive. Techniques for low-temperature wafer-scale integration "above silicon" are being developed. They allow for the "reuse" of a silicon wafer, and for adding passive components and MEMS on top of them, at low cost. Figure 7.14 shows above-IC processing for depositing high-quality RF passives

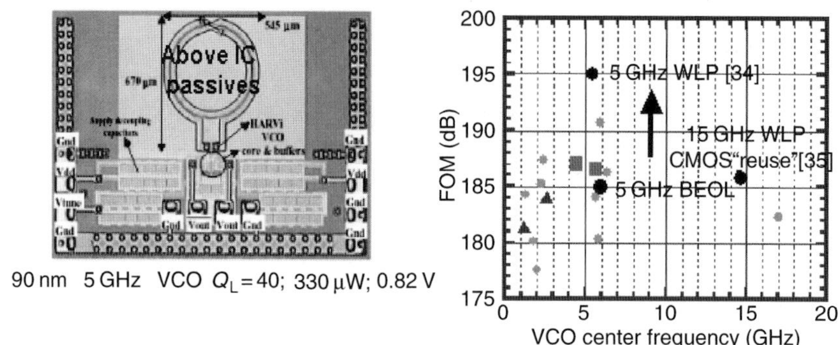

Fig. 7.14. Above-IC processing of passive components leads to higher performing circuits

(Linten et al. 2004). This allows a 90 nm CMOS VCO to run at 0.82 V for 330 μ W power and $-155\,\mathrm{dBc\,Hz}^{-1}$ at 1 MHz phase noise. Carchon et al. (2005) show how "reuse" of the 5 GHz CMOS part with another above-IC inductor leads to a 15 GHz oscillator with a better figure-of-merit than with state-of-the-art full-CMOS integration.

Healthcare and wellness will be an important application domain for Ambient Intelligence, and interfacing between electronics and living bio-tissues will be of crucial importance. Fromherz (2005) and Lee et al. (2005) present new breakthroughs in coupling bionics to electronics and in cell manipulation. For more information about this subject we also refer to Chap. 4.7.4 of this volume.

7.4 Conclusions

Ambient Intelligence is the next wave of information technology for enhancing human experience and improving the quality of life. It implies a consumer-oriented industry, driven by software from the top, and enabled and constrained alike, by nano-scale physics at the atomic level.

"More-Moore" will be needed to deliver Giga-ops computation and Giga-Hz communication capabilities for stationary and wearable devices. The grand challenge will be to design flexible multi-processor platforms with two-orders-of-magnitude lower power dissipation than today's microprocessors at one twentieth of the cost, while coping with the realities of nano-scale physics. We have presented a number of emerging techniques to cope with this challenge. Design of such systems will depend on our ability to create multi-disciplinary alliances capable of covering the huge span between AmI dreams and their implementation in the interaction of billions of nano-scale devices.

On the other hand, "more-than-Moore" technology is needed for the design of autonomous wireless sensor networks. The complexity is not in the number of transistors, but in clever combinations of technologies, circuits, and system architectures to design ultra-low-power, ultra-low-cost, ultra-simple sensor nodes for Ambient Intelligence.

8

Software Platforms

N. Georgantas, P. Inverardi, and V. Issarny

"Object-oriented, component-oriented, service-oriented. Is "human-oriented" next?"

8.1 Introduction

Ambient Intelligence is an emerging *user-centric* service provision paradigm that aims at enhancing the quality of life by seamlessly offering relevant information and services to the *mobile* user, anywhere and at anytime. The *ubiquitous* property applied to both computing and networking implies a useful, pleasant, and unobtrusive presence of the system everywhere – at home, en route, in public spaces, at work. Omnipresence of the system entails further its capacity to integrate diverse, *heterogeneous* computing and networking facilities, and to provide service in an ad hoc way, where the users and the system have no mutual a priori knowledge. While a number of base enablers such as wearable and handheld computers, wireless communication, and sensing mechanisms are already commercially available for deploying base infrastructures supporting the AmI vision, the development of AmI software systems still raises numerous scientific and technical challenges due to the distinguishing features of Ambient Intelligence, which may be encapsulated in the above introduced notions: *mobility*, *heterogeneity*, *ad hoc nature*, *ubiquity*, and *user-centrism*, where *dynamics* shall be added as a global property encompassed by all the others.

These notions and their special meaning in Ambient Intelligence are extensively discussed in Sect. 8.2 of this chapter from the standpoint of *software systems*. More specifically, two major paradigms in the software domain, namely *software architectures* (SA) and *middleware*, are identified as most suitable instruments towards mastering the distinguishing features of Ambient Intelligence. Specific requirements for SA and middleware towards Ambient Intelligence are then elicited out of a brief overview of the state of the art in these two fields. Further, drawing from fundamental principles of SA and middleware and the identified requirements, a generic *architectural framework* for AmI software systems is outlined; it comprises a number of architectural entities abstracting essential features of AmI infrastructures, and a set of conceptual viewpoints allowing the methodical study of these infrastructures.

The realization of the AmI vision or the relative pervasive computing paradigm has been the objective of numerous research efforts in the software engineering domain. In Sects. 8.3 to 8.5, we survey three comprehensive research attempts, under the names of *Aura*, *Gaia*, and *WSAMI*, aimed at providing feasible solutions to (part of) the requirements identified in Sect. 8.2. We study these three systems on the basis of our generic architectural framework, considering them as distinct instantiations of this framework. This enables aligning the observations coming from the three systems, which leads us, in Sect. 8.6, to evaluate the degree to which these systems cover the identified distinguishing features of Ambient Intelligence. This comparative assessment allows us to draw conclusions about ongoing and future research in the area of software platforms for Ambient Intelligence.

8.2 Software Systems for Ambient Intelligence

Software engineering for Ambient Intelligence benefits from recent developments in the software domain, especially those that address distributed software systems. A number of software infrastructures were introduced in the early 1990s to facilitate the development of distributed systems, based on the *object-oriented* paradigm (e.g., CORBA, Java-RMI, DCOM). Using an object-based software infrastructure, distributed applications and higher-level system functionalities may be developed using a core execution environment, which offers basic functions for managing objects life cycle and remote interactions (Emmerich 2000). Although this significantly eases development of distributed applications, application developers still have to devise solutions to the enforcement of non-functional properties like dependability and persistence management. This has thus led to the emergence of *component-oriented* (Szyperski et al. 2002) technologies at the end of the 1990s, which introduce the notion of container. A container is an object host that enforces key non-functional properties for business applications like transaction and security management (Emmerich 2002). Software technologies have further evolved towards *service-oriented* computing (Papazoglou and Georgakopoulos 2003) in the early 2000s, so as to support the development of open distributed applications over the Internet. This allows packaging applications as autonomous services for access and possible composition with other applications over the Internet.

These advances in the software domain provide the base for AmI software systems. Nevertheless, addressing the distinguishing features of Ambient Intelligence requires further dedicated software engineering research. Thus, research in the area of distributed systems shall result in generic solutions to ubiquitous computing and networking on top of heterogeneous infrastructures for AmI environments. To this end, the *middleware* paradigm – key element in all the software technologies outlined above – provides the most promising architectural choice for AmI software systems. Furthermore, software engineering research shall lead to a solid foundation for the thorough development of AmI software systems. To this extent, the *software architecture* paradigm (Shaw and Garlan 1996; Bernardo and Inverardi 2003) appears to be the right abstraction, around which the entire software life cycle can be organized. In the following two sections, we thoroughly discuss the requirements for both SA and middleware posed by Ambient Intelligence. Based on the elicited requirements, we then introduce a generic architectural framework for AmI software systems.

8.2.1 Software Architectures

Software architectures are the stage of software development in which the structure of the system is outlined by identifying: (1) the decomposition level of the system, i.e., the *components*, and (2) a first model of the system's dynamic behavior, i.e., how components will interact through *connectors*. SA are design artifacts, used to improve the comprehension of the system, and to clearly document the development of the system in order to better face maintenance and evolution (Garlan 2000). Recently, due to the increasing use of component-based technologies and supporting middleware to develop applications, SA are being promoted as run time artifacts, playing the role of dynamic system representation in order to guarantee some degree of consistency of the system structure in the presence of evolution during system execution (Aldrich 2003; Morrison et al. 2004).

By allowing static and dynamic system description, SA support predictive – both qualitative and quantitative – analysis, verification, and validation of software systems, thus improving the confidence on the system behavior, and enhancing the dependability of the developed systems. In the literature, several approaches for the description and analysis of SA have been proposed (Issarny et al. 2002; Bernardo and Inverardi 2003; Balsamo et al. 2004). However, current notations and techniques are inadequate to address the new requirements that Ambient Intelligence poses on SA. This is in particular due to the dynamics of AmI applications with respect to both user-centric and computer-centric contexts.

Referring to the distinguishing features of Ambient Intelligence, from the SA perspective, *dynamics* amounts to coping with applications whose structure and behavior can change, i.e., components can appear/disappear during the computation. *Mobility*, whether logical or physical, also implies dynamism, since it changes in general the behavior of a system. *Heterogeneity* might require specific components that allow the application to uniformly run on the infrastructure. *Ad hoc nature* forces the adoption of specific topologies (e.g., peer-to-peer) and interaction modalities (e.g., service discovery). *Ubiquity* induces the choice of highly scalable SA. *User-centrism*, targeting minimum user distraction and proactiveness, puts more behavioral burden on the components that support the user and requires more sophisticated interaction and coordination modalities among components; it might also require self-adaptability of the application.

The state of the art in the field of SA for distributed systems that address the above properties is much preliminary and fragmented, focusing on a small set of system attributes. SA for self-adaptive systems can be found in the literature as, e.g., illustrated in Sect. 8.3 with the Aura system. However, proposed approaches lack associated design and validation methodologies. Furthermore, they address very specific adaptation, i.e., with respect to computer-centric context-awareness. Comprehensive design and validation of self-adaptive systems require addressing both functional and non-functional properties and the highly dynamic environment. In addition, adaptation applies to both the application and middleware layers.

Other relevant work has been carried out on architectural paradigms and middleware that support communication between distributed, mobile components (Murphy et al. 2000; Bernardo and Inverardi 2003). These solutions are interesting from an architectural viewpoint. In Murphy et al. (1999), a middleware is introduced where communication between components is mediated by a virtual common workspace (Carriero and Gelernter 1989). The peer-to-peer architectural paradigm applied in Sailhan and Issarny (2003) is much suited for ad hoc (Blair and Campbeel 2000) provides mechanisms for dynamic self-reconfiguration based on a meta-model of the middleware itself. Finally, dynamic service discovery and composition are an emerging research area primarily addressed in the context of middleware systems that also impacts the SA domain. This raises a number of challenges related to security, dynamic type checking, service profile publication and retrieval, service interference, new binding schemes that cater for quality of service (QoS), and integration of heterogeneous components. Research in analysis and verification of SA of evolvable and composable services is quite immature, though some approaches are emerging (Baresi et al. 2003; Issarny et al. 2004).

8.2.2 Middleware

Middleware is a software layer that stands between the networked operating system and the application (Bernstein 1996), and provides well-known, reusable solutions to frequently encountered problems like heterogeneity, interoperability, security, dependability. As suggested in the previous section, middleware constitutes a key architectural building block of AmI systems. Nevertheless, a number of challenges remain in addressing the distinguishing features of Ambient Intelligence.

A significant enabler of systems' *ubiquity* is availability of networked computing resources in all situations. This can only be made possible by the combined exploitation of both ad hoc and infrastructure-based wireless network technologies, and of the diversity of wireless devices. Relevant solutions include networked operating systems for tiny-scale devices, e.g., (Levis et al. 2004), together with the availability of an all-IP network over most network technologies. However, integration of the various networked devices cannot be fully solved at the networked operating system layer only. Higher-level abstractions need to be provided at the middleware layer to allow AmI applications to be distributed over various nodes. A simple example is enabling access to rich multimedia content within the networked home and its display on the most relevant devices according to the location of interested users. Further, by their very nature, AmI systems must be open, and ease access and sharing of resources to users. Therefore, ubiquity additionally poses the problem of adequate management of the user's identity to ensure security and privacy.

User-centrism in AmI systems requires adaptation of applications to their environment and users, which is often called context-awareness. Many middleware systems have concentrated on abstracting the complexity of context-awareness (Dey et al. 1999; Kidd et al. 1999). However, most systems that have been introduced focus on adapting applications to system resource-related context, as opposed to user-related context.

Resource-aware adaptation includes supporting the dynamic composition of AmI applications according to networked resources, so that the resulting composition realizes the target functional and non-functional behavior, which reveals the *dynamic* and *ad hoc nature* of such applications. This functionality is facilitated by resource discovery protocols that allow dynamically advertising, discovering, and accessing resources that are available in the environment. However, existing resource discovery protocols allow seeking resources according to the target functional behavior, but hardly take account of the QoS that is to be delivered.

Dealing with the dynamics of AmI applications is closely related to device *mobility* management at the middleware layer. In particular, there has been extensive study on middleware systems managing the distribution of content over mobile nodes. Results in this area include solutions to the handling of disconnected operations (Kuenning and Popek 1997), the caching of nomadic data on untrusted servers, the sharing of data in ad hoc networks (Mascolo et al. 2001; Sailhan and Issarny 2003), and the location of mobile content (Castro et al. 2001).

Furthermore, ad hoc composition of AmI applications in the presence of user mobility requires dealing with the *heterogeneity* of the software infrastructure and in particular of the middleware (Grace et al. 2003). Specifically, an application implemented upon a given middleware cannot interoperate with services developed upon another. Middleware heterogeneity manifests itself in the two major middleware functionalities, i.e., service discovery (Bettstetter and Renner 2000) and service interaction. Thus, middleware shall overcome two heterogeneity issues to provide interoperability in the AmI environment, i.e., heterogeneity of service discovery protocols (Bromberg and Issarny 2004), and heterogeneity of service interaction protocols. Additionally, heterogeneity may be due to versioning. Different deployed versions of service discovery and interaction protocols hinder device interoperation.

8.2.3 Architectural Framework for Software Systems

Following our discussion on SA- and middleware-related AmI requirements, we introduce the *architectural framework* for AmI software systems depicted in

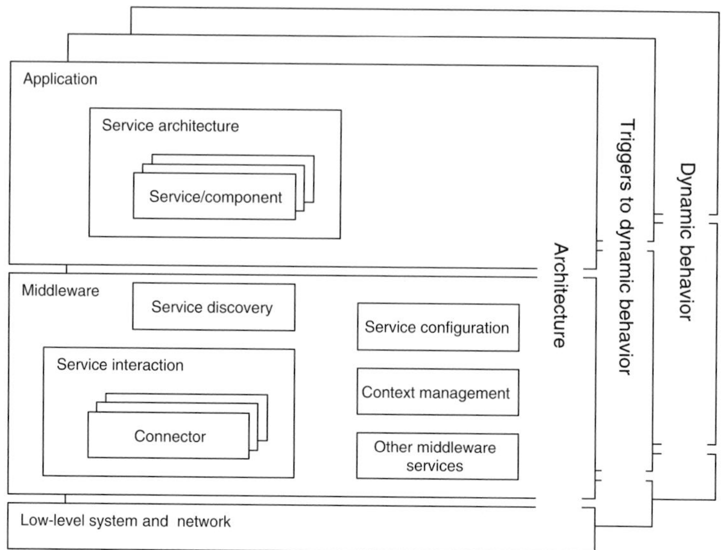

Fig. 8.1. Generic architectural framework

Fig. 8.1. We apply a typical three-layer structure, i.e., Application–Middleware–Low-level system and network, which is followed by all middleware-based distributed software systems.

In the *Application* layer, applications are formed from networked *services–components* in SA terminology – which conform to a specific *Service Architecture*; the latter prescribes the way in which service implementation details are abstracted to enable service discovery and invocation.

The *Middleware* layer incorporates a number of essential functionalities already discussed in the two previous sections, such as *Service Discovery* and *Service Interaction*, the latter based on one or more interaction mechanisms represented by *connectors* in SA. Further, *Service Configuration* integrates all mechanisms that support dynamic composition, configuration and reconfiguration of distributed applications from networked services and related connectors in the AmI environment. The above middleware functionalities are complemented by *Context Management* and a number of *other middleware services*, e.g., supporting data management and non-functional properties such as security and QoS.

Finally, the *Low-level system and network* layer integrates the networked operating system, device drivers, and software libraries providing base system and network functionalities on which the middleware executes.

In the sections that follow, we survey Aura, Gaia, and WSAMI as instantiations of the generic architectural framework. We follow three viewpoints in the study of each system. Besides *Architecture* of the system, *Triggers to dynamic behavior* concerns the different dimensions of input information that the system is able to perceive and that stimulate the system's activity, such as target functional properties, QoS and resource-control requirements, context data, and user requirements and preferences. Then, *Dynamic behavior* concerns the dynamic discovery and retrieval of application services, resources and middleware services; and the dynamic composition, (re)configuration and self-adaptation of applications. Certainly, there are other important viewpoints, such as data management, dependability or performance, which we had to omit due to space limitations. The goal of our study is to review how Aura, Gaia, and WSAMI respond to the requirements identified for both SA and middleware, under the light of the distinguishing features of Ambient Intelligence.

8.3 Aura: User Task-Driven Environment Configuration

The key objective of Aura is to allow users to focus on their real tasks rather than being distracted by dealing with the configuration and reconfiguration of computer systems to support these tasks (Garlan et al. 2002). The Aura infrastructure performs automatic configuration and reconfiguration of pervasive computing environments according to the user's task and intent.

8.3.1 Architecture

The Aura infrastructure (Sousa and Garlan 2002; Sousa and Garlan 2003) comprises a number of architectural components, superposed on a layered architecture, as depicted in Fig. 8.2. The *Task Management* layer captures knowledge about user's tasks and associated user's intent and context. Based on this information, the *Task Manager* (TM) coordinates the configuration and reconfiguration of the environment to best serve the user. The *Environment Management* layer provides a level of indirection between the environment-independent requests made by the Task Management and the concrete applications and devices of the *Environment* layer. *Suppliers* abstract these applications and devices; they are employed by the *Environment Manager* (EM) to support a user's task. Finally, the *Context Observer* collects and provides physical context information to TM, EM, and context-aware applications. An environment has one instance of TM, EM, and Context Observer, and several instances of suppliers. Thus, at the level of a self-contained environment, the Aura infrastructure is centralized, meaning that a number of central entities concentrate knowledge and coordination of the functionality of the environment. Communication among all Aura components is asynchronous, employing XML-based messages.

Suppliers are implemented by wrapping existing applications to conform to the infrastructure's configuration APIs. For example, *MS Word* and *Notepad* can each be wrapped to provide a *text editing* service. Further, suppliers take care of service interconnection by employing the existing communication infrastructure of the environment (networks and middleware), according to the specific characteristics of the encapsulated application.

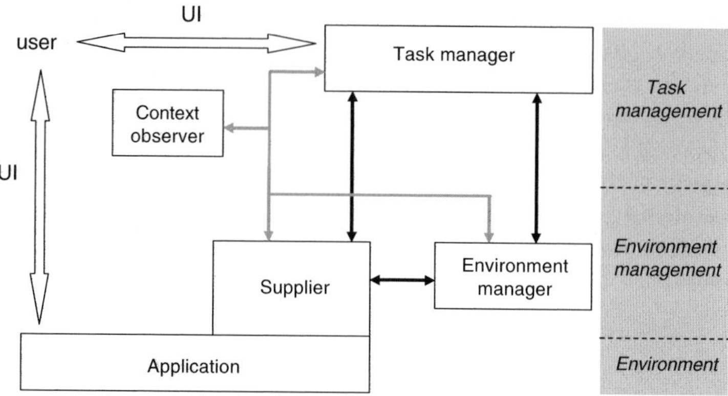

Fig. 8.2. Aura architecture

Aura puts emphasis on the dynamic configuration of the environment carried out jointly by the TM and the EM, which embody the Service Configuration functionality of the generic architectural framework. Then, service components and connectors of the framework are not prescribed by Aura: existing applications and communication mechanisms are employed, while suppliers wrap both for configuration purposes.

8.3.2 Triggers to Dynamic Behavior

Aura's dynamic behavior aims at optimally matching user tasks to environment functionalities. Different understanding about tasks is established at the different levels of the Aura architecture. Thus, at TM level, the user has a high-level view about the abstract services (e.g., *language translation*) and service interconnections (e.g., *text in, text out*) that he/she requires for his/her task. At Environment Management level, there is a deeper knowledge: EM maps services to concrete suppliers (e.g., *babel fish translation*) and service interconnections to concrete architectural connectors (e.g., *pipe over TCP*).

Focusing on task description at Task Management level, this is expressed as a number of alternative sets of services and service interconnections that can support the task. This description is complemented with a set of user preferences, which comprise: (1) *configuration preferences*, which capture the preferred services and service interconnections to support a task, (2) *supplier preferences*, which capture the preferred components to supply the required services, and (3) *QoS preferences*, which capture the acceptable QoS levels and preferred tradeoffs.

Aura integrates a mathematical model for expressing each type of user preferences and their aggregation in the form of a *utility function* (Poladian et al. 2004). Thus, a utility value between 0 and 1 may be calculated for each type of user preferences, where a value close to 1 means that preferences are satisfied. Computing the optimal matching between user tasks and environment functionalities corresponds to maximizing the aggregate utility function.

The configuration preferences model reflects how happy the user is with each possible set of services and service interconnections that can support the user's task. The supplier preferences model reflects: (1) how happy the user is with the choice of a specific supplier for a specific service, and how much he/she cares, (2) how much the user is willing to wait for suppliers to be set up, for a specific assignment of suppliers, and how much he/she cares, and (3) how happy the user will be if the supplier for a specific service is replaced during execution. The QoS preferences model reflects for each QoS dimension of a service or connection, how happy the user is with each supported level of the specific QoS dimension, and how much he/she cares about this dimension.

8.3.3 Dynamic Behavior

Aura's utility model is used to guide both the initial configuration and the ongoing reconfiguration of the environment. Further, a service registering and matching mechanism is employed. Suppliers register with EM a description of the services they offer, as well as of the ways these services are reachable through connectors. Thus, upon a task request, EM identifies specific suppliers as able to serve a user task. Further, it checks suppliers for interconnection compatibility and interconnects them through specific connectors.

Dynamic task configuration is carried out as follows. The user describes to TM the requested task and his/her configuration preferences. TM determines the alternative sets of services and connections that can support the task, and communicates them to EM. For each set of services and connections, EM – using the supplier and QoS preferences models – determines the optimal supplier and connector assignment as well as the QoS levels and resource consumption bounds at which each supplier and connector should run, considering the QoS and resource profile of each possible supplier. For this assignment, a utility value is calculated. EM returns this information to TM. Finally, TM – using the utility values for all alternative sets of services and connections and the configuration preferences model – determines the optimal set of services and connections for supporting the user's task and asks EM to set them up. Furthermore, the employed suppliers use determined QoS levels and resource consumption bounds to determine appropriate resource-adaptation policies for the wrapped applications and connectors.

Dynamic task reconfiguration may be carried out at different levels:

- At Task Management level, a service or connection currently used may be replaced by another service or connection. These are changes initiated by the user or by some context change.
- At Environment Management level, a supplier or a connector currently used may be replaced by another supplier or connector, or new resource consumption bounds may be established for a supplier. These are changes

initiated by EM, when the employed suppliers or connectors no longer offer the best utility for the requested services and connections. Causes for this may be: components failing or becoming disconnected, new components becoming available, or a significant resource variation in the employed components. To avoid user distraction in case of task upgrading, context information or the user himself/herself may be consulted through TM before performing a reconfiguration.
- At Environment level, resource-adaptive applications may handle resource variations locally, according to resource-adaptation policies determined by the associated suppliers.

Aura's dynamic behavior is further supported by the capability of saving the user-level state of a task, which concerns properties such as the services and connections employed, or user-interaction parameters like window size, cursor position, etc. TM captures the user-level state of a task by polling each supplier. Capturing and recovering the user-level state is done: (1) when the user moves from one environment to another or when the user swaps one task for another, in order to enable task suspension and resumption, or (2) upon reconfiguration of a task. State recapturing is triggered by some change in the user's context or intent that indicates that the task is about to be suspended. It may also be done periodically to ensure recovery of an almost up-to-date state in the case of a failure.

Finally, Aura supports composite services. A composite service is provided by a *composite supplier*, which coordinates the suppliers of the component services. The structure of a composite service is exposed to both EM and TM, which control the composition.

8.4 Gaia: Programmable Active Spaces

Gaia is built upon a meta-operating system developed as a distributed middleware infrastructure that coordinates networked devices and software components contained in a physical space, thus, rendering this space an integrated programmable environment, called an *active space* (Roman et al. 2002). By complementing the core meta-operating system with an application framework, Gaia enables the development, deployment, and execution of distributed, multi-device, mobile applications within active spaces. Gaia builds upon a CORBA platform.

8.4.1 Architecture

The Gaia infrastructure comprises three major components: *applications*, the *application framework*, and the *kernel*, as depicted in Fig. 8.4. The Gaia application framework (Hess et al. 2002; Roman and Campbell 2003 consists of:

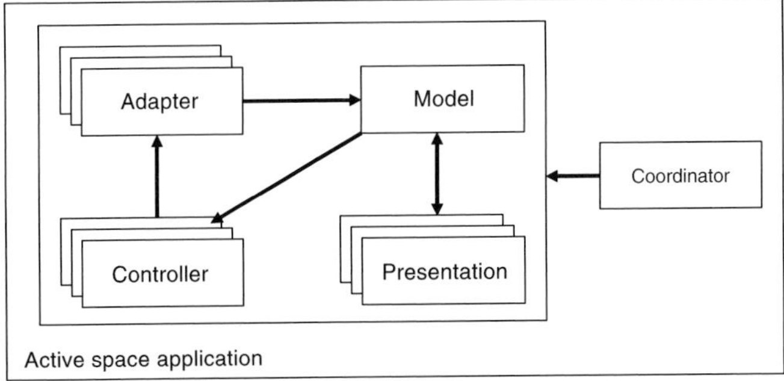

Fig. 8.3. Gaia application model

(1) a component-based application model, (2) application management (AM) functionality, which performs tasks such as dynamic configuration or reconfiguration of applications, and (3) a set of policies, which customize AM tasks. The Gaia application partitioning model adopts and extends the traditional Model–View–Controller (Krasner and Pope 1988) model, defining five components, as depicted in Fig. 8.3: (1) *model*, which implements the application logic; it can be a single component or a collection of distributed components, (2) *presentation*, which transforms the application's state into a perceivable output representation, such as a graphical or audible representation; multiple presentation components may be attached to a model component, (3) *controller*, which is an input entity (e.g., GUI, sensor) that can alter the state of the application; multiple controller components may be attached to a model component, (4) *adapter*, which maps events generated by controllers into requests to the application model, therefore decoupling controllers from specific models, and (5) *coordinator*, which encapsulates information about the application components' composition and enables altering this composition.

The Gaia kernel (Roman et al. 2002) comprises: (1) a *component management core* (CMC), which allows creating and destroying components on any execution node in the active space, and (2) a set of basic services employed by applications (Fig. 8.4):

- The *event manager* implements an event-based communication mechanism based on decoupled suppliers, consumers, and channels. The event manager manages all channels, providing a single, centralized entry point for events.
- The *context service* infrastructure may comprise several *context providers*, which are specialized to different context information. A registry supports retrieval of available context providers.
- The *presence service* detects – proactively or reactively – and maintains presence information about all active space entities, i.e., software components, devices, and people. Diverse mechanisms are used depending on the

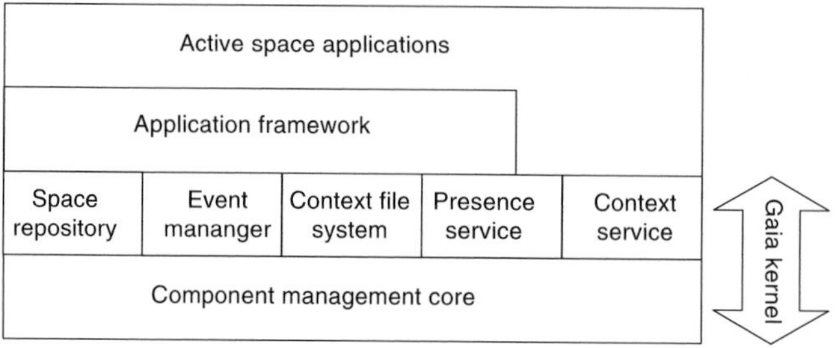

Fig. 8.4. Gaia architecture

target entity, such as: (1) heartbeats on a default event channel sent periodically by software components and devices to notify the service that they are in the active space, (2) RF active badges carried by users, (3) a beacon broadcasted by the service, publishing references to the Gaia kernel services.

- The *space repository* stores information about all active space entities and allows browsing and retrieving entities on the basis of specific attributes. All entities are associated with an XML description that contains their properties; users have user profiles.
- The *context file system* is a distributed, context-aware file system that uses application-defined properties and environmental context information to proactively organize and make available user's data.

The Gaia kernel shall be hosted by every node of the active space. However, most of the kernel services include centralized entities that concentrate knowledge and coordination of the functionality of the environment. These entities are hosted by specific nodes of an active space. Thus, at the level of an active space, the Gaia infrastructure is centralized.

In Gaia, event-based communication is used for notifying about changes in the active space or in the state of components, for both kernel and application components. Further, Gaia supports synchronous (RPC) communication. Application components may as well use other communication mechanisms available at nodes of the active space, such as streaming.

The middleware-oriented Gaia architecture follows pretty closely the generic architectural framework by clearly prescribing a Service Architecture (the Gaia application model), and a Middleware layer with rich functionalities, such as an advanced Service Discovery mechanism jointly supported by the presence service and the space repository. Two Service Interaction mechanisms are provided, based on CORBA connectors.

8.4.2 Triggers to Dynamic Behavior

Gaia's applications are associated to users. Gaia's dynamic behavior aims at mapping applications to available resources of a specific active space. To this end, applications are described in a generic way and instantiated for specific active spaces. An *application generic description* (AGD) is an active space-independent configuration description, listing generic application components and their requirements, e.g., use of a specific OS. An AGD is created by the application developer. Application instantiation and other AM tasks can be customized according to application framework policies. Policies can be default, provided by the application framework, or defined by application developers or users.

Furthermore, Gaia's dynamic behavior may depend on context. Context-aware applications are enabled by incorporating context controllers and by relying on the context service. Context controllers receive context information from specific context providers and synthesize specific context events that alter the application's state. Further, the presence service captures and provides a specific type of context information, which can be used by both applications and kernel services.

8.4.3 Dynamic Behavior

Gaia supports a mechanism for registering and discovering active space entities. A device entering an active space receives the beacon broadcasted by the presence service, resolves the event manager, and initiates the heartbeat mechanism. The presence service receives the heartbeat event and sends a presence event to notify the rest of the space about the new device. The space repository receives the presence event, contacts the device to retrieve its XML description, and stores the information. In a similar way, a user entering an active space, carrying an RF active badge, is detected by the presence service, and has his/her profile stored in the space repository.

Dynamic application instantiation is carried out as follows. AM uses the related AGD and the space repository to map an application to available resources of a specific active space, thus generating an *application-customized description* (ACD). This mapping can be assisted by the user, or be automatic following application framework policies. An ACD is a concrete configuration description, listing specific application components and their execution nodes. Then, AM uses the ACD and CMC to instantiate and assemble the application components. All components that were not already active initiate the heartbeat mechanism, and are thus registered with the active space in the same way as described above for a device.

Gaia allows changing the composition of an application dynamically upon user's request. The user may indicate a new device providing an application component that should replace a component currently used. AM uses CMC to create the new component – if not already active – and possibly destroy the old one and updates the application coordinator component with the new application composition.

Gaia supports further application mobility between active spaces (Roman et al. 2002) through application suspension and resumption. The application framework allows saving the state of the application at two levels. At the model component level, the functional state of the application can be saved in some application-dependent format. At the coordinator component level, the configuration of the application can be saved in the form of an ACD. Both forms of application state can be saved persistently in the Gaia context file system, thus, being accessible from different active spaces. Saving and retrieving application state is used to support application suspension and resumption. AM learns about users entering and leaving the active space through the presence service. When a user leaves, AM obtains the list of applications associated to the user and suspends them. Then, when the user enters an active space, it resumes the suspended applications.

8.5 WSAMI: Ad hoc, Decentralized AmI Environments

WSAMI is a lightweight middleware infrastructure, efficiently deployable on wireless, resource-constrained devices, which enables ad hoc, totally decentralized AmI environments (Issarny et al. 2005). WSAMI builds upon the Web services paradigm (WebService, online), whose generality and pervasiveness due to the ubiquity of the Web allow for dealing with the interoperability requirement associated with provisioning AmI services. WSAMI exploits further Web services to support service provisioning both in the local and in the wide area through the extension of the local wireless environment to the wide-area Internet.

8.5.1 Architecture

The fundamental architectural element of the WSAMI infrastructure is the *Core Middleware* that provides essential functionalities, enabling interaction among mobile Web services. Building upon the generic Core Middleware, two application architectural models have been elaborated along with supporting middleware services. The WSAMI architecture is depicted in Fig. 8.5.

The Core Middleware (Issarny et al. 2005) comprises: (1) the *WSAMI language* for specifying mobile composite Web services, (2) a lightweight, efficient middleware *Core Broker* realizing SOAP (SOAP, online), and (3) the *Naming and Discovery* (ND) service for dynamically locating requested services.

The WSAMI language is an XML-based language enriching the WSDL-based description of Web services (WSDL, online) in order to enable: (1) specification of composite Web services, (2) specification of QoS properties of Web services, and (3) specification of additional, middleware-level QoS properties enforced by customizing the communication path between interacting services. The minimal Core Middleware shall be deployed on every node of the environment. Additional WSAMI middleware-level services provide advanced features, but do not have to be supported by all nodes.

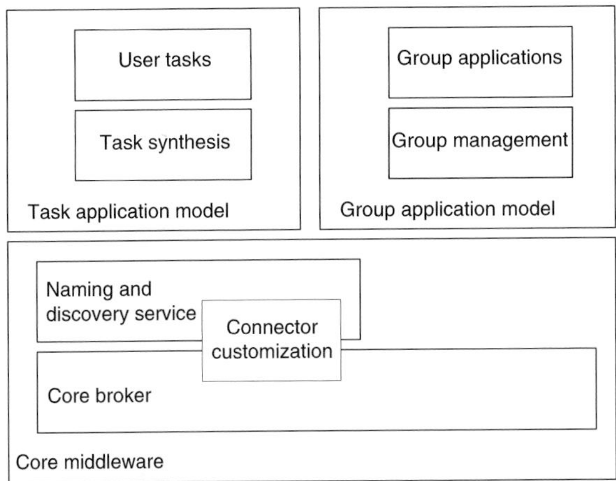

Fig. 8.5. WSAMI architecture

The *task application model* enables structuring complex user tasks that are performed by employing the services and resources of the environment (Georgantas and Issarny 2004). This model identifies a number of component classes:

- An *elementary service/resource* component models an atomic service or resource of the environment, including I/O devices.
- A *generic composite service/resource* component composes elementary service/resource components by employing a *coordination* component that encloses the stateful workflow of the composition, and a *computation* component that encloses stateless processing functionality.
- A *generic task* component models end-use functionality by employing a coordination component, a computation component, and several *user interface* (UI) components. A UI component decomposes into a *UI front-end* component i.e., a generic I/O component, and a *UI back-end* component that encloses task-specific functionality as well as adaptation functionality specific to the UI front-end. In the same way, adaptation functionality through a back-end component is provided for any I/O component. A generic task component composes a number of elementary or composite service/resource components to structure a task, as depicted in Fig. 8.6.

Elementary service/resource components, generic composite service/resource components, and generic task components are atomically deployed either on the user's portable device or on any node of the environment; each component has a WSAMI description. The task application model integrates the *Task Synthesis* (TS) middleware service that enables dynamic composition of user tasks.

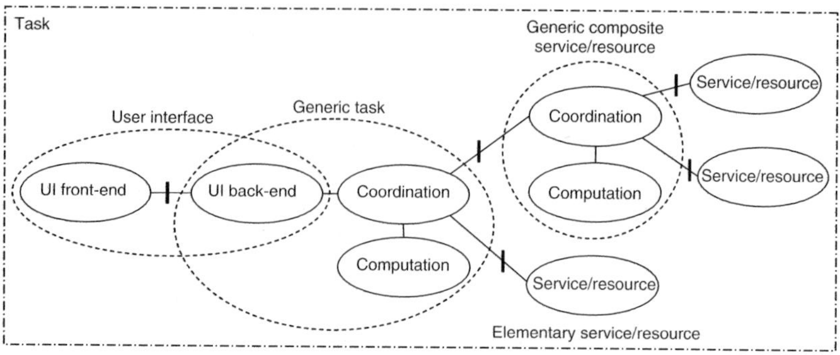

Fig. 8.6. WSAMI task application model

The *group application model* enables structuring group applications, such as content sharing and collaborative editing, executed by ad hoc, dynamic groups of peer-to-peer interacting wireless devices. Managing dynamic groups poses tremendous challenges for the development of applications due to the high complexity exacerbated by the changing environment. Therefore, the group application model amounts to a generic *Group Management* (GM) middleware service, which manages the dynamic group, relieving applications from such complexity, and can be customized to support diverse applications (Boulkenafed et al. 2004).

WSAMI supports the essential middleware functionalities identified in the generic architectural framework, i.e., Service Discovery and Interaction, the latter based on Web services SOAP connectors. It further enables two alternative Service Architectures (the task and group application models), each one along with a dedicated Service Configuration mechanism (the TS and GM services).

8.5.2 Triggers to Dynamic Behavior

In the group application model, WSAMI's dynamic behavior is based on the precise, parameterized specification of the group membership property. The group membership is primarily defined with respect to a given functional property that shall be offered by any node in the group. The group membership is further enriched to include: (1) the relative location of nodes expressed, e.g., in distance units or in number of hops, (2) authentication of nodes with respect to a security domain, (3) connectivity of nodes by using the underlying network protocols, (4) QoS-awareness, concerning QoS attributes that shall be offered by all nodes in the group, such as reliability, security, performance, and transactional behavior, and (5) resource-awareness, concerning resources and corresponding resource levels that shall be offered by all nodes in the group, such as CPU load, memory, bandwidth, and battery.

8.5.3 Dynamic Behavior

In the decentralized WSAMI environment, dynamic behavior is either shared among peer devices, or coordinated by a device that assumes this role dynamically. Dynamic behavior amounts to dynamically discovering services and composing them into applications.

The ND service is based upon the WSAMI description of services and a simple matching relationship. It is a decentralized discovery service based on maintaining repositories of information on deployed service instances on each node of the local environment. ND permits a prioritized service search within the local environment and on the Internet – the latter by contacting universal UDDI (online) repositories – taking into consideration whether a node is power-plugged or battery-powered, and adapting to the infrastructure-based and ad hoc modes of the local wireless network.

The Core Middleware provides one more basic capability for dynamic behavior. Based on the WSAMI description of a service, WSAMI supports customizing the SOAP connector between interacting services by interposingappropriate middleware-level components (customizers) that adhere to the pipe and filter architectural style. In this way, QoS properties such as security, performance and reliability can be dynamically enforced at middleware level. The Core Broker supports connector customization together with the ND service: customizers are retrieved and interposed, when the service itself is retrieved.

Building on the basic capabilities for dynamic behavior offered by the Core Middleware, dynamics of the AmI environment is further treated within the two application models.

In the task application model, the TS service is dynamically composed of two components: a *TS front-end* component, which mainly supports interaction with the user, and a *TS back-end* component, which provides task synthesis functionality. TS front-end resides on the user's device, while TS back-end resides on the node hosting the related generic task component, which may not be the user's device; this minimizes message exchange overhead of task composition over the network. A user wishing to carry out a task launches the TS front-end on his/her device. TS front-end employs ND to retrieve all available generic task components in the local environment, including the ones deployed on his/her device. Then the user selects a specific task. TS front-end contacts the appropriate TS back-end, which then employs ND to retrieve all the composite or elementary service/resource components to be integrated in the task, in hierarchical order, along with the required connector customizers. Finally, TS back-end assembles the retrieved components.

In the group application model, the GM service takes care of assembling mobile nodes that together allow meeting target functional and non-functional properties, and of further making transparent failures due to the mobility of nodes. Building on the multi-parametric group membership allows GM to be most generic and configurable. GM assumes three functions: First, it discovers nodes that are eligible for group membership. This function is carried out by each node in its vicinity by employing ND. Second, it establishes the group. This function is managed by a single node – in order to minimize message exchange – called the *leader* which is periodically changed within the group. And third, it manages the group's dynamics, i.e., it updates group membership according to the dynamics of the network's topology. This function amounts to periodically repeating the discovery and establishment phases in order to update the group. The period is dynamically adapted according to the relative update statistics of the last two periods.

8.6 Assessment and Research Challenges

The three surveyed systems are aimed at substantiating the distinguishing features of Ambient Intelligence. In the following, we assess the extent to which this goal has been reached, and identify research challenges for SA and middleware that are still to be addressed.

With regard to *mobility*, Aura and Gaia introduce the concept of application mobility: applications are associated to users; when users move between environments, the fixed infrastructure takes care of saving the current application state, suspend the user's applications in the current environment, and resume them later in the new environment. However, connection between environments has been realized only in Gaia through a commonly accessible file system, while Aura envisages a more complex functional cooperation between infrastructure instances of distinct environments, which has not yet been elaborated. WSAMI incorporates a more traditional view of user/device combined mobility. An application is either a distributed task or a group application, i.e., (co-)set up and possibly (co-)executed on the mobile user's portable device, according to the services/devices available in the vicinity, which in general are also mobile. This makes the WSAMI environment much more "mobile" than the Aura and Gaia environments: not only users and their personal devices, but also the infrastructure may be mobile, i.e., the infrastructure may be assembled from mobile devices. Mobility is a recurring issue in AmI research. Unrestricted user, device, and computation mobility shall be supported by middleware, while computation mobility is further related to logical mobility, which also concerns SA.

Ubiquity within Aura and Gaia is pursued through deployment of multiple, rich in functionality, centralized environments, interconnected through file sharing and possibly functionally as well. However, this richness limits the range of deployment of these environments. WSAMI relies on the universality of the Web to provide an indeed ubiquitous, however less rich, AmI environment. Integrating the worldwide base of computing platforms, services and data in a way that makes it easy to access them at anytime, anywhere, and from different devices is a major challenge for AmI middleware.

Heterogeneity is treated in all three systems by imposing a homogeneous system infrastructure that exposes specific APIs, and well-defined service abstractions. In Aura, this infrastructure supports discovery of services and configuration of applications, while in Gaia and WSAMI, the infrastructure additionally supports application interaction. Thus, by assuming homogeneity, the issue of middleware heterogeneity is not treated. Nevertheless, WSAMI's reliance on standardized Web services makes it more pervasive than Aura and Gaia, which introduce proprietary middleware. Further, regarding application interaction, Aura integrates diverse connectors, exploiting any existing communication infrastructure of the environment, while WSAMI and Gaia impose specific connectors. Apparently, interconnection of devices and services in AmI environments is an open research issue. Research, spanning both SA and middleware, shall be aimed at offering the right paradigms and programming abstractions for distributed component discovery and interaction. The issue of middleware heterogeneity shall orient this research towards a higher level of abstraction, where discovery and interaction semantics shall be defined independently of diverse syntactic realizations.

The *ad hoc nature* of AmI environments is well treated in all three systems by employing service – and resource – discovery, and application mapping to available services. However, service components may be composed into applications as long as they implement functionally and syntactically compatible – therefore, accordingly designed and developed – interfaces. Thus, another case of heterogeneity, between service interfaces of semantically compatible services, is an issue yet to be dealt with.

User-centrism is manifested in Aura by system's reactiveness or proactiveness in terms of interpreting user's preferences, context, and intent, and configuring accordingly computing environments to support user's tasks. Nevertheless, no elaboration has yet been done on proactively exploiting user's context and intent. Gaia applies user-centrism through: proactiveness, triggered by user's presence, in terms of starting, suspending, and resuming applications, and making available user's data files; support for user-defined policies customizing application management tasks; and support for context-aware applications. Further, as discussed above, in both Aura and Gaia, applications are associated to users and not to platforms – based on an environment-independent description – and follow them while they move. Similarly, in WSAMI, user-driven tasks have abstract descriptions and are instantiated according to the current environment. User-centrism is the most typical among the identified features of Ambient Intelligence, and probably the one that is most open to new approaches. Research challenges concern modeling, retrieving, and reasoning on user's context and intent in order to reactively or proactively adapt to user's needs and serve user's requests.

Finally, *dynamics* is present in all the above features. In the survey of the three systems, we have focused on the system's response to dynamics in terms of dynamic configuration and reconfiguration of both application and system services. All three systems support dynamic application configuration, each time specific to the employed application architectural model. Aura enables QoS- and resource-aware assembly and configuration of services and connectors. WSAMI supports SOAP connector customization, limited to the pipe and filter style. Additionally, WSAMI enables support for advanced non-functional properties for group applications through parameterized group management. With regard to dynamic self-adaptation, Aura supports QoS- and resource-aware application reconfiguration. By capturing and recovering – periodically for the case of failure or upon voluntary reconfiguration – the user-level state of a task, Aura can ensure the consistency of the task in terms of high-level state and user's data. Further, WSAMI supports reconfiguration of group applications by managing the underlying group dynamics.

Nevertheless, application consistency upon reconfiguration shall additionally be ensured with respect to functional and non-functional properties, and this is a challenge for both SA and middleware. A key issue is the identification of an appropriate description level of the SA that will allow dynamicity and self-adaptability to be adequately expressed and verified. Obviously, in the dynamic context of AmI applications, verification cannot be carried out entirely statically. Therefore, a further challenging research issue involves the relationship between the SA and the middleware infrastructure, which provides the appropriate middleware support for dynamic reconfiguration and adaptation.

It is evident that the open research issues in the software domain towards Ambient Intelligence are numerous and challenging. Dealing with them is hard, and is further exacerbated by the requirements for acceptable response times, dependability, security, privacy, and trust in the open, not centrally administrated, highly variable in resources in space and in time AmI environment. Among AmI challenges, user-centrism is probably the most fascinating one, evoking even philosophic and futuristic questions about "human–machine integration". Then, after object-oriented, component-oriented, and service-oriented, will "human-oriented" be the next software engineering and computing paradigm?

9

Mobile Computing

B. Svendsen

"Innovators don't change the world. The users of their innovations do."

Michael Schrage, MIT Media Lab

9.1 Computing Everywhere and Anywhere

Mobile computing is understood as the capability of exchange of information and/or data between devices at any remote locations. It is one of the technology enablers for realising the AmI vision.In a manner of speaking, the mobile telephone opened the opportunity. It has been embraced by the public and business and has changed the way people communicate and interact and how business and work life is organised. It has also demonstrated possibilities that innovative people have developed further.

The other important technology development is the Internet, which ideas originally were to provide, *distributed computing* in order to facilitate better utilisation of computing resources. Another Internet driver was communications safety in providing robust network topologies. Not the least, a third aspect was extending the human communication space as promoted by J.C.R. Licklider (Spilling and Lundh 2004) who already in 1960 developed a vision of an "On-line community of people".

When we also add the tremendous development of computing power and data storage capabilities, mobile computing is thus a natural development from these two major communications changes in the last two to three decades.

But mobile computing is not only about people directly accessing network resources; it is also about machine networking, often referred to as *Machine-to-Machine* (M2M) communications. The Massachusetts Institute of Technology launched their initiative "Things that think"; in order to study the possibilities and impact of embedded intelligence in every day things and gadgets.

Advanced, powerful, easy-to-use devices combined with increasingly fast and efficient voice/data/video transport networks will allow people, both in a private and work situation at any site immediate access to a wide range of information that was formerly stored, managed, and retrieved at various remote locations.

This gives several opportunities. A building entrepreneur can use mobile computing devices and applications to perform several tasks directly from the construction site. This includes job costing, materials receiving, electronic work orders, control and alarm maintenance, and communications with headquarters or other jobsites.

The interaction between commercial actors and the community services have become easier by using mobile computing devices to connect to the databases at city or county authorities, freeing workers, and supervisors from trudging to a city hall for approvals.

Mobile computing is out there, but at the moment as more or less isolated applications, and one challenge of the domain is to bring them together to support the realising of the AmI vision.

9.2 Applications of Mobile Computing

Endless possibilities open up if we only are able to grasp the potential of people and devices being networked – anywhere and anytime. It is difficult to foresee future killer applications, and what we see today often sets limits to our imagination. But if we are going to create the future, we have to try to imagine what can be possible and what could be taken up, taking all our current technological and humanitarian knowledge into account. Asking the question: "What kind of service do you want?" seldom gives very good answers, because usually you cannot want something that you have never seen or experienced. Thus, technology development must be targeted to create a playground for actors: the citizens, the society, the corporations, and not the least people with new ideas that can bring the society forward.

But it is also important to be specific and to identify potential applications and possibilities which we already see the start of and use that as guidelines for our efforts. Thus, in this section, some applications of mobile computing are described.

9.2.1 Your Personal Economy Will Be in Your Pocket

Mobile commerce (m-commerce) is the possibility of performing payments using a mobile device. It involves a financial transaction and a mobile wireless device.

Payments via a mobile telephone provide convenience to mobile users, cost savings to merchants and revenue growth for mobile operators. M-commerce has already been introduced in some European countries and is an emerging opportunity over the next years. Types of existing and upcoming m-commerce services include the following aspects:

- Prepaid recharge for mobile users.
- Parking payment.
- Payment of goodsfrom vending machines.
- Payment of toll road charges.

- Purchase of transport ticket (city systems, railway, air transport).
- Purchase of event ticket (theatre, sports, cinemas).
- Bill payments (net banking, charging credit cards).

Services that have proved to be popular, initially with young people, are purchase of ring tones, icons, and games.

The use of the mobile device for m-commerce applications is likely to grow. The inclusion of proximity technology, like e.g. RFID or NFC (see later), opens up more applications as the user interactions become easier, and it opens up virtually endless possibilities. Some examples are listed below:

- A youngster could download a coupon from a rock music poster to her cell phone and "beam" it to a payment terminal for a discount when buying the musician's CD.
- A commuter could search the Web for the most convenient train leaving for his home, pay for the ticket and download it to his computer, then use the computer to pass it to his mobile phone. He could also search and download it directly on his mobile phone. On board the train, he could bring the phone near the conductor's reader to pay his fare.

The latter example becomes reality very soon, as the public transport operator in greater Frankfurt, Germany already has been testing this and is launching a project in 2005.

9.2.2 Electronic Tags Replace Bar Codes and Add More to It

Electronic tags have a huge potential in different logistics applications. Equipping, e.g. retail goods with electronic tags, may ease both handling and tracking of products all from production, via storage all the way to the end consumer. It is likely to reduce unnecessary storage and also to reduce theft. Another benefit is the possibility to recover lost items, which also provides a gain for the manufacturers.

Luggage handling at, e.g. airports and by freight carriers using electronic tags may lead to a more secure handling and less misplaced items.

For these purposes optical bar codes are used today. These are limited in the amount of information that can be stored (printed) and also subject to damage and loss. Electronic tags can contain more information about the item in place, e.g. ownership, origin, certificates, start and destination, and expiry date for food, just to mention some. They are also more robust in the handling because they do not require "line of sight" between tag and reader.

As an example, Wal-Mart, USA's biggest retail chain is launching electronic tags on cases and pallets from their top 100 suppliers in 2005.

9.2.3 The House Knows What You Are Up To

A number of applications could radically change the daily life in the whole society, in the home sphere, at work and in social life. Almost any everyday object may be equipped with a tiny microcontroller, fixed to or even printed on the device, with storage and computing capacity sufficient to exchange data messages with the external environments when connected to a network.

Demonstrations of "smart homes" have highlighted how "smart devices" connected to the local home network may both ease the daily life as well as improve quality of life in the home sphere.

The alarm clock can initiate routine tasks in the morning, like turning the lights on, activating the coffee machine, and bringing other household devices from a "dormant" and power saving state into ready-to-use mode.

The intelligent bathroom can keep inventory of medicine bottles (equipped with an electronic tag) and remind to take their pills or renew their prescriptions. People with need for medical control/observation could take their tests in the bathroom and the test result be immediately transferred to the medical doctor. The status of the home can also be controlled with at mobile unit from any remote locations.

Home safety is another aspect of the smart home concept. Alarms due to unauthorised intrusion, fire detection, water supply leakages, etc. can today be transferred to local emergency centrals via the normal telephone line. In the future, we should see the addition of remote control of critical functions as well as more active and detailed surveillance that can be controlled by the owner and by an authorised party like, e.g. a guard company or the local fire brigade and police.

Energy efficient homes can be utilised by networking heating and cooling sources. A home server may set up a programme for directing the energy consumption according to the time of day and actual use of the different rooms in the home. Current systems exist which are able to monitor the temperature in each room and controlling, e.g. electric ovens to maintain a predefined temperature according to the time of day and day of week as preferred by the owner. In areas of the world where electricity is not the main heating source, microcontrollers can control, e.g. gas heaters or central heaters based on bioenergy. Cooling systems become more widespread and are also a part of this.

We could even see that rooms or parts in a house could be equipped with proximity detectors to human presence and dynamically change the preferred indoor climatic settings according to this, however the time delay constant must be taken into account. The smart home could learn from user behaviour in order to adapt the energy use.

9.2.4 Attentive Vehicles – Road and Transport Safety

In the European Union countries, road traffic accidents claim more than 40,000 lives and they leave more than 1.7 million people injured every year (CARE, online). Inattentive drivers represent a serious danger in road traffic. In USA police reports show that 20–30% of crashes are caused at least in part from inattentive drivers.

Incorporating active attention management systems in cars can detect, e.g. hard manoeuvres and initiate messages from an onboard phone, navigation systems, and warning lights. Cars can be equipped with sensors that can detect if it is about to leave the lane unexpectedly and alarm the driver. This can also be taken further to make the roads themselves intelligent. Then, information about off-road driving can be automatically routed to the emergency central and coupled with possible alarm messages from the car itself, containing, e.g. data about speed, position, owner's identity, number of passengers, and their identities (there may be sensors in the seats), etc. Emergency teams may then be quickly launched and also tailored for the situation.

The potential benefits in road transport with respect to safety are high. In Europe the current accident level represents estimated costs, both direct and indirect, of 160 billion euro, not to mention the human cost.

Intelligent Transport Systems (ITS) will incorporate such safety systems as described above. ITS will additionally provide means to make transport of people and goods more efficient, as well as enabling enhanced services to transport users. Entertainment together with location and context aware information will make the travel experience better.

Examples of such information are of course about traffic conditions for car travellers, expected arrival times, delays, and other irregularities for users of public transport, whether it is job commuting or long distance travelling. Other types of services are tourist and service information.

9.2.5 Public Services for Citizens

Introducing *electronic identity cards* will be basic for implementing enhanced access to community services for the citizens. It is also possible to foresee that this "card" is virtual and embedded into the smart card (SIM) of the mobile phone.

This card will contain information about the person which can ease access to rightful services like, e.g. voting, social services, and also contain adequate information which legal authorities are entitled to have access to, like tax-paying information and criminal records. It could also contain a health record that can be valuable in an emergency situation, and also providing medical prescriptions entered by the GP, and validated by the pharmacy.

An electronic identity card must imply a high security level, which may, e.g. reduce fraud and virtually eliminate false identities. It should of course become the *e-passport*.

9.2.6 Monitoring Product Quality Improves Food Safety

Safer production and safer products can be achieved by adding intelligent tags to the products and facilities in the process from manufacturing via storage to the end consumer. Providing safe and healthy food is an example. An emerging trend is also that citizens want to know the origin of fresh food.

One step on the way is to provide, e.g. refrigerated counters with wireless temperature sensors. The sensors can report status and deviations from the ideal conditions and shopkeepers as well as food manufacturers can act upon it. Too high storage temperatures are often registered by the local watchdog agencies when they perform their spot tests. The benefit of the wireless approach is that it makes shop refurnishing and reconfiguration easy.

9.3 In the Bottom Lies the Technology

How are we going to provide the carriers for the above-mentioned applications? This section gives a very brief overview of the current wireless technologies, which can be used in order to build a mobile computing network supporting the AmI vision.

This is neither a complete listing nor comprehensive overview of the technology situation. For this purpose the reader should approach the different standard organisations or manufacturers.

Basically two main technological components are essential for realising the mobile computing in an efficient manner. This is *wireless communications* and *wireless location*. Both will be shortly discussed below. In order to exchange information between devices, a communication channel must be available. Location is not necessarily a must, but in order to provide better services and more efficient use of the communication systems, user location is beneficial. It is a must in order to simplify user interaction and advanced M2M communication. In addition we will present trends in user terminal equipment as well as proximity detection concepts like electronic tags and similar.

9.3.1 Wireless Communication

Modern, digital wireless communication available to the public started for real with GSM, designed in the 1980s and deployed in Europe in the 1990s. Now, some 10 years later many countries have achieved a penetration of more than 80%. GSM is not only confined to Europe, and as of January 1, 2005, *GSM had more than 1.2 billion subscribers in more than 200 countries worldwide* (GSM Association, online). This makes GSM, together with other cellular technologies the most important communication channel for mobile computing applications.

GSM was essentially designed to deliver mobile voice service, allowing users to initiate and receive calls everywhere, even in fast moving vehicles. The Short Message Service (SMS) was originally thought of as an add-on that might increase the attraction of the GSM system, but later became a success of its own. The data extension service GPRS with the latest enhancement,

EDGE, typically provides data rates up to 230 kb/s. GPRS and EDGE allow mobile phone users to browse the web, download games, and send and receive picture messages (MMS).

The 3G standard UMTS has been deployed in several areas in Europe, and will from the start provide higher data rates, up to 384 kb/s, and a richer service environment. Already enhancements are being specified which will enable even higher data rates. At the start of 2005, there were already 16 million global UMTS subscribers (UMTS Forum 2005).

Today's deployed 3G systems provide services such as video telephony and high quality video streaming in addition to services known from 2G and 2.5G. Other multimedia communication services, such as video sharing and video chat, will be enabled through the IP multimedia Subsystem (IMS). IMS complements 2.5G/3G systems with a standard IP based multimedia communication platform.

The increasing popularity of Internet and the World Wide Web has led to demand for access to Internet also from mobile terminals. As a result there has been a development of new wireless access systems designed for hot spots and other high capacity access systems, in parallel with development of built-in wireless access cards in mobile terminals. Products based on the Wireless Local Area Network (WLAN) standards from IEEE have become extremely successful and through the work of the WiFi Alliance product interoperability has been secured to the benefit of the users. This is the wireless Ethernet and WiFi communications are now integrated in virtually any new laptop computer on the market and becomes more and more common also in palmtop computers, or PDAs (Personal Digital Assistant).

Personal Area Networks (PAN) is the term used for technologies like Bluetooth or similar. Bluetooth has been there for some time and virtually any mobile phone encompasses this standard. Often used to provide a voice channel to a wireless headset, it can also be used as a cable replacement between the phone and a laptop. Bluetooth devices can form ad hoc networks and one important feature is the discovery function, which enables two devices to automatically communicate when they are in each other's range.

9.3.2 Wireless Location

Locating objects and people is one of the prerequisites of Ambient Intelligence. There is a need for systems to obtain accurate, reliable, and updated positions data of mobile terminals and other objects. It may be absolute geographic positions or relative positions between objects, terminals, or users.

Location can be provided by the wireless communication network itself or by additional technology.

Positioning by satellite is well known as GPS (Global Positioning System) terminals have become common consumer market products. Basic accuracy of GPS for public use is 30 m, thus add-ons have been made to improve this.

Differential GPS (DGPS) uses extra information of the system accuracy provided by a radio transmitter placed at a known position in order to correct the GPS read out. In this way, the position accuracy can be brought down to less than 1 m, in some cases down to cm-level.

A similar system is the Russian GLONASS. GPS is owned and operated by the US Department of Defence, which is one reason for the upcoming European satellite navigation system GALILEO. This is expected to be operational in 2008.

Wireless communications systems with fixed base stations or access points have an inherited location capability by identifying the fixed point, the base station, which the mobile terminal communicates with. Such location by proximity is used today in mobile cell networks, Bluetooth and WLAN environments. Enhancement of the proximity concept may be obtained by measuring the signal strength.

Position can also be determined by employing triangulation techniques in wireless networks, basically in the same manner as the satellite systems do, however currently these technologies are expensive for operators to install.

9.3.3 Multistandard and Flexible – Software Radio

Products that integrate different technologies have been on the market for some time. Today mobiles that offer several capabilities in the same unit overtake pure GSM phones. Modern mobiles and "light-weight" PCs like PDAs (Personal Data Assistant) provide accessibilities to several network types making them very flexible.

Software defined radio (SDR) is an emerging technology making it possible to adaptively reconfigure both network and terminal capabilities, even configure new radio concepts in the devices.

According to the *Software Defined Radio Forum* (SDR Forum), SDR is a collection of hardware and software technologies that enable reconfigurable systems architectures for wireless networks and user terminals (SDR Forum, online). SDR is one of the technologies that will make future devices much more flexible and powerful.

9.3.4 Wireless Proximity Technologies – Electronics Tags

Radio Frequency IDentification, or RFID, is a generic term for technologies that use radio waves to automatically identify people or objects. There are several methods of identification, but the most common is to store a serial number that identifies a person or object, and perhaps other information, on a microchip that is attached to an antenna. The chip and the antenna together are called an RFID transponder or an RFID tag.

RFID based systems are already well established in toll roads, for access control and payment. During later years the following developments have sped up the adoption of the RFID technologies:

- The RFID mandate from Wal-Mart, as described later.
- The EPC (Electronic Product Code) Standard.
- The NFC (Near Field Communications) technology.
- The development of contactless smart cards.

Electronic Product Code Standard – EPC – is an emerging RFID standard developed by the Auto-ID Center, a former academic research project of MIT. It is now overtaken and maintained by EPCglobal, Inc, a joint venture between EAN International and the Uniform Code Council (EPC, online).

EPC is the RFID version of the UPC barcode standard. Like UPC, EPC is intended to be used for specific product identification as well as case and pallet identification. EPC goes beyond UPC by not only identifying the product as, but also providing access to additional data about the origin and history of the specific units. It potentially allows you to track the specific unit's history as it moves through the supply chain.

The EPC network takes an important role, as all data related to EPC will exist in the network. The historic data and other information about a unit are not stored in the RFID tag, but in the network. A pointer giving access to the data is stored in the RFID tag, working in the same way as an Internet address does.

Near Field Communication technology, or NFC, is a wireless protocol targeted towards consumer electronics. It is said to enable secure means of communication between various devices without the user having to exert much intellectual effort in configuring their "network". When two NFC-devices are brought close enough they are linked up in a peer-to-peer network. Thereafter, the devices can be set up and continue the communications using longer range technologies like, e.g., Bluetooth or WiFi (ECMA 2004). The goal of the NFC Forum is to establish the technology as an open platform offering the best benefits for everyday consumers.

NFC adds intelligence and networking capabilities to the phone and creates new opportunities to add product and service capabilities to the handset like digital transactions and sharing in close proximities.

The NFC chip can act both as a reader and as storage card, and it is compatible with the contactless card standards (see below). A mobile phone equipped with these facilities is an ideal device for making payments since the interaction between the card and the mobile phone is under the user's control and the card can be remotely updated or cancelled by the mobile operator.

The introduction of the NFC technology has raised the attention of many industries like mobile phone manufacturers, credit card companies, content providers, and network operators.

The public transport authority in Frankfurt has announced the use of NFC technology to offer a ticketing solution based on mobile phones with access to their existing contactless smart card ticketing infrastructure (EC–IDA 2004).

Contactless smart cards have reached a mature state with more than 540 million cards in the market. Several major transport systems in Asia have been in public service for more than a year. Other cities are in the process of switching to this new technology primarily to reduce maintenance cost, increase system reliability and security and to provide greater convenience for users. Contactless-smart cards may also be used to substitute magnetic stripe cards for payment and authentication. They provide all the security features of smart cards in addition to the advantage of being contactless. The MIFARE standard is the industry standard for contactless smart cards, and is also conforming to contactless standards from ISO.

9.3.5 Wireless Concepts and Technologies for Novel Applications

A whole range of potential users cannot access telecommunications services based on traditional terminals. Such users include people, objects, machines or even animals, and the application needs may be surveillance, monitoring, location services or any other measurable parameter. Opposed to traditional thinking about terminals, such usage may require low terminal costs vs. value of object, usage patterns, device lifetime and terminal sizes. These requirements introduce a need for infrastructure and terminals with quite different properties than the traditional systems.

Such systems have existed for some time, partly as academic projects, and partly as proprietary and fairly expensive technology tailored for special purposes. However, recently industry actors have showed interest in the area. As a consequence, we are now seeing advanced integrated components that are critical for the success of such *lightweight, cost effective, mobile narrowband radio based network with location and sensing capabilities.*

Two kinds of mobility apply to the user. *Intermobility* refers to absolute mobility and implies the need for roaming techniques and a contiguous set of base stations along the route of movement. *Intramobility* refers to mobility within acell, or relative mobility where objects may move, but without need for roaming as the whole set of related nodes are moving, and even the network access is moving together with the nodes.

Systems with the properties as described above are basically small, cheap off-the-shelf embedded computers, limited in terms of computation and storage capabilities, but with local narrowband radio communications. They are easy to deploy and since the components are low cost one may envision a large-scale deployment within many different user segments. Since there exist basic computation and storage capabilities within every node of the system, the system itself can be made relatively “smart” by combining sensing and local programmable logic.

The technology has matured over the last couple of years. The two most important trends are *miniaturisation* and *reduced production cost.* Prices of small low cost radio chips with embedded microprocessors with computation capabilities of 1994 personal computers are as low as 3 USD.

Sensor networking based on advanced communication systems and sensing technologies are emerging in the market. Key properties for such systems are low cost, very long component lifetime, robustness and in some cases advanced ad hoc multihop routing protocols. The systems are often one-time deployment systems, which means that broken components or even components that have run out of power, simply are thrown away and replaced with new ones.

9.4 Challenges Imposed on the Domain by Ambient Intelligence

How will people react to and meet the new technical facilities that enable Ambient Intelligence? The biggest challenge may be to define and select the services and applications that will be accepted in the market. Although the technology and the services doubtless have the capability to improve the quality of daily life for most citizens, concern has been voiced about venturing privacy. There is a challenge for regulatory authorities to develop and implement regulatory framework that will guarantee maintenance of privacy protection. Service providers and network operators are also challenged by this to provide methods and concepts that take care of both safety and security and at the same time ensure ease of access for the citizens.

One of the most useful possibilities of Ambient Intelligence may be the citizen's access to public information and public databases. Open access to such information and databases is needed to obtain full effect of this option. Authorities are challenged to develop systems and routines that will secure any citizen access to public information of interest to her/him.

On the technical side a number of improvements are required to ensure seamless connection and smooth operation across different networks and different technologies. Development and agreement on open interface standards is a key prerequisite for the success of Ambient Intelligence. Agreement on global protocol standards is another challenge. A success in this area, followed by implementation in the different networks, will have a great impact on the user friendliness of AmI services. Migration to IP in network infrastructure is another issue under consideration or implementation in many networks. IP-based networks will lead to more efficient procedures and simplify problems related to network security, reliability and quality of service.

Development of user terminals and software dedicated for upcoming new services is another challenging area. Power consumption of user terminals and stand-alone wireless terminals represents a problem today. The challenge is to develop new types of batteries or alternative power feeding systems.

The growth of wireless systems will increase the pressure and demand for radio frequency resources and will require a more efficient utilisation of the frequency spectrum. There seems to be a need both for allocating new frequencies for upcoming wireless systems and a general reallocation of radio frequencies. There is a challenge to frequency authorities in Europe to define new methods for a fair and efficient frequency spectrum management. Furthermore, there is a challenge to the industry to develop modulation and coding systems that with improved efficiency of the spectrum utilisation. New spectrum management methods can ease the pressure and increase the utilisation. Dynamic spectrum trading principles and cognitive radio technologies should be further developed.

9.5 How Far Have We Come?

So far we have shown how mobile computing is one of the technology basics for realising the AmI vision, as well as giving examples of applications and key technologies, which makes it possible. Now, let us look at some examples from the last years which show how far we have come with respect to implementing mass market applications based on mobile computing technologies.

9.5.1 Buying and Paying by the Mobile Phone

Bank services via the Internet are already well established in most European countries. These services can also be accessed from mobile units over wireless access networks. Services like payments via the bank can be approved or carried out from a mobile unit. Electronic cash is another payment mechanism. Basically this is a replacement for coins and notes, and the payment itself does not involve a bank transaction. The mobile phone may serve as purse or wallet, which the user may fill up by using a net-bank account or similar, and thereafter using the mobile phone to pay articles from a vending machine or from an Internet-shop, or simply to fill up the user's account for prepaid mobile service. For security reasons payment transactions require electronic signature by using Public Key Infrastructure (PKI).

The public transport operator of Frankfurt's greater area in Germany, Rhein-Main Verkehrsverbund (RMV), announced that it would launch a trial on m-ticketing in early 2005 based on NFC technology developed by Philips and Sony. It has been tested and the project will be run in the town of Hanau, using Nokia mobile phones equipped with NFC shells. RMV has already used contactless- smart card tickets for some time, and the new m-ticketing solution will be compatible with the existing infrastructure (EC–IDA 2004).

9.5.2 Retail Chains Is Starting to Use Electronic Tags

RFID has for many years been extensively used (in Norway) for access control at toll roads. The user's car must be equipped with an electronic card containing a radio communication chip and data storage capabilities. The data are exchanged with the corresponding terminal at the tollbooth when the car is passing, and the user's account with the toll company is charged with the cost of passing.

Wal-Mart is USA's biggest retail chain with almost 8,500 stores in the United States and globally and with a global revenue of 256.3 billion USD in 2003 (Wal-Mart 2004). In 2003, Wal-Mart announced its plans for rolling out RFID technology. The company is planning to implement RFID tags step by step both in the supply chain and in the retail business. From January 1, 2005, Wal-Mart requires their top 100 suppliers to have their cases and pallets "chipped" by using RFID-technology carrying EPCs. By doing this, commodity trade logistics and security will benefit. Following a successful pilot test carried out in 2004 the company has announced that it will proceed with the implementation in compliance with the original plans.

It is expected that the suppliers will use something like 8 billion electronic tags a year, thus this implementation has a significant impact on RFID becoming mature and mainstream. The UK retail chain Tesco has done similar tests and plans to roll out RFID technology through its supply chain by 2008. In January 2005, Tesco signed a contract for delivery of 4,000 readers and 16,000 antennas during the year. The Wal-Mart implementation plan has had a significant impact on the interest for RFID in different applications, and a number of demonstrations, tests, pilot implementations and different applications have been reported. Thus Gillette has demonstrated the use and benefits of RFID in the supply chain, where a misplaced case was identified and directed to its intended destination. Furthermore, Gillette has demonstrated how RFID and the EPCglobal Network could enable manufacturers to see product quantities in the retail supply chain.

9.5.3 Electronic Luggage Tags at Airports

Hong Kong International Airport is deploying the largest RFID network undertaken so far in Asia, using RFID equipment for tagging and tracking baggage. The project is set to go live on January 1, 2005, alongside the existing bar code system. Delta Air Lines has carried out two successful RFID trials in 2004 and is planning to roll out an RFID system to track all luggage it handles at US airports. Delta plans to use disposable RFID baggage tags that will be attached to passenger luggage at check-in at every US airport the company serves. The labels will enable Delta to track each item throughout the carrier's baggage-sorting operation and onto the plane through any transfer airports for connecting flights and finally onto the baggage carousel at the passenger destination.

9.5.4 United States Introduces Electronic Passports

The US Government Printing Office has in 2004 tested electronic passports (e-passports) embedded with RFID chips with the intention to standardise the chip. The US State Department plans to begin distributing limited numbers of electronic passports to American travellers at the end of 2004 and to make them fully available in the first quarter of 2005. It is intended that all American passports issued annually will be e-passports by late 2005.

9.5.5 Electronic Shepherds and Surveillance

In Norway, Telenor R&D has designed a low cost, low bandwidth, wireless network system with the capability to realise the requirement for data/information exchange independent of location. The objective is to build proof-of-concepts, based on the ideas described in the overview section of this chapter. The following three examples of field trial systems solve specific tasks for users by augmenting the intelligence of existing systems and tools used within a process by deploying "invisible gadgets" on objects and persons. These field trials are:

- Monitoring refrigerated units.
- Monitoring livestock.
- Monitoring soldiers.

Monitoring temperature in refrigerated units adds significant value to grocery stores. Collected data are used for optimising energy consumption, resulting in typically 5–10% savings. Such a system also offers a cost-effective way of collecting data based on wireless technology. In addition, temperature data are included as a part of the industry's internal control programme. New regulations force the industry to log the temperature several times a day. Until now temperature readings have to be done manually. Telenor and partners have conducted a trial with one of the largest retailers in Norway.

Temperature sensors track data from every refrigerated unit. The data are routed from the end node to the user based on the network bridge that may be connected to any available network access. The nodes and their sensors continuously monitor every important point of interest in the shop area and notify the staff only when something is on the verge of happening. They function most of the time as basic loggers, but turn "smart" when needed. Also, there is a high degree of intermobility in the system, all nodes move freely within its associatedcell. This means that the staff can rearrange the shop without worrying about existing infrastructure

Another example is *monitoring movements of livestock* in grazing land, adding value to farmers. Only in a small country as Norway, every year 2 millions of sheep are sent out to grazing land. And every year more than 100,000 sheep are lost during the grazing season. The reasons are many, but the most usual ones are predator attacks and accidents in the terrain. This annual animal loss has severe economical consequences for many farmers. The current way of dealing with this problem is good old herding like they did it 2,000 years ago. Already in 1995, The "Electronic Shepherd"-project was established in North Norway. A cooperation between Telenor R&D, MIT, and Solvik Gård, later became part of MIT's field laboratories in the FabLab-project (Thorstensen et al. 2004).

By adding communication and logic capabilities to animals with reporting facilities on the whereabouts of the animal, and also capability to monitor the state of the animal and its environment, the system works as an invisible agent for the farmer, giving him information when he needs it. This application also utilises the mobility aspect of the system, by providing all the animal nodes connectivity to mobile networks. That means the extended network access follows the individual.

Telenor R&D and The Norwegian Army have recently started a trial for *monitoring position and condition of soldiers.* This will add great value to the Army's Operation Command Centre. The system shall generate on-line information of military units' position. This information, visualised in a digital map, will give the Centre a better overview. The information chain will be less time-consuming and this enables a more effective reaction chain. It can also help to resolve information stubs, and by collaboration with the other units in the system, generate a more informative picture of what is happening at the moment. The information may be used to make decisions locally or to inform decision makers.

9.5.6 Mobile Phone as Tourist Guide

In the cathedral (Nidarosdomen) in Trondheim, Norway, the mobile phone as a tourist guide has been demonstrated. A trial service developed by Telenor R&D and partners was offered in 2004. The service utilises positioning information combined with visual and spoken tourist information stored in a local network. Tourist information was available in several languages and delivered to the user either on request or pushed to the user at certain "hot spots" inside the cathedral.

A WLAN network with position capabilities was installed in the cathedral. The tourist information was stored in a server connected to the network. Users of the service would need to register and have an admittance tag with WLAN transmitters connected to the mobile phone. Selected information might then be pushed to the user's terminal, based on the location of the terminal.

9.5.7 Attentive Vehicles Seek to Prevent Road Accidents

In 2003, Volvo added an attention management system to its S40 sedans. Sensors pick up steering actions, accelerator position, and other vehicle dynamics and feed data into a computer, which looks for evidence of side-slipping, overtaking, or hard braking. When it notices such demanding manoeuvres, the system initiates messages from the onboard phone, navigation system, and warning lights.

Motorola and DaimlerChrysler demonstrated a minivan outfitted with a similar system. More recently, Volvo has been testing cameras that can detect drowsy eyelids and suspicious land crossings. In March 2004, the European Union launched a four-year, 12.5 million euro project to develop industry wide standards for adaptive driver–vehicle interfaces by 2008.

A "simpler" approach, which demonstrates the benefits of attentive vehicles, comes from the French car manufacturer Citroën, which in its C5-model from 2004 incorporates optical sensors under the car. The sensors of this "Lane Departure Warning System" (LWDS) are picking up the reflection from white and yellow road marking in order to detect if the car is driving outside the markings. A vibration in the driver's car seat is activated, unless the manoeuvre is accompanied by the use of the indicator lights.

9.6 Concluding Remarks

Initially, MIT's initiative on "Things that think" was mentioned. Some of the visions on embedded intelligence are drawn up in the book "When things start to think" (Gershenfeld 1999). Part of "thinking" is "talking" and mobile computing becomes a reality. It is one of the enablers for realising the vision of Ambient Intelligence. This chapter has only shown some of the possible applications, enabling technologies and examples of initiatives in the domain. The real challenge for the domain imposed by Ambient Intelligence is to provide mobile *ubiquitous* computing, and to provide applications and services that are truly beneficial to the society and citizens. Technology scepticism is significant, and to gain trust and confidence from the public is crucial. If we succeed in that, there are few limits to the possibilities.

10

Broadband Communication

P. Lagasse and I. Moerman

"I do not fear computers. I fear the lack of them."

Isaac Asimov

10.1 Vision

Mobile connectivity is a prerequisite for the deployment of true AmI systems. The concept of Ambient Intelligence requires the interaction of a variety of devices, appliances, sensors, and processors with persons who should not feel constrained in their movements when using the system. In order to best understand the goals and requirements of the broadband communications that underpin the AmI system let us imagine the existence of a "connectivity ether." Pervasive from a personal to a global scale, present anywhere, anytime, this connectivity ether makes us a part of a global electronic nervous system that connects all people and devices with processing or sensing capabilities.

The essential properties of this connectivity ether are that it is scalable, adaptable to changing circumstances, and self-healing. Above all it must be trustworthy and must keep us secure while being respectful of our privacy. As we do not want to wait, it must have a very low latency for whatever service or application we want to use. Finally it must protect us against unwanted data and against information overload by intelligent filtering of the exponentially growing amount of data sent over the network.

In order to achieve this vision there is no need to start from scratch. A wide variety of performant wireless technologies are already available or are currently being developed. The challenge we face is how to build this connectivity ether from the myriad of ICT devices, systems, and networks that are currently available or under development. A number of issues that need to be resolved are quite obvious:

- The various mobile and wireless networks and protocols must seamlessly converge with the fiber-optic feeder and backbone network.
- The complexity of interconnecting all those heterogeneous networks must be completely hidden so as to transform the complexity of the technology into simplicity of use.

- The restoration and self-healing capabilities should guarantee 100% availability. In view of the underlying complexity of the network, a self-healing and graceful degradation capacity appears to be the only solution to achieve the availability required by life-critical applications, which are bound to exist in such a pervasive network.
- The usual technical specifications of network bandwidth must be translated into a requirement for low latency of service.
- Regarding trust and security, leaving the user stranded to protect himself against all forms of e-garbage and cyber-crime is simply not an option.

10.2 Ongoing Evolutions in Broadband Communication

Nowadays, a lot of broadband applications are taken for granted in fixed networks. These applications require a high level of Quality of Service (QoS) and are generally characterized by high bandwidth requirements and low latencies, which can currently only be offered by fixed broadband access technologies (such as DSL, cable, and FTTH). The challenge future telecom operators and service providers are facing is to examine how these bandwidth hungry and QoS demanding services can also be provided in wireless access networks.

The number of Internet users is still expanding rapidly (from 688 million users, end 2003 (ITU-D, online) to 889 million users in March 2005 (Internet-World Stats, online)), while the number of fixed broadband Internet users is increasing even more rapidly. End 2004 there were about 150 million broadband users (Point Topic Ltd., online), which is 17% of all Internet users. However, when looking at the mobile subscriber statistics revealing 1,688 million subscribers in December 2004 (GSM World Statistics, online), we may conclude that the mobile terminal is without any doubt the most popular terminal. These numbers clearly indicate the great potential of broadband communications, not only for fixed, but also for mobile broadband services. In future users will take it for granted to have access to the same broadband services from their mobile terminal as they have now from their fixed terminal. The unprecedented growth of mobile communications further triggers the emergence of novel mobile applications and services, of which an overview is given in Chap. 9 (Mobile Computing).

In this chapter we first give an overview of the various wireless technologies, leading to the question whether we really need all of them. Then we will describe the concept of personal networking that is currently under development and that will allow to integrate the variety of wireless technologies into a network that comes close to the realization of the ideal situation of the connectivity ether.

10.3 Too Many Wireless Technologies Today?

To address the question whether or not there are too many different technologies for wireless communication we first present an overview of some of the existing technologies.

10.3.1 Overview of Existing Wireless Technologies

During the past decade, many technological and experimental advances have been observed in mobile and wireless networks. Table 10.1 gives an overview of the main wireless technology standards available today and their main characteristics (theoretical bit rate, spectral frequency, coverage area), and application domains.

The IEEE has developed several standards, which complement each other in terms of application domain and coverage area. The 802.15 standards (IEEE 802.15, online) define short-range technologies for Wireless Personal Area Networks (WPAN). A WPAN, often just called PAN, is a small wireless network, in which a limited number of personal devices (such as PCs, mobile phone, PDAs, etc.) communicate with each other through short-range wireless connections. The different 802.15 standards are targeted at different data rates. The 802.15.1 standard (Bluetooth) (Haartsen 2000; Bluetooth, online) has a medium data rate (1 Mbps) and was originally meant as a low-cost wireless technology to connect relatively cheap electronic devices hereby avoiding the need for a cable. Currently Bluetooth is however gaining interest as a WPAN technology. The 802.15.3 standard (Ultra-Wide Band or UWB) enables very high date rates (up to 480 Mbps), while 802.15.4 (IEEE 802.14.1, online), better known as the global standard for ZigBee (ZigBee, online) technology is a platform for low data rate, low-cost, and power-efficient wireless communication.

The group of 802.11 standards (IEEE 802.11, online), also known as Wi-Fi (Wireless Fidelity), aims at home and business wireless LANs and commercial hotspots and has a medium coverage area from 10 m (indoor) to 500 m (outdoor). The most popular standards, available in most laptops and PDAs nowadays, are 802.11b and its higher bit rate successor 802.11g, both operating at 2.4 GHz and interoperable. The IEEE 802.11a, operating at 5 GHz was originally only allowed in the United States, but is recently also permitted in Europe for indoor communication only (outdoor communication is forbidden because of interference with RADAR). In Europe there is HiperLAN/2, the WLAN standard developed by ETSI (European Telecommunication Standard Institute) (ETSI HIPERLAN/2, online). 802.11a and HiperLAN/2 have been the competing standards for a long time. Although HiperLAN/2 has proven to perform better in terms of throughput, the 802.11a standard currently has a definite lead in the worldwide market as the top choice for WLAN deployments. However, more recently the 802.11h standard (not shown in Table 10.1) had been proposed to satisfy regulatory requirements for operation in the

Table 10.1. Wireless technology standards

technology	theoretical bit rate	frequency	range	main application
HomeRF	1 Mbps (v1.0), 10 Mbps (v2.0)	2.4 GHz	∼ 50 m	wireless network for home and small office
IEEE 802.11b	1, 2, 5.5, and 11 Mbps	2.4 GHz	25–100 m (indoor), 100–500 m (outdoor)	WLAN, hotspot
IEEE 802.11g	up to 54 Mbps	2.4 GHz	25–50 m (indoor)	WLAN, hotspot
IEEE 802.11a	6, 9, 12, 24, 36, 49, and 54 Mbps	5 GHz	10–40 m (indoor)	WLAN, hotspot
Bluetooth (IEEE802.15.1)	1 Mbps (v1.1)	2.4 GHz	10 m (up to 100 m)	cable replacement, WPAN
UWB (IEEE 802.15.3)	110–480 Mbps	mostly 3–10 GHz	∼ 10 m	digital imaging and multimedia applications in WPAN
IEEE 802.15.4 (e.g., ZigBee)	20, 40, or 250 kbps	868 MHz, 915 MHz, or 2.4 GHz	10–100 m	home automation, remote monitoring and control
HiperLAN2	up to 54 Mbps	5 GHz	30–150 m	WLAN
IrDA	up to 4 Mbps	infrared (850 nm)	max. 1 m (line of sight)	instant transfer of data or digital images
IEEE 802.16	32–134 Mbps	10–66 GHz	2–5 km	last mile fixed broadband access
IEEE 802.16a	up to 75 Mbps	< 11 GHz	7–10 km (max. 50 km)	
IEEE 802.16e (broadband wireless)	up to 15 Mbps	< 6 GHz	2–5 km	

GSM	9.6 kbps	900 MHz	300 m (urban)	speech, SMS
		1.8 GHz 1.9 GHz	5–10 km (rural)	
GPRS	max. 171.2 kbps (typically 20–50 kbps)	900 MHz, 1.8 GHz, 1.9 GHz	as GSM	data, MMS, video
UMTS	144 kbps (rural) 384 kbps (urban) 2048 kbps (indoor)	2 GHz range	1 km (macro-cell) 400 m (micro-cell) 75 m (pico-cell)	speech, SMS, MMS, data, video

5 GHz band in Europe. 802.11h is essentially 802.11a with additional European features and actually incorporates several HiperLAN/2 functionalities.

The IEEE 802.16 family (IEEE 802.16, online), better known as the wide-range WiMax technology, is designed to provide wireless last-mile broadband access in the Metropolitan Area Network (MAN), delivering performance comparable to traditional cable and DSL. The main advantages of systems based on 802.16 are the cheap installation cost and the ability to quickly provision broadband services in scarcely populated areas, which are difficult to reach with wired infrastructure.

We further mention another medium-range standard, HomeRF (Negus et al. 2000). This is a wireless standard developed mainly for consumer use. Although a technically strong solution, it lacks strong vendor support.

A final (very) short-range wireless technology, shown in Table 10.1, is IrDA (Williams 2000). IrDA is short for Infrared Data Association, a group of device manufacturers that developed a standard for transmitting data via infrared light waves enabling transfer from one device to another without any cables. Many devices such as PCs, PDAs, and printers are equipped with IrDA ports. IrDA ports support roughly the same transmission rates as traditional parallel ports. The only restrictions on their use are that the two devices data must be very close to each other and that there must be a clear line of sight between them.

The technologies discussed so far are rather fixed wireless communication or access technologies (offering wireless communication or wireless access, but not really supporting mobility or only low mobility), while the last three technologies presented in Table 10.1 are wireless technologies supporting high user mobility.

GSM or Global System for Mobile communications represents the second generation (2G) mobile phone standard and is world's most widely used mobile system (GSM World (online); Rahnema 1993). GSM was originally developed for Europe, but has now excess of 75% of the world market. GSM was initially designed for operation in the 900 MHz band and subsequently modified for the 1,800 and 1,900 MHz bands. The 900 and 1,800 MHz frequency bands are used in Europe, Asia, and Australia, while the 1,900 MHz frequency band is primarily deployed in America. GSM is designed for voice and is operated in circuit switched mode.

GPRS (General Packet Radio Service) is a packet-based communication service for mobile devices that allow data to be sent and received across the GSM network (Samjani 2002). GPRS enables always-on wireless Internet access over the GSM network without the need for a dial-up modem. GPRS can theoretically achieve 171.2 kbps using multi-slot techniques, but the available bit rate today is generally much less (typically between 20 and 50 kbps). GPRS is an upgrade to the existing GSM network. The GSM network still provides voice and the GPRS extension handles data, hereby allowing voice and data to be sent and received at the same time. GPRS is a step toward 3G and is therefore often referred to as 2.5G.

UMTS or Universal Mobile Telecommunications System is the European implementation of the third generation (3G) mobile telephone standard (UMTS, online). The 3G mobile system is standardized under 3GPP (Third Generation Partnership Project), a collaboration between ETSI and other regional telecommunication standard bodies. The goal of UMTS is to enable networks that offer true global roaming and can support a wide range of voice, data, and multimedia services on rapidly moving wireless devices. UMTS provides multimedia services in the 2 GHz band and offers data rates of 144 kbps at vehicular speed in rural areas (macro-cell), 384 kbps at pedestrian speed in urban areas (micro-cell) and 2 Mbps indoor (pico-cell). Unlike GPRS, UMTS is based on a completely different technology than GSM (UMTS is based on Wideband Code Division Multiple Access or WCDMA), while GSM is based on Time Division Multiple Access or TDMA, implying that a completely different access network has to be built from scratch. This explains why UMTS shows a much more costly and incremental deployment than GPRS.

10.3.2 Do We Really Need All Those Technologies?

From the overview in previous section, it is clear that there exist many wireless technologies. Unfortunately none of the wireless technologies is powerful enough to support all professional and private services of a mobile user anytime and anywhere. Each wireless technology is designed for some specific application or user context. When the user is moving at vehicular speed, then a wireless technology that supports high user mobility (such as GSM, GPRS, or UMTS) is required. However such technologies are bandwidth constrained, and hence, when the user mobility is low, other technologies, such as a high-bandwidth WLAN may be preferred. Low bit rate power-constrained wireless devices, such as low-cost battery-empowered sensors or small home appliances, do not require a relatively expensive, high bit rate, high power-consuming Wi-Fi radio interface, but will benefit from cheap wireless technologies like ZigBee. It is obvious that the coexistence of different heterogeneous wireless technologies is indispensable for future ubiquitous wireless communications. In future even more advanced, higher bit rate and more power-efficient wireless technologies will emerge, and will have to coexist with current wireless standards. Interworking between heterogeneous wireless networks is hence getting increasingly important.

Some interworking mechanisms are already being developed in the framework of "beyond 3G" systems. "Beyond 3G" systems are considered to be heterogeneous networks with multiple Radio Access Technologies (RATs) as well as reconfigurable user terminals in order to allow mobile users to enjoy seamless wireless services irrespective of their location, speed, or time of the day. They will allow users to choose their access technology according to his or her needs. This concept of "Always Best Connected" is shared by the 3GPP and it has laid the groundwork for a UMTS/WLAN interworking specification (Gustafsson and Jonsson 2003). While WLANs can offer high bandwidth, cellular networks provide a (nearly) full coverage. Cellular operators no longer see WLAN networks as a competitive RAT, but consider WLAN as a true complementary technology to the 3G cellular networks. Cellular operators embrace the benefits of WLAN to build Internet access hotspots and offer roaming agreements for their mobile customers. UMTS/WLAN integration is not only advantageous for the mobile users, but also for the mobile network. As soon as WLAN coverage is available (typically in town or business centers), a mobile user should switch seamlessly from the ubiquitous low bandwidth UMTS network to a high bandwidth WLAN. This will not only improve the QoS, but will also increase the capacity in the cellular network.

WLAN/UMTS interworking can be achieved by means of different degrees of coupling (Apostolis et al. 2002; Pinto et al. 2004): from very tight coupling, where the WLAN access network is considered as a generic UMTS radio access network, to loose coupling, where WLAN access network is connected to the UTRAN (UMTS Terrestrial Radio Access Network) core network via the Internet. The higher the degree of coupling, the faster the vertical UMTS/WLAN handover can be achieved, but the larger the technical (and economical) challenges.

Interworking between heterogeneous wireless access technologies enables a mobile user to enjoy seamless Internet connectivity irrespective of location, speed, or time of the day. However, wireless Internet access anytime and anywhere is not enough to support all a person's professional and private activities. Nowadays an average user has several devices at several distant locations, such as the mobile devices he carries with him, devices at home, in the office, in the car, etc. A user wants to have a trusted communication between his many local and remote devices, hereby avoiding complex configuration and authentication steps. A more global approach will be needed to provide a global seamless connectivity of heterogeneous devices over heterogeneous communication networks in all circumstances.

10.4 How to Deal with All These Wireless Technologies?

10.4.1 The Concept of Personal Networking

The many technological advances and market demands for mobile and wireless networks have triggered the introduction of a diversity of terminals and wireless devices (notebook, PDA, mobile phone, GPS-terminal, camera, sensors, home appliances, etc.) and the development of various new multimedia services (text and multimedia messaging, real-time video broadcast/multicast, video-conferencing, video-on-demand, wireless payment, gaming, chatting, remote monitoring, etc.). Nowadays an average user carries several personal wireless devices with him, while having more devices at home, in the office, and probably also in his car. The number of personal devices is expected to increase during the next years. Although many devices, wireless technologies, and advanced services are already available today, they are not yet able to adapt autonomously to the user context and the user's preferences. Trusted communication between heterogeneous personal devices may be possible today, but is very complex as a lot of manual configuration and authentication steps are involved. Many newly introduced services have a great potential, but are often perceived as a burden as they do not take into account the user's expectations, the ease-of-use, the limitations of the user's terminal, or the user's network environment.

From PAN to PN

The introduction of the Personal Area Network (PAN) by the IEEE 802.15 working group (see previous section on wireless technologies) is already a first step toward communication between personal devices. However, a PAN only consists of personal devices in the close vicinity of the person and hence only offers a solution at the local scale, while the resources or services needed by a person are not necessarily in the close vicinity of the person. The concept of personal networking, which is studied for example in the IST Integrated Project MAGNET (IST MAGNET, online) (My personal Adaptive Global NET), can bring a global solution to satisfy a (mobile) person's professional and private needs, by providing trusted communication between the many local and remote personal devices. A Personal Network (PN) is a dynamic collection of interconnected heterogeneous personal devices, not only the local devices centered around the person, but also personal devices on remote locations such as devices in the home network, the office network, and the car network (Niemegeers and Heemstra de Groot 2003). The PN is organized in ad hoc interconnected clusters, consisting of heterogeneous devices, and linked by various suitable interconnection mechanisms (e.g., the Internet, WLAN, GSM, UMTS, PSTN, etc.), as shown in Fig. 10.1. The different clusters are interconnected by secure tunnels between devices with gateway capabilities. The different devices in a cluster may be heterogeneous in terms of processing power, data storage capacity, radio interface(s), battery resources, display size, resolution, etc.

The PN concept can be very well illustrated with the virtual home truck scenario presented in Fig. 10.2. Transportation and logistics represent a major business industry employing millions of truck drivers. Each day, these people spend hours in their vehicle while driving, waiting, or sleeping, and they are often several days away from home. By creating a virtual home environment, these people can be offered the ability to stay in touch with their family, to stay connected with their company and clients, or the possibility to contact their colleague truck drivers, which could have great commercial potential taking into account the large number of truck drivers worldwide.

Consider a truck equipped with a mobile phone, broadband Internet access, LCD display, headset, etc., forming a cluster of cooperating devices. When finished working, a truck driver could set up an Internet connection to his home. At home, a cluster of cooperating cameras, speakers, headsets, provides the truck driver with a virtual home environment. Through this environment, he can virtually walk around, seeing his family, talking with them, watching together a movie, playing games, etc.

When driving, the truck driver can listen to his digital music collection by streaming it from a server in his network at home. When the truck driver stops at a parking, he can read his e-mail, search for colleagues, play a game with other truck drivers, etc. When the truck driver arrives at a client, his

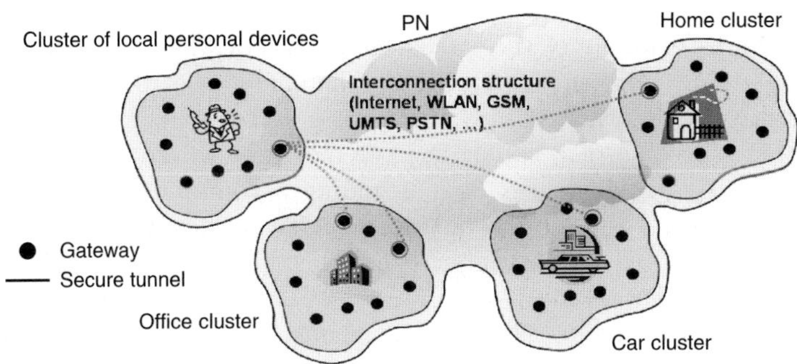

Fig. 10.1. Personal Network concept

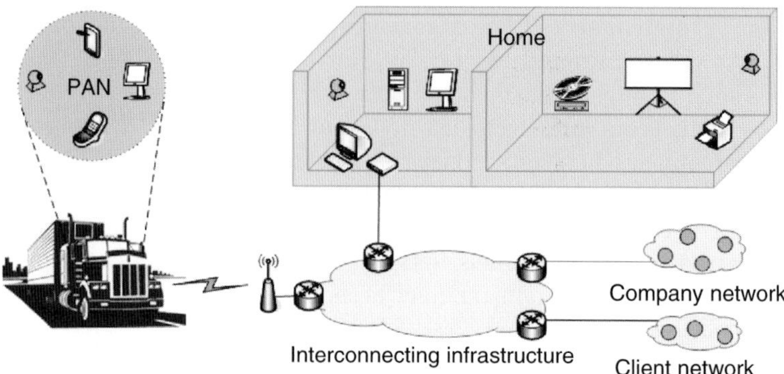

Fig. 10.2. Virtual home truck

PAN can connect to the client's company network and download the necessary documents. The documents can be digitally signed, handed over to the client and a copy can be uploaded to the truck driver's company, reducing the administrative burden. It should be noted here that the client network is not part of the PN of the truck driver. A PN should hence not only be able to support its own personal services, but also services offered by foreign devices or foreign networks.

From PN to Multi-PN

A PN has been defined earlier as a distributed networking solution to support a person's professional and private services by integrating all of a person's devices capable of network connectivity, whether in his or her (wireless) vicinity or at remote locations such as office and home. However, interaction and communication between multiple persons may also be of prime importance in some private or professional scenarios and should therefore also be supported by the PN concept. Multi-PN communication is actually a special case of communication of a PN with foreign nodes, where the foreign nodes belong to another person's PN. The multi-PN communication has already been touched in the previous example of the virtual home truck, when addressing the interactive gaming between truck drivers. The conference/meeting scenario in Fig. 10.3 presents another, more illustrative example of multi-PN.

Consider the case of a conference (or meeting) where multiple persons, each representing a single PN, are attending a conference, taking place in different locations. Each room is equipped with a wireless network printer, an access point providing wireless Internet access, a large display, and a file server to temporary store relevant files. A number of participants also have Internet access through the use of their mobile phone. As not everyone is directly within each other's send range, multi-hop routing is used in order to enable person-to-person connections. If someone has interesting information for other

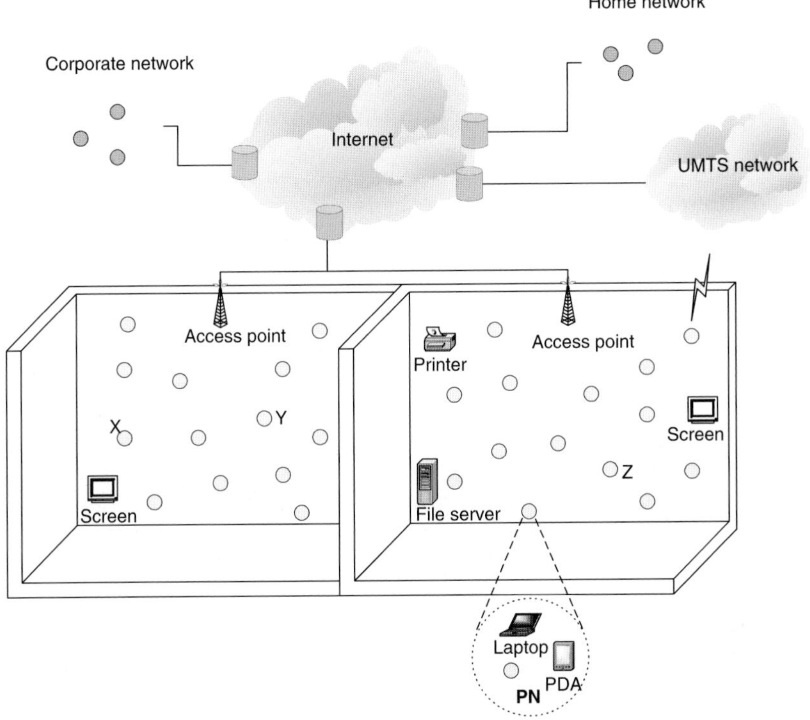

Fig. 10.3. Conference/meeting scenario

persons, he can upload his files to the file server, hereby setting the necessary protection mechanism. The subgroup of persons for whom the information is destined, is automatically informed and can download the information at their convenience. Upon download, some authentication step is required in order to prevent misuse of information. If needed, files can be printed on the network printer. At a given moment person X would like to give person Y some information about another interesting project. Unfortunately this information is stored on the laptop of a colleague Z attending a meeting in a remote room. The colleague is searched for over the multi-PN and the requested file is transferred through multi-hop routing. If needed, a person can access his personal or corporate files by using one of the access points (directly if the person is within the range of an access point or via multiple hops when the access point is out of reach) or even by borrowing the uplink (e.g., a UMTS link) of another person, if this person grants access to the first person. If someone quickly wants to present some extra information to the other attendees in the meeting, it can be sent to the large display. During a meeting multiple persons can work on the same file, which is displayed on the large screen.

For more examples of potential PN and multi-PN scenarios we refer to Chap. 9 (Mobile Computing).

10.4.2 Main Challenges of Personal Networks

Some existing technologies may offer solutions to a specific aspect of the PN scenario. However, in order to realize the full PN scenario meeting a person's communication needs, many (existing and new) technologies need to be combined into an integrated solution. The main PN research challenges are listed below.

Self-organization

A PN needs to be easy to use, setup, configure, and maintain as well as fast and secure. The different heterogeneous personal devices have to automatically form clusters. Once a personal device is switched on, it should be immediately recognized by the other local personal devices and automatically integrated in the cluster, without user intervention. A cluster has to be capable to deal with different link technologies. Not all devices necessarily have the same radio interface; some devices may have multiple radio interfaces and will have to forward data between devices with different radio interfaces.

The clusters have to automatically discover potential gateways, which can give access to interconnection structures. If a device in a cluster needs access to a device or service in another cluster, the most appropriate gateway has to be selected in each cluster and a secure tunnel has to be created between the gateways.

Support of Dynamics and Mobility

A PN consisting of several clusters interconnected with each other over fixed infrastructure can be very dynamic: devices and clusters may only intermittently be accessible due to mobility, energy constraints, and radio link characteristics. Clusters may change their point of attachment to the fixed infrastructure (e.g., when the user together with the devices he carries is moving.), clusters may merge and split (e.g., the cluster around the user can merge with or split from the home or office network), devices may enter or leave (e.g., by switching on or off, or because of a bad radio link). This dynamic behavior will strongly influence the communication in PNs, and one of the challenges is to provide connectivity and to allow service continuity in a dynamic PN environment. The user mobility and network dynamics may have a serious impact on many other mechanisms such as addressing, routing, gateway discovery, service discovery, and context discovery.

Naming, Addressing, and Routing

Each service, device, cluster, and PN should have a unique name and all network interfaces a unique address. The reason to have names is to hide irrelevant information from users and to facilitate user access to devices and services. The user does not need to be aware of the actual location/directory or any other detailed information of the subject. As the clusters and the PN are continuously dynamically changing, the addressing of different devices and interfaces may be altered. The user should not worry which dynamically assigned address corresponds to which device, but only use the same name for the same device. The addressing should further allow efficient routing.

Ad hoc Networking

Multi-hop aspects and ad hoc networking techniques can play an important role in the realization of the PN concept, not only in scenarios where multiple PANs (clusters) come together and form an ad hoc network (such as in the conference scenario when no fixed infrastructure is involved), but also for PN-wide routing (cluster to cluster over fixed infrastructures). A PN may not be seen as a pure ad hoc network as defined in the IETF MANET context: "massive network of hundreds or thousands of possibly highly mobile nodes in which routing plays a leading role." In the PN concept, the focus will be much broader and much more challenging than what is called "MANET routing." Context awareness can be exploited to incorporate additional intelligence in existing ad hoc routing protocols. Further, the possible existence of multiple radio interfaces within one cluster (Bluetooth, 802.11b, 802.11a, Zigbee, etc.) imposes additional challenges on the ad hoc routing protocols. In the PN vision, an ad hoc network of multiple clusters will normally not operate as a stand-alone network (as opposed to pure ad hoc networks). Clusters will interact with other clusters or foreign nodes via Internet access points over infrastructure-based networks.

Context Awareness and Context Discovery

Context awareness is a prerequisite for realizing the AmI vision. A PN has to adapt as quickly as possible to context. Hereby context means time information (day, hour), location (geographical position and environment), the person's agenda and schedules, the person's activities, network resources (e.g., available bandwidth), network conditions (e.g., quality of the link, latency), availability of personal devices, device properties (e.g., type of display, resolution, battery capacity), presence of other persons and their available devices and services, user profile (e.g., preferences, subscriptions for Internet access, and charging policies), etc.

A context discovery framework is required for collecting and updating context information and for offering relevant context information to any context-aware application or service. If a context discovery framework is available, a context-aware application or service does not have to care about gathering of the context information, but can just rely on the context discovery framework. Below we present a few examples of context-aware services.

- Gateway discovery and gateway selection: if multiple gateways are available, the gateway selection may be based on the bandwidth and latency requirements of the required service, on battery capacity, on available Internet subscriptions and charging limits, etc.
- Ad hoc routing: routing can be more efficient if constraints like battery capacity and quality of the links are taken into account. Routing can further benefit from the presence of other person's devices (provided that the other person grants access to his devices and the necessary security precautions are taken).
- Service discovery: the result of a service request may depend on the user preferences, the location, the time information, the devices properties, the available bandwidth, etc.
- QoS support: through interaction with the context (and service) discovery framework, the best network, terminal, and service parameters can be automatically selected for a particular service and a given context. As the context discovery frameworks not only keep information about the network conditions and terminal capabilities, but also are aware about the user preferences, a PN is offering more than just QoS. A PN is able to translate the user's (subjective) Quality of Experience requirements into (objective) QoS parameters.

Service Discovery

A PN contains many services that must be discovered and used. A service discovery mechanism is needed to locate the place, where the service can be found, to specify what the service provides and to describe the procedure how the service can be accessed and controlled. The service discovery mechanism should not be limited to local service discovery in the vicinity of the user, but should also be able to detect remote services in remote clusters of the PN or services offered by other persons or service providers. The service discovery mechanism has to further take into account the dynamics of the PN and has to be aware that there is no permanent connectivity between nodes and between clusters. It can hence not (always) rely on infrastructure-based servers.

Security and Privacy

Secure communication must be guaranteed in the PN through adequate key establishment, encryption, and authentication mechanisms. Security is required at different layers: at the link layer for secure connectivity between devices within the same radio domain, at network layer for secure routing and secure tunneling over insecure network infrastructures (such as the Internet), at the service level for secure service and context discovery and services access. All data needs to be secured: not only application data, but also control and management information. Different levels of trust relationships between communication entities should be possible in terms of duration (from ephemeral to permanent trust relations) and in terms of cryptographic keys (from light-weight keys to highly robust keys). Keys and/or passwords used in authentication and encryption procedures must be installed in each personal device so that they are not accessible to unauthorized people. The PN must provide a simple management tool to the end-user for establishing and managing trust relationships.

10.4.3 Some Personal Network Solutions

PN Architecture

Figure 10.4 shows the PN architecture, proposed by the IST MAGNET project. The architecture is composed of three abstraction levels (ALs): the connectivity, the network, and the service AL.

The connectivity AL consists of various wired and wireless link layer technologies, organized in radio domains, including infrastructure links.

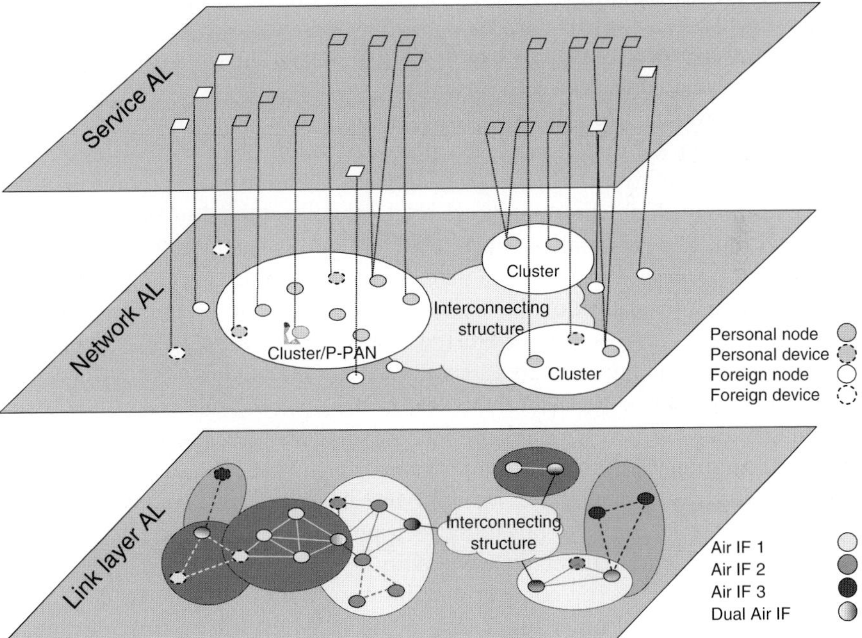

Fig. 10.4. PN architecture: view of three abstraction levels

A distinction is made between node and device, where a node is a device that implements IP. The link layer will allow two nodes or devices implementing the same radio technology to communicate if they are within radio range.

To allow any two devices within a PN to communicate, a network AL is needed. This level divides the nodes and devices into personal and foreign nodes and devices, based on the trust relationship between each other. Only devices that are able to establish a permanent or long term trust relation can be part of the user's PN. A long term trust relation implies that sharing and borrowing devices is also taken into account (e.g., family devices, devices from work). Personal devices that have such a long term common trust relation form clusters and these clusters can communicate with other clusters via interconnection structures. The P-PAN or Private Personal Area Network is a special cluster and represents the collection of personal devices around the person.

The highest level in this architecture is the service AL, which incorporates two types of services; public and private services. Public services are offered to anyone, while private services are restricted to the owner or trusted persons by means of access control and authentication.

The rest of this chapter will mainly focus on architectural concepts at the network AL.

Cluster Organization

The first step toward cluster organization is the initial assignment of a long term (or permanent) key. This procedure only occurs when a new personal device is introduced for the first time in the P-PAN. The long term key establishment can be understood as a personalization process.

At the connectivity AL, personal devices that have direct radio connectivity and that share the same trust relation will establish a trusted communication link. To this end, a short term key will be generated starting from the long term key. When more personal devices are discovered with which a trusted communication link can be established, a cluster of personal devices is formed that only consists of trusted communication links. The short term key establishment happens frequently, each time the cluster is formed (e.g., every morning when the user switches on his personal devices around him), while the long term key establishment only occurs once (when a new device is introduced for the first time).

Once secure link layer connectivity has been realized, communication at the network level can take place between personal devices, without using devices that have different or no ownership relation. To this end, the network layer needs access to the connectivity information at each node. Not all devices in a cluster will have direct links to all other nodes, because of different link layer technologies or radio-range limitations. This implies that a cluster might be a multi-hop network.

All personal nodes within one cluster can securely communicate with each other using IP as the common protocol. The use of IP as a common language makes it possible to have a network layer architecture that is independent from the heterogeneity of the underlying link layer. The cluster can also consist of limited devices that have no IP capabilities. As these personal devices can offer important services to the cluster, they should be connected somehow. To this end, a personal node that is able to communicate with this IP-incapable device will serve as its contact point (proxy), making its services accessible to the other nodes.

As the cluster network architecture is an IP-based multi-hop network, it should provide addressing and routing functionalities in order to enable efficient and secure intra-cluster communication. In order to cope with the specific characteristics and dynamics of clusters (nodes can join and leave the cluster, clusters can merge or split, nodes can move, nodes can have multiple wired/wireless interfaces), a distributed architecture is preferred. In such an architecture, all nodes in the cluster will cooperate in an ad hoc manner to provide the necessary network functionality (cluster formation and management, routing, addressing) and security functionality.

Finally, a cluster will not only operate as a stand-alone network, but it will also interact with its immediate environment, such as local foreign nodes or interconnecting structures. Nodes in the cluster that provide connectivity to nodes and devices outside the cluster are called gateway nodes. Gateway nodes will require some special functions such as address translation, set up and maintenance of tunnels, filtering incoming traffic, etc. These tasks might be quite heavy for some personal nodes, so it is useful to select powerful personal nodes as gateway nodes if possible. The process of finding capable gateway nodes with links to foreign nodes or interconnecting structures is another network function, which has to be provided by the cluster.

PN Organization

A PN can have multiple clusters that are geographically dispersed, but that have access via gateway nodes to the interconnecting structure. In order to form a PN and realize inter-cluster communication, two requirements need to be fulfilled. First of all, the clusters need to be capable of locating each other in order to establish tunnels between them. Secondly, once the PN has been formed, it should be able to maintain itself regardless of changes in gateway and node mobility. For these requirements to be fulfilled, we introduce the concept of a PN agent, a management framework that can be either centralized, under the control of a single provider or in a fixed cluster, or distributed over multiple providers or operators. Clusters that have obtained access to the interconnecting infrastructure announce their presence to this PN agent. This presence information should at least include the name of the PN the cluster belongs to, the point of attachment to the interconnecting structure, i.e., the IP address of the gateway through which the cluster can be reached and some

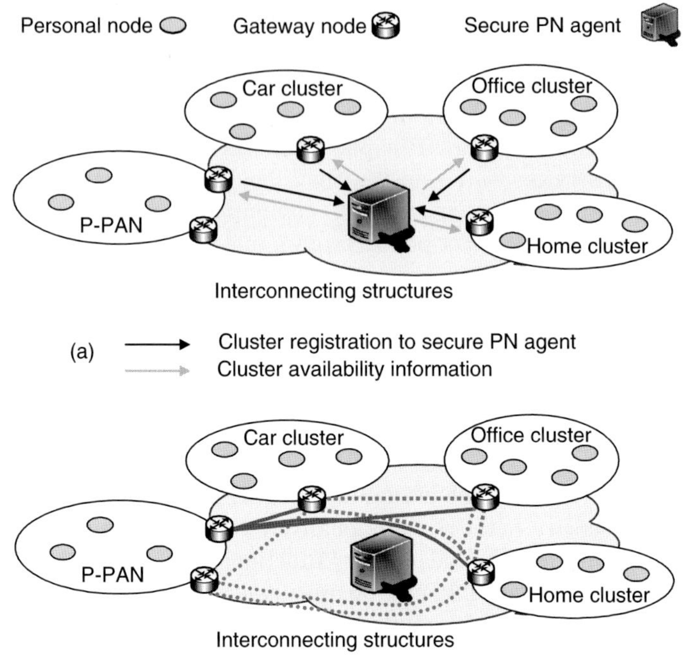

Fig. 10.5. PN formation and maintenance

credentials to verify this information. The PN agent will communicate this information to the cluster gateway nodes and this information will trigger the creation of secure tunnels between the clusters, as shown in Fig. 10.5 For simplicity the PN agent is presented as a central entity, but it could be as well distributed. The purpose of the tunnels is twofold. First, they provide secure inter-cluster communication by shielding the intra-PN communication from the outside world. Secondly, these tunnels are established dynamically. When clusters move, the information in the PN agent will be kept up-to-date. As a consequence, the PN agent will function as a secure database that tracks the PN clusters.

The information in the PN agent is tightly coupled to other components of the architecture such as the naming system, resource and service discovery framework and tunnel management mechanism. The PN agent can also be used by foreign nodes that wish to communicate with the PN. The only thing foreign nodes need to know in order to communicate with the PN is the contact information of the PN agent. The PN agent will find the appropriate service or node to connect with.

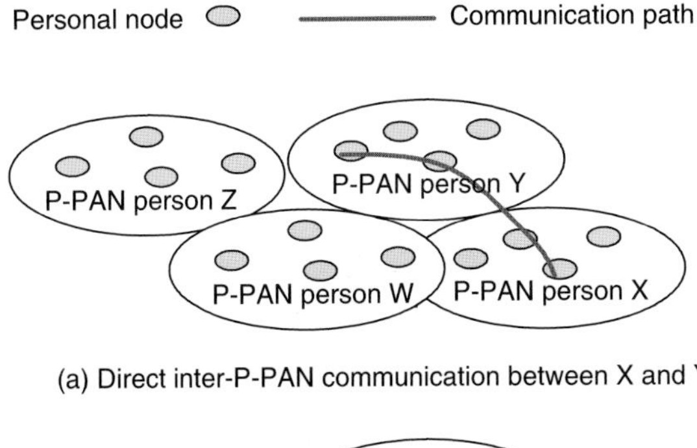

(a) Direct inter-P-PAN communication between X and Y

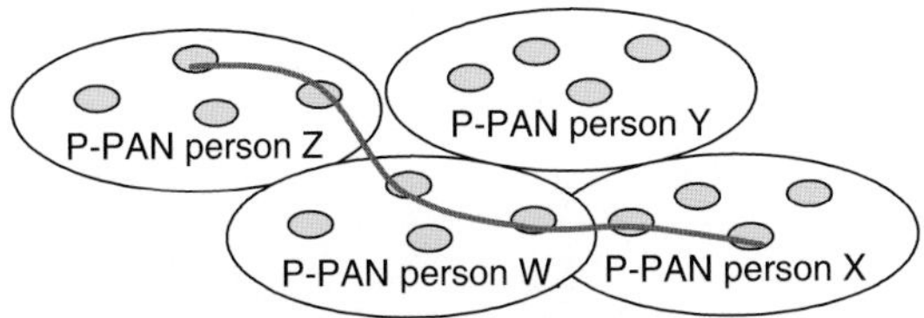

(b) Indirect inter-P-PAN communication between X and Z

Fig. 10.6. Inter-P-PAN communication

Multi-PN Communication

A first type of multi-PN communication is inter-P-PAN communication. Inter-P-PAN communication is the communication between personal devices of two persons that are in each other's neighborhood. This communication can be direct or indirect. Direct means that the communication paths do not include devices belonging to a third party. Indirect means that the communication paths include nodes that belong to other persons. Figure 10.6a, b illustrates direct and indirect inter-P-PAN communication.

Inter-P-PAN communication does not rely on the fixed infrastructure, resulting in multi-hop wireless communication paths. This implies that all network mechanisms such as resource and service discovery, naming, addressing, and routing should be able to operate in such an infrastructureless environment. For instance, nodes need to have unique addresses for communication, and intra-cluster routing should be extendable to neighboring P-PANs in a secure way, etc. Of course, access to the fixed infrastructure could be available, but will probably not be the most efficient way of networking.

Inter-PN communication is the communication between personal devices of two persons that are at remote locations. Figure 10.7 illustrates this type of communication.

When one PN wants to communicate with another PN at a remote location, it first has to find out how to reach this PN. To this end, naming and the PN agent concept will play a major role. By identifying the PN by its name and contacting the PN agent, the agent can provide the necessary information to establish communication with the remote PN, as it stores the IP addresses

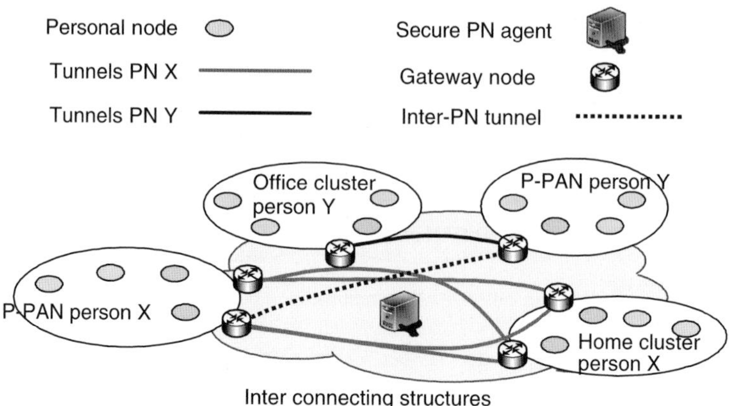

Fig. 10.7. Inter-PN communication

of the gateway nodes. This means that the central agent has to function as a kind of broker for establishing communication between PNs. Based on the information provided by the PN agent, a secure tunnel between the PNs can be established. The addressing solutions developed for intra-PN addressing should be extensible to inter-PN communication. In addition, routing protocols should be able to create connections, over the inter-PN tunnel, to the remote PN based on the assigned PN addresses.

10.5 Conclusions

In this chapter we have shown how the concept of personal networking, as studied within the IST MAGNET project, may bring a solution for the realization of the connectivity ether. The PN concept creates a personal distributed environment where persons can interact with various devices not only in the close vicinity, but potentially anywhere. This general PN architecture bridges heterogeneous (wired and wireless) networks and different wireless technologies, and hence hides the complexity of the connectivity to the end-user. The network layer is the glue that binds all a person's devices together into one PN. The network layer is based on a long term trust relationship that can offer communication between all a person's devices in a secure way. In this chapter, some solutions were presented to some of the most important networking issues needed to realize intra-PN and multi-PN communications that are context aware, and further support the dynamics of the PN and the user mobility. Those solutions based on results from the IST MAGNET project represent one of the possible implementations of AmI networking.

Within the European context, the IST project Ambient Networks (IST Ambient Networks, online), also develops complete and coherent solutions for ambient networking, enabling the easy and dynamic composition of disparate networks amid an ever-increasing heterogeneity of technologies and provider structures. The main objectives of the project are to define a set of adaptive and self-configuring mobile network components, which will reduce planning, deployment, configuration, and network maintenance costs and a comprehensive, integrated security framework, preserving end-to-end network protection and robustness against attacks.

Another related IST project is WINNER (Wireless World Initiative New Radio). This project is rather focusing on providing new radio technologies to make mobile communication systems more adaptable to user needs anytime and anywhere. WINNER is working toward enhancing the performance of mobile communication systems through improvements of radio transmission.

A third related IST project is E2R (End-to-End Reconfigurability) (IST E2R, online). The key objective of the E2R project is to devise, develop, and trial architectural design of reconfigurable devices and supporting system functions to offer an expanded set of operational choices to the users, applications and service providers, operators, regulators in the context of heterogeneous mobile radio systems.

The Ambient Networks, WINNER, and E2R projects are also part of the Wireless World Initiative (WWI) (WWI, online).

In the framework of AmI networking, two more IST projects should be mentioned here – although there may be more IST projects dealing with some aspects of Pervasive Computing and Ambient Intelligence – are IST VESPER (Virtual Home Environment for Service Personalization and Roaming Users) (IST VESPER, online) and IST RUNES (Reconfigurable Ubiquitous Networked Embedded Systems) (IST VESPER, online). The purpose of the VESPER project is to develop a service architecture for provision of a Virtual Home Environment (VHE) across a multi-provider, heterogeneous network, and system infrastructure. The VHE (Bougant et al. 2003) is defined as a concept for personal service environment portability across network boundaries and between terminals. This concept includes that users are consistently presented with the same personalized features, user interface customizations, and services in any network and any terminal, wherever the user may be located.

The vision within the RUNES project is to enable the creation of large-scale, widely distributed, heterogeneous networked embedded systems that interoperate and adapt to their environments. The inherent complexity of such systems must be simplified for programmers if the full potential for networked embedded systems is to be realized. The widespread use of network embedded systems requires a standardized architecture that allows self-organization to suit a changeable environment. RUNES aims to provide an adaptive middleware platform and corresponding application development tools, which provide programmers the flexibility to interact with the environment where necessary, whilst affording a level of abstraction that facilitates ease of application construction and use. This will allow for a dramatic cut in the cost of new application development and a much faster time to market.

Another European initiative is the Eurescom project Personal Nets (Eurescom, online). Personal Nets are a generic concept for providing an individualized, user-centric solution to the longer term integration of all communication and information services. The Personal Nets project is a feasibility study to test the concept of Personal Nets. The main objectives of this study are to identify and clarify from a customer point of view, the technical and business models, and concepts of Personal Nets bearing the changing paradigm of the digital economy in mind (i.e., different models for cooperation in service provisioning) and to investigate the underlying technologies, services and applications, and the business model of the Personal Nets concept.

In the United States there are also several research initiatives on the connectivity and cooperation between personal devices, such as the MOPED Project at University of Illinois at Urbana-Champaign (Kravets et al. 2001) and the Oxygen project at MIT (Oxygen, online). The goal of the MOPED Project is to enable cooperation between personal devices. The project presents a networking model that treats a user's set of personal devices as a MOPED, an autonomous set of MObile grouPEd Devices, which appears as a single entity to the rest of the Internet. All traffic for a MOPED user is delivered to the MOPED, where the final disposition of traffic is determined. To the outside world, this MOPED appears as a single device with a single interface or identifier. In reality, the group of devices cooperates to provide better services to the user. A MOPED component however needs constant contact with the infrastructure to function properly. A MOPED component cannot facilitate further ad hoc communication with MOPED components from other persons.

The vision of the Oxygen project, which is a project about pervasive human-centered computing, is that in the future computation will be human-centered and will be freely available everywhere, like batteries and power sockets, or oxygen in the air we breathe. Oxygen enables pervasive, human-centered computing through a combination of specific user and system technologies. Oxygen's user technologies directly address human needs. Speech and vision technologies enable humans to communicate with Oxygen as if they are interacting with another person, saving much time and effort. Automation, individualized knowledge access, and collaboration technologies help humans to perform a wide variety of tasks that they want to do in the ways they like to do them. Computational devices embedded in homes, offices, and cars sense and affect the immediate environment. Handheld devices empower humans to communicate and compute no matter where they are. Dynamic, self-configuring networks help machines locate each other as well as the people, services, and resources they want to reach. Software that adapts to changes in the environment or in user requirements helps humans to do what they want when they want to do it.

It is clear that the "connectivity ether" described in the introduction is no longer a distant dream. Various research projects are rapidly developing architectures that will transform the current wireless and wire-line communication networks into this connectivity ether.

11

e-Infrastructure and e-Science

T. Hey, D. De Roure, and A.E. Trefethen

"e-Science is about global collaboration in key areas of science, and the next generation of infrastructure that will enable it."

11.1 Introduction

The digital world of the Grid meets the physical world through a variety of sensors, instruments, and interfaces. These two significant trends in computing technology – more devices around us, more integration and power behind the scenes – have a symbiotic relationship. Grid applications, as exemplified by many projects in the UK e-Science program, have increasingly become aware of the need to focus on this digital–physical interface of the Grid, and the pervasive and ubiquitous computing community is looking towards the Grid for aspects of processing, data handling, integration, and access. Meanwhile there is also interest in applying middleware techniques across these distributed computing domains.

Both trends stand to benefit from a third significant movement in computing – the move towards machine-processable explicit knowledge as in the Semantic Web. This enables the automation and interoperability, which is increasingly necessary in these open, distributed systems and is essential to realize their full ambitions. This is illustrated through the adoption of Semantic Web technologies within the practice of Grid computing, a field of endeavor known as Semantic Grid (De Roure et al. 2005)]. These technologies are also being adopted within pervasive computing, for example in representing context information and device ontologies. Significantly, they also facilitate a capability for automated inference – an aspect of "intelligent" behavior.

Ambient Intelligence is more than the notion of pervasive computing, i.e., devices everywhere – it is the notion of intelligence in the surrounding environment supporting the activities and interactions of the users. It is the Grid technologies, and indeed the Semantic Grid, which shift us from sets of deployed devices into a genuine AmI environment. Ambient Intelligence is the meeting of the Semantic Grid and the physical world.

In this chapter, we will provide a background overview of the technologies and their applications – how the needs of e-Science applications are addressed by a Grid infrastructure, the nature of pervasive and AmI applications, and the role of Semantic Web technologies. We will then discuss some of the challenges that need to be met to realize this vision and provide illustrative case studies.

11.2 e-Science and the Grid

The international particle physics community is used to working in large collaborations with physicists from many institutions distributed around the globe, using large remote experimental facilities. Given the distributed nature of modern particle physics experiments, Tim Berners-Lee, who was working at CERN – the particle physics accelerator laboratory in Geneva – in the early 1990s, recognized that the particle physics community needed a tool for exchanging information within these large collaborations. From such origins the World Wide Web was born.

After a slow start, the international particle physics community enthusiastically adopted the Web for information exchange within their experimental collaborations – the first Web site in the USA was at the Stanford Linear Accelerator Center. Now, just a decade later, scientists are attempting to develop a new generation of collaboration tools. Besides being able to access information at different sites, they want to be able to integrate, federate, and analyze information from many disparate and distributed data sources – these data sources include data archives as well as networks of sensors and Radio Frequency Identification Devices (RFIDs) and to access and control computing resources and experimental equipment at remote sites.

11.2.1 The e-Science Vision

At the end of the 1990s, John Taylor became Director General of Research Councils at the Office of Science and Technology (OST) in the UK – roughly equivalent to Director of the National Science Foundation (NSF) in the USA. Before his appointment to the OST, Taylor had been the Director of HP Laboratories in Europe and HP as a company has long had a vision of computing and IT resources as a "utility". Rather than purchase expensive IT infrastructure outright, users in the future will be able to pay for IT services as they require them, in the same way as we use the conventional utilities such as electricity, gas, and water or as in "pay as you go" telephone billing. Taylor recognized the trends for scientific collaborations summarized above and realized that many areas of science could benefit from a common IT infrastructure to support multidisciplinary and distributed collaborations. He therefore put together a bid to the UK Government for a significant new initiative in this area and articulated a vision for this type of collaborative "e-Science" (Taylor, online).

> e-Science is about global collaboration in key areas of science, and the next generation of infrastructure that will enable it.

Taylor's bid to the Government was successful and in 2001 the UK initiated a £250M, 5-year e-Science program to develop the tools, technologies, and infrastructure to support such multidisciplinary and collaborative science. It is important to emphasize that e-Science is not a new scientific discipline – rather, the e-Science infrastructure developed by the program should allow scientists to do "faster, better, or different" research.

Such an e-Science infrastructure is in fact very close to the vision that J.C.R. Licklider (Lick) took with him to DARPA when he initiated the research project that led to the ARPANET. Larry Roberts, one of his successors at DARPA and principal architect of the ARPANET, describes this vision as follows (Segaller 1998).

> Lick had this concept of the intergalactic network which he believed was everybody could use computers anywhere and get at data anywhere in the world. He didn't envision the number of computers we have today by any means, but he had the same concept – all of the stuff linked together throughout the world, that you can use a remote computer, get data from a remote computer, or use lots of computers in your job. The vision was really Lick's originally.

The ARPANET led to the present day Internet – but the killer applications have so far been email and the Web rather than the distributed computing vision described above. Of course, in the early 1960s, Licklider only envisaged connecting a small number of scarce, expensive computers at relatively few sites. However, the relentless improvements in silicon technology over the past 40 years summarized in Moore's Law – Gordon Moore's prediction that the number of transistors on a chip would double about every 18 months and that the price-performance would be halved – has led to an explosion in the number of supercomputers, mainframes, workstations, personal computers, and other devices that are connected to the Internet.

11.2.2 The e-Science Infrastructure

The high-speed national research networks that constitute the underlying fabric of the academic Internet have long connected scientific collaborations. Under the banner of "e-Science", scientists and computer scientists around the world are now collaborating to construct a set of software tools and services to be deployed on top of these physical networks. The goal is a core set of middleware services that will allow scientists to set up secure, controlled environments for collaborative sharing of distributed resources for their research. Collectively, these middleware services and the global high-speed research networks will constitute a new global e-Infrastructure – Cyber-infrastructure in the USA – for collaborative scientific research.

How does this e-Science vision relate to the Grid? The term 'Grid' was first used in the mid 1990s to denote a proposed distributed computing infrastructure for advanced science and engineering. At the time, the intent was to use distributed computing resources as a "meta-computer," and the name was taken from the electricity power Grid – from a loose analogy with the idea that computing power would be made available for anyone, anywhere to use in the same way as we use electrical power. However, it became clear that there are very few problems for which the computational workload can usefully be distributed across multiple supercomputers. Similarly, merely connecting many individual processors across a wide area network is no substitute for a genuine supercomputer with its high performance low latency network. Users with an appropriate "supercomputer problem" to solve just need access to an available supercomputer. At the other extreme, users requiring to run many independent, single processor computational tasks – such as "ensemble" or "parameter search" Monte Carlo simulations – can take advantage of the type of distributed computing enabled by the Grid. However, the e-Science vision is much broader than such computational paradigms and is intimately related to data and information sharing. In this connection, Foster et al. (2001) have formulated the following re-definition of the Grid in terms of a middleware infrastructure to facilitate collaboration.

> The Grid is a software infrastructure that enables flexible, secure, coordinated resource sharing among dynamic collections of individuals, institutions and resources.

This vision of the Grid in terms of the middleware necessary to establish the virtual organizations needed by scientists is surprisingly close to Licklider's original vision. Unfortunately, present-day versions of Grid middleware provide only a small part of the functionality required for e-Science collaborations. Nevertheless, the vision of a set of middleware services that will allow scientists to set up "Virtual Organizations" (VOs) tailored to the needs of their specific e-Science communities has proved to have universal appeal. This vision is at the heart of the UK's e-Science program (Hey and Trefethen 2002) and a similar vision is embodied in the "Atkins" report on Cyber-infrastructure for the NSF (Atkins 2004).

11.2.3 Web Service Grids

Web Services are the distributed computing technology that the IT industry is uniting around to be the building blocks for interoperable, distributed IT systems (W3C, online). Web Services are a specific realization of a Service Oriented Architecture (SOA, online) in which services interact by exchanging messages in SOAP (Simple Object Access Protocol) format while the contracts for the message exchanges that implement those interactions are described in WSDL (Web Service Definition Language). By encapsulating internal resources within the service and providing a layer of application logic between those resources and the consumers, the owners of the service are free to evolve its internal structure over time (for example, to improve its performance or dependability), without making changes to the message exchange patterns used by existing service consumers. This encourages loose-coupling between consumers and service providers which is important for building robust interenterprise IT systems since no one party is in complete control of all parts of the distributed application.

The standards body for Grid, the Global Grid Forum (GGF), is developing an Open Grid Services Architecture (OGSA) based on Web Services (GGF, online). By leveraging developments in Web Services technologies, e-Science application developers will be able to exploit the tools, documentation, educational materials, and experience from the Web Services community when building their applications. The e-Science community can focus on building the higher-level services specific to the application domain while responsibility for design of the basic components of a reliable underlying infrastructure is left to the IT industry. The GGF will soon publish standards for basic services such as information services, execution management, data access and integration, resource management, and security. These basic services together with standards for portal technology and visualization services will enable scientists to use generic middleware infrastructure services to build their application specific VOs. This is the rationale for the UK e-Science "plug and play composable services" vision for Grid middleware.

11.2.4 The "Data Deluge" as a Driver for e-Science

One of the key drivers underpinning the e-Science movement is the imminent deluge of data from the new generations of scientific experiments and surveys (Hey and Trefethen 2003). In order to exploit and explore the Petabytes of scientific data that will arise from these high throughput experiments, supercomputer simulations, sensor networks, and satellite surveys, scientists will need assistance from specialized search engines and data mining tools. To create such tools, the data will need to be annotated with relevant metadata giving information as to provenance, content, conditions and so on and, in many instances, the sheer volume of data will dictate that this process is automated. In the next few years, scientists will create vast distributed digital repositories of scientific data that will require management services similar to those of more conventional digital libraries as well as other data-specific services. The ability to search, access, move, manipulate, and mine such data will be a central requirement for this new generation of collaborative science applications. Some examples are given below to substantiate these claims.

There are a relatively small number of centers around the world that act as major repositories of a variety of scientific data. Bioinformatics, with its development of gene and protein archives, is an obvious example. For example, the Sanger Centre near Cambridge in the UK (Sanger, online) currently hosts 20 Terabytes of key genomic data and has a cumulative installed processing power (in clusters – not in a single supercomputer) of around a Teraflop(s). Sanger estimates that genome sequence data are increasing at a rate of four times each year but that the associated computer power required to analyze these data will "only" increase at a rate of two times per year – still significantly faster than Moore's Law. A second example from the field of bioinformatics will serve to underline the point (Stuart 2002). It is estimated that human genome DNA contains around 3.2 Gbases, which translates to only about a Gigabyte of information. However, when we add to this gene sequence data, data on the 100,000 or so translated proteins and the 32,000,000 amino acids, the relevant data volume expands to the order of 200 Gigabytes. If, in addition, we include x-ray structure measurements of these proteins, the data volume required expands dramatically to several Petabytes, assuming only one structure per protein. This volume expands yet again when we include data about the possible drug targets for each protein, to possibly as many as 1,000 data sets per protein. And there is still another dimension of data required when genetic variations of the human genome are explored. To illustrate this bioinformatic data problem in another way, let us look at just one of the technologies involved in generating such data generation. Consider the production of x-ray data by the present generation of electron synchrotron accelerators. At 3 seconds per image and 1,200 images per hour, each experimental station generates about 1 Terabyte of x-ray data per day. At the next generation "DIAMOND" synchrotron currently nearing completion in the UK (DIAMOND, online), the planned "day 1" beam lines will generate many Petabytes of data per year, most of which will need to be shipped, annotated, analyzed, and curated.

A different data/computing paradigm is apparent for the particle physics and astronomy communities. In the next decade we will see new experimental facilities coming online that will generate data sets ranging in size from 100s of Terabytes to 10s of Petabytes per year. Such enormous volumes of data exceed the largest commercial databases currently available by one or two orders of magnitude (Gray and Hey 2001). Particle physicists are energetically assisting in building Grid middleware that will not only allow them not only to distribute Petabytes of experimental data amongst the 100 or so sites and 1,000 or so physicists collaborating in each experiment, but also to generate vast amounts of simulated data to understand their huge detectors. They are also developing environments and tools that will allow them to perform sophisticated distributed analysis, computation, and visualization on all or subsets of the data (Chin et al. 2003; NERC, online; GENIE, online; GeWiTTS, online; Integrative Biology, online).

As a final example from the field of engineering, consider the problem of health monitoring of industrial equipment. The UK e-Science program has funded the DAME project (DAME, online) – a consortium analyzing sensor data generated by Rolls Royce aero-engines. It is estimated that there are around 100,000 Rolls Royce engines currently in service. Each trans-Atlantic flight made by each engine, for example, generates about a Gigabyte of data per engine – from pressure, temperature, and vibration sensors. The goal of the project is to transmit a small subset of this primary data for analysis and comparison with engine data stored in three data centers around the world. By identifying the early onset of problems, Rolls Royce will be able to lengthen the period between scheduled maintenance periods and increasing profitability. The engine sensors will generate many Petabytes of data per year and decisions need to be taken in real-time as to how much data to analyze, how much to transmit for further analysis and how much to archive. Similar (or larger) data volumes will be generated by other high-throughput sensor experiments in fields such as environmental and earth observation, and of course human health-care monitoring.

From all these examples it is evident that e-Science data generated from sensors, satellites, high-performance computer simulations, high-throughput devices, scientific images and so on will soon dwarf all of the scientific data collected in the whole history of scientific exploration. Until very recently, commercial databases have been the largest data collections stored electronically for archiving and analysis. Such commercial data are usually stored in Relational Database Management Systems (RDBMS) such as Oracle, DB2, or SQLServer. As of today, the largest commercial databases range from 10s of Tbytes up to 100 Tbytes. In the coming years, we expect that this situation will change dramatically in that the volume of data in scientific data archives will vastly exceed that of commercial systems. Inevitably this watershed will bring with it both challenges and opportunities. It is for this reason we believe that the data access, integration and federation capabilities of the next generation of e-Science middleware will play a key role for both e-Science and e-Business.

With this imminent deluge of scientific data, the issue of how scientists can manage these vast datasets becomes of paramount importance. Up to now, scientists have generally been able to manually manage the process of examining the experimental data to identify potentially interesting features and discover significant relationships between them. In the future, when we consider the massive amounts of data being created by simulations, experiments, and sensors, it is clear that in many fields they will no longer have this luxury. The discovery process – from data to information to knowledge – needs to be automated as far as possible. At the lowest level, this requires automation of data management with the storage and organization of digital entities. At the next level, we require automatic annotation of scientific data with metadata describing both interesting features of the data and of the storage and organization of the resulting information. Finally, we need tools to enable scientists to progress beyond the generation of mere structured information towards the automated knowledge management of our scientific data.

11.3 Pervasive Computing and Ambient Intelligence

Enabled by technology advances and driven by customer demand for portable devices which are smaller, lighter and which run for longer on batteries, pervasive computing devices are now becoming increasingly prevalent in our everyday life – for example, the mobile phones, PDAs (portable digital assistants), digital cameras, global positioning systems, and other electronic accessories that we carry on our person and place in our everyday environment. These portable devices communicate using a range of wireless technologies including Bluetooth, ZigBee, wireless Ethernet, and the Global System for Mobile Communications (GSM). Internet protocols are evolving, for example through IPv6, to accommodate increasing numbers of devices, mobility, and automatic configuration. Noting Moore's Law, the clear technology trend is towards massive deployment of low-cost and low-power devices.

These pervasive devices form the intersection between the physical world and the digital world. Like the e-Science examples above, they provide a deluge of data that can demand sophisticated processing – effectively we are immersing ourselves in a vast, highly heterogeneous sensor network. Significantly, they also provide the means of interaction with the digital world.

11.3.1 Sensor Networks

Pervasive computing systems have the potential to deliver environmental and social benefits, through improved monitoring, providing access to information to create decision making capabilities and alert mechanisms. There are many application areas, particularly involving wireless sensor networks, including environmental monitoring in such applications as pollution control, disaster recovery, in medical sensors and patient care, and in business applications such as building the smart supply chain (Gaynor et al. 2004). For example, the NASA/Jet Propulsion Laboratory conceived the "Sensor Web" to take advantage of inexpensive mass consumer-market chips to create platforms that act together as a single instrument and which may be embedded into an environment to monitor and control it (Delin et al. 2005). As the technologies continue to improve we will see smaller, lower-power and lower-cost devices. The latest research has created novel methods for harvesting energy from the environment to provide self-powered micro-systems, giving a glimpse of the degree of pervasiveness the future may hold (Glynne-Jones and White 2001).

Prior to sensor networks, measurements would typically be taken manually, infrequently and at a small number of locations. Using pervasive devices, we can achieve a considerably higher spatial and temporal density of measurements. Not only does this increase data volume but the computational task implied by this may be quite considerably higher than before – for example, where the data feed into models and simulations performing numerical calculation, or where there is a demand for real-time data processing. With increasing numbers of deployments, the data integration task also becomes substantial.

Some applications effectively make use of everyday devices as a sensor network. For example, the more information that can be obtained about the current context of a person, the better able the system is to adapt to their requirements and provide appropriate services. This includes determining location and movement. Other applications focus on using multiple information sources in order to track the movements of individuals, raising a set of privacy issues and a significant technical challenge in drawing reliable inferences from multiple information sources of variable quality.

11.3.2 Interaction

Pervasive devices enable users to interact with the Grid. Within e-Science projects, we see examples of the use of specialized devices to collect data as well as general-purpose devices – such as mobile phones – being used to monitor and control experiments. Context is important for information delivery – providing the right information on the right device at the right time and with the right level of intrusiveness. In fact this rich notion of context becomes the query: using context knowledge, devices can act proactively and autonomously in order to provide the user with the information they require (NGG Expert Group 2004). Context is also crucial to information acquisition where it provides important metadata to aid the subsequent interpretation of the data and to record its provenance.

From the perspective of pervasive applications, the availability of Grid services makes it possible to achieve computational tasks that would not be possible on the devices themselves. For example, a digital camera can produce a volume of data that would benefit from remote Grid processing for tasks such as data compression or feature extraction – even more so with a digital video camera. Many interactive pervasive applications demand real-time processing and therefore require significant computational power, to be delivered on demand – one example of this is mobile mixed reality. Hence we envisage an important role for the Grid in support of new applications, especially as the devices generate (and store) larger quantities of data.

11.3.3 Intelligence

The data processing tasks in support of pervasive applications require sophisticated data mining, data integration, and the use of services such as feature extraction, pattern matching, language translation, and gesture classification. Sometimes they involve working with multiple sources of information, such as in situation assessment. It is these aspects, coupled with the context-awareness described above, that take us from a world of pervasive devices into a world of Ambient Intelligence. Many of these tasks stand to benefit from Grid processing and can be delivered as Grid services.

The applications also require sophistication in the behavior of the infrastructure – dynamic discovery and composition of services and devices in order to meet context-specific requirements. This contains many challenges since it is necessary for it to occur dynamically and there may be many applications competing for the same resources, which may not always be available. This is, for example, the territory of intelligent agents, which conduct negotiations in order to form coalitions to meet current requirements. As Grid services become increasingly available, closely related issues will arise there.

11.4 The Semantic Web

The applications discussed in the previous sections demonstrate that there is a very considerable degree of automatic processing, interoperation, and integration that is demanded by both Grid computing and Ambient Intelligence. The key to achieving this is to provide the necessary information in a standard, machine-processable form. For this we turn to another significant trend in contemporary computing: the Semantic Web (Berners-Lee et al. 2001). The Semantic Web is described in the World Wide Web Consortium (W3C) Semantic Web Activity Statement as the following initiative.

> ...to create a universal medium for the exchange of data. It is envisaged to smoothly interconnect personal information management, enterprise application integration, and the global sharing of commercial, scientific and cultural data. Facilities to put machine-understandable data on the Web are quickly becoming a high priority for many organizations, individuals and communities. The Web can reach its full potential only if it becomes a place where data can be shared and processed by automated tools as well as by people. For the Web to scale, tomorrow's programs must be able to share and process data even when these programs have been designed totally independently.

The Semantic Web technologies are based on the Resource Description Framework (RDF), which provides a standard way to represent metadata – which could for example be data about documents, objects, devices, Grid resources or people. The shared vocabularies that are used – and which are the key to interoperability – are called ontologies and can be represented in the OWL Web Ontology Language. Currently rule languages are being developed to operate alongside OWL. Tools for RDF are readily available, including for example the "RDF triplestores" that provide a means of working with large volumes of RDF metadata and conducting queries upon it.

RDF can be used to describe the various entities in our Ambient Intelligence and Grid systems, for example, resources, services, devices, context, and user profiles. The creation of ontologies for these entities are topics of current research efforts, as are the techniques of "semantic matching" which help discover appropriate entities. As well as describing entities, RDF also enables us to model the relationships between things in the physical world. As metadata are recorded in different times and places about a specific object with a unique identifier, it is effectively linked together by that common identifier. Hence the metadata accumulate, forming a rich, machine-processable body of knowledge.

11.4.1 Towards a Semantic Grid

A few years ago De Roure et al. (2001) introduced the notion of the Semantic Grid which advocated "the application of Semantic Web technologies both *on* and *in* the Grid." From the requirements derived from the diverse set of UK e-Science applications they identified a need for maximum reuse of software, services, information, and knowledge. Although the basic Grid middleware was originally conceived for hiding the heterogeneity of distributed computing, the authors contended that users now required "interoperability across time as well as space" to cope with both anticipated and unanticipated reuse of services, information, and knowledge.

The best practice in Semantic Grid is emerging through a series of e-Science projects, which have applied Semantic Web technologies to Grid applications in a variety of ways. The myGrid e-Science project builds on Semantic Web technologies and is researching high-level middleware to support personalized in silico experiments in biology (MyGrid, online). These in silico experiments use databases and computational analysis rather than laboratory investigations to test hypotheses. In myGrid, the emphasis is on data intensive experiments that combine the use of applications and database queries. The biologist user is helped to create complex workflows with which they can interact and that can also interact with workflows of other researchers. Intermediate workflows and data are kept, notes and thoughts recorded, and different experiments linked together to form a network of evidence as is currently done in bench laboratory notebooks.

The computer scientists and biologists in the project have together developed a detailed set of scenarios for investigation of the genetics of Graves' disease, an immune disorder causing hyperthyroidism, and of Williams-Beuren syndrome, a gene deletion disorder that affects multiple human systems and also causes mental retardation. To implement its ideas, the project has built a prototype electronic workbench based on Web Services. They have identified four categories of service: (1) external third party services such as databases, computational analyses and simulations, wrapped as Web Services; (2) services for forming and executing experiments such as workflows, information management and distributed database query processing; (3) services for supporting the e-Science methodology such as provenance and notification; (4) semantic services, such as service registries, ontologies, and ontology management, that enable the user to discover services and workflows and to manage several different types of metadata. Some or all of these services are then used to support applications and build application services. The project has developed a suite of ontologies – roughly speaking, agreed vocabularies of terms or concepts – to represent metadata associated with the different middleware services. Semantic Web technologies such as DAML+OIL (DAML, online) and standards body W3C's Web ontology language OWL (W3C OWL, online) then allow the prototype myGrid workbench to operate, interoperate, and reason over these services intelligently. The project has demonstrated the potential of such an approach to in silico bioinformatics experiments and is now attempting to produce more robust semantic components that will allow users to personalize their own research environment (Stevens et al. 2004; Wroe et al. 2004).

In a new paper, the same authors have revisited the e-Science program 3 years on from their original analysis to examine if their expectations have been realized (Miles et al. 2003). They now see the e-Science requirements as a spectrum, with one end characterized by automation, virtual organizations of services and the digital world, and the other end characterized by interaction, virtual organizations of people and the physical world. From experience with projects such as myGrid they have abstracted 12 key requirements for the Semantic Grid (1) resource description, discovery and use; (2) process description and enactment; (3) autonomic behavior; (4) security and trust; (5) annotation; (6) information integration; (7) synchronous information streams and fusion; (8) context-aware decision support; (9) support for communities; (10) smart environments; (11) ease of configuration and deployment; (12) integration with legacy IT systems. They also identify five key technologies that are being used to address these requirements in some of the UK e-Science projects: (1) Web Services; (2) software sgents; (3) metadata; (4) ontologies and reasoning; (5) Semantic Web Services. The Semantic Grid is not yet a reality but the UK e-Science projects are providing a valuable test-bed for Semantic Web technologies.

11.4.2 Semantic Web and Pervasive Computing

Semantic Web technologies are also being adopted within pervasive computing, for example in representing context information (De Roure et al. 2005) and device ontologies (Strang and Linnhoff-Popien 2004), allowing device capabilities and service functionality to be explicitly represented. This allows devices to advertise their services and permits a vision where a device might extend its functionality by contracting the use of a service from another device. The "task computing" activity at Fujitsu Laboratories of America and University of Maryland illustrates the combination of Semantic Web and pervasive computing (Matsuoka et al. 2003), encouraging device manufacturers to incorporate Semantic Web technologies into their devices in order to give end-users easier and more flexible use of the features of the devices. The Task Computing environment includes Web Services as well as RDF, OWL, and Universal Plug and Play (UPnP).

These technologies are being used today, but they are typically applied in the context of information on the Web, which evolves slowly. In the context of pervasive computing, vast amounts of metadata may be generated quickly and in a distributed fashion. This poses engineering challenges for the RDF triplestores and in handling distributed knowledge. It is also interesting to note that pervasive computing can assist with automatic metadata capture and therefore in building the Semantic Web – the value of the Semantic Web relies on good metadata being available. Hence Semantic Web technologies can help build the AmI infrastructure and Ambient Intelligence can help build the Semantic Web.

11.5 The Symbiosis of Grid and Ambient Intelligent Computing

The digital world of the Grid meets the physical world through a variety of sensors, instruments, and interfaces. These two significant trends in computing technology – more devices around us, more integration and power behind the scenes – have a symbiotic relationship. Grid applications, as exemplified by the projects described above, have increasingly become aware of the need to focus on this digital–physical interface of the Grid, and the ubiquitous computing community is looking towards the Grid for aspects of processing, data handling, integration, and access. Below we consider seven aspects of the intersection between the Grid and the physical world. In the next section, we will then consider three case studies of projects working at this intersection.

11.5.1 Devices Need the Grid for Computation

As sensors and sensor arrays evolve, our AmI environments are acquiring data with considerably higher temporal and/or spatial resolution than before. This data deluge demands considerably greater computational power to process and analyze it, especially where there are also real-time processing requirements. Currently, many pervasive deployments are relatively small scale, due to small numbers of devices or small numbers of users, but they will demand more Grid processing as numbers inevitably scale up – and, in particular, as the data arecombined with other sources which are also scaling up as this gives rise to a very significant increase in complexity. Grid services also provide computational augmentation of devices to support AmI applications – for example gesture recognition, language translation, and the computational tasks demanded by augmented reality.

11.5.2 Devices Need the Grid for Integration

Research in Pervasive Computing and Ambient Intelligence tends to focus on individual devices and deployments of these, with an interoperable infrastructure within a deployment – there is no established common distributed systems infrastructure standard to handle the interworking of multiple diverse sets of devices. Storz argues that, through Grid technologies to link sets of devices together, significant potential exists for building ubiquitous computing applications on a hitherto unprecedented scale (Matsuoka et al. 2003). This wider-scale integration is necessary to realize the full benefits of the AmI vision.

11.5.3 The Grid Needs Devices to Interface with the Physical World

There are many e-Science applications that use devices for data acquisition, interaction, or notification. There are some familiar "Grid-enabled" devices in a laboratory – the pieces of scientific equipment, instruments, and "Grid appliances" connected directly to the Grid. At one end of the spectrum, we have instruments such as telescopes and x-ray diffractometers, but our interest here is in the multitude of AmI devices in the surroundings of the Grid user. Traditionally the human user interface to the Grid has been through the graphical user interfaces of applications and portals. In the context of Ambient Intelligence, the interface becomes the devices in the user's environment. These devices may be used manually or they may be part of the environment, and they may be collecting data or providing notification and information display/visualization.

11.5.4 Virtualization

Grid computing and Ambient Intelligence are each about large numbers of distributed processing elements. At an appropriate layer of abstraction, they both involve similar computer science challenges in distributed systems. Specifically, these include service description, discovery, and composition, issues of availability and mobility of resources, negotiation of quality of service, autonomic behavior, and of course charging, security, authentication, and trust. Both need ease of dynamic assembly of components, and both rely on interoperability to achieve their goals. The peer-to-peer paradigm is also relevant across the picture.

Both Grid and pervasive middleware infrastructures stand to benefit from Semantic Web technologies. Again this is about semantic interoperability, but in the middleware, we need service description, discovery, and composition, and indeed research areas such as Semantic Web Services are being applied both to Grid and to pervasive computing. Hence the semantic approach sits above the large-scale distributed systems of pervasive and Grid computing.

11.5.5 The Information Systems Perspective

Much of Ambient Intelligence and Grid Computing is actually about information. As parts of these distributed systems come together, interoperability of the information is essential. Kindberg notes "Too often, we only investigate interoperability mechanisms within environments instead of looking at truly spontaneous interoperation between environments. We suggest exploring content-oriented programing – data-oriented programing using content that is standard across boundaries – as a promising way forward". (Stortz et al. 2003).

It is not just the content but also the metadata that are essential to achieving interoperability. The techniques of the Semantic Web are quite appropriate here; particularly as they lend themselves to automated processing. In particular, some inference capability at this level helps to put the "Intelligence" into Ambient Intelligence.

11.5.6 Grid Computation on Networks of Devices

Though the area is less well developed at this time, some researchers are interested in running Grid computations over networks of devices. One case for this is in intelligent sensor networks where it would be advantageous to shift some of the back-end computation out into the field in order to perform some processing on the fly. Another is the future vision of large numbers of devices being available – either through Ambient Intelligence or, one day, through fabrication of large numbers of processing elements that are embedded in everyday materials, as explored in the Amorphous Computing project at MIT (Kindberg and Fox 2002).

11.5.7 Self-organization

Self-configuration, self-management, self-optimization, and self-healing are all desirable properties in both Grid and pervasive systems – sometimes described as "autonomic" behavior. There is evidence of this sort of behavior in ad hoc networks, while Grids are typically more highly managed and controlled. The vision is of adding new devices or Grid nodes in an arbitrary manner and having the system adapt to them, and similarly adapting to change and failure.

Semantic Web technologies offer a means for the machine-processable knowledge, which is necessary for this level of automation, and we can envisage a self-organizing Semantic Grid (Abelson et al. 2000). Agent-based computing has also been proposed as an appropriate technology in both Grid and pervasive middleware. In addition to a service-oriented approach, agents bring an important notion of autonomous behavior, and this could be crucial to the necessary degree of automation in these large-scale systems. Agents are also able to respond to their dynamic circumstances using techniques of service negotiation, a capability needed in both the Grid and pervasive contexts.

11.6 Case Studies

Below we present three examples of projects that build on the relation between Grids and Ambient Intelligence. The examples are taken from various domains including materials science, health care, and sensor applications.

11.6.1 CombeChem

The CombeChem e-Science project aims to enhance the correlation and prediction of chemical structures and properties, using technologies for automation, semantics, and Grid computing (CombeChem, online; De Roure 2003; Frey et al. 2003). A key driver for the project is the fact that large volumes of new chemical data are being created by new high throughput technologies, such as combinatorial chemistry in which large numbers of new chemical compounds are synthesized simultaneously. The need for assistance in organizing, annotating, and searching these data is becoming acute. The multidisciplinary CombeChem team has therefore developed a prototype test-bed that integrates chemical structure–property data resources with a Grid-based computing environment. The project has explored automated procedures for finding similarities in solid-state crystal structures across families of compounds and evaluated new statistical design concepts to improve the efficiency of combinatorial experiments in the search for new enzymes and pharmaceutical salts for improved drug delivery.

One of the key concepts of the CombeChem project is "Publication@Source" by which there is a complete end-to-end connection between the results obtained at the laboratory bench and the final published analyses (Hughes et al. 2004). This starts with Ambient Intelligence in the smart laboratory and Grid-enabled instrumentation. By studying chemists within the laboratory, handheld technology has been introduced to facilitate the information capture at this earliest stage; see the SmartTea project (Frey et al. 2002; Schraefel et al. 2004; SmartTea, online). Using tablet PCs, the system has been successfully trialled in a synthetic organic chemistry laboratory and linked to a flexible back-end storage system. A key finding was that users needed to feel in control and this necessitated a high degree of flexibility in the lab book user interface. The computer scientists on the team investigated the representation and storage of human-scale experiment metadata and introduced an ontology to describe the record of an experiment and a novel storage system for the data from the electronic lab book. In the same way that the interfaces needed to be flexible to cope with whatever chemists wished to record, the back end solutions also needed to be similarly flexible to store any metadata that might be created. Additionally, pervasive computing devices are used to capture laboratory conditions, and chemists are notified in real time about the progress of their experiment using pervasive devices.

These data then feed into the scientific data processing. All usage of the data through the chain of processing is effectively an annotation upon it, and the provenance is explicit. The creation of original data is accompanied by information about the experimental conditions in which it is created. There then follows a chain of processing such as aggregation of experimental data, selection of a particular data subset, statistical analysis, or modeling and simulation. The handling of this information may include explicit annotation of a diagram or editing of a digital image. All of this generates secondary data, accompanied by the information that describes the process that produced it.

In a sister project called eBank, raw crystallographic data are annotated with metadata and "published" by archiving in the UK National Data Store as a "Crystallographic e-Print" (Frey et al. 2002). Publications can then be linked back to the raw data for other researchers to access.

Crucially, this digital record is enriched and interlinked by a variety of annotations be they data from sensors, records of use, or explicit interaction. The annotations need to be machine processable, and useful for both their anticipated purpose and interoperable to facilitate subsequent unanticipated reuse. This is achieved by the deployment of Semantic Web technologies; RDF is used through the system. At the time of writing there are 70 million RDF triples in the CombeChem triplestore. The target is 200 million, making this a substantial Semantic Web deployment.

Another e-Science project, CoAKTinG (Bachler et al. 2004), has created tools to enhance the collaboration between e-Scientists, and has used CombeChem as a case study. This effectively extends the digital record by including the meetings between the e-Scientists, fully interlinked using semantic annotation. This is a further step towards the complete digital record. Originally conceived to provide real-time "presence" information about other people in a distributed meeting, the tools have been extended to integrate with devices so that device status is also communicated. Hence the tools support the broader smart environment and can be seen as part of the AmI infrastructure.

11.6.2 Grid Based Medical Devices for Everyday Health

Devices for health monitoring are an important AmI application. Grid and pervasive computing come together in the Grid Based Medical Devices for Everyday Health project, in which patients who have left hospital are monitored using wearable computing technology. Since the patient is mobile, position and motion information is gathered (using devices such as accelerometers) to provide the necessary contextual information in which to interpret the physiological signals. The signal processing occurs on the Grid and medics are alerted – by pervasive computing – when the patients experience episodes that need attention.

The interesting infrastructure research question is to what extent the Grid services paradigm can be deployed in the direction of the devices. The project has been conducted using Globus Toolkit 3. The devices and sensors typically have limited computational power and storage and only in some cases may be capable of hosting Grid Services, generally interfacing via a Grid service proxy instead. Intermittent network availability had been found to be a critical problem, and the project has developed a framework for supporting both mobile Grid clients and services in an intermittently connected network environment – it has produced a standard Web Services interface and associated software toolkit which provides a generic mechanism for exposing mobile/remote sensing devices as Grid services.

The additional contextual information provided by the wearables – GPS and accelerometer data – is essential to interpretation of the physiological signals. Modeling of context, reasoning about it, and managing it call once again upon Semantic Web technologies. The project also features an information portal that can be accessed by pervasive devices, and can be used to access physiological data but also for monitoring and diagnosis of the deployed system.

11.6.3 FloodNet

The FloodNet project (FloodNet, online) is an example of an intelligent sensor network making environmental measurements. Deployment is facilitated by wireless technologies but there are significant issues of power for the devices and the need for very long unattended periods of operation. Grid processing on the back end has significant benefits to the operation in the field.

FloodNet has deployed a set of intelligent sensor nodes around a stretch of river, chosen for its tidal behavior so that, for test purposes, there are regular variations in water level. The nodes are powered by solar cells in conjunction with batteries and each node communicates with its neighbors using wireless Ethernet. A special node relays the data back using GPRS. This is an ad hoc network, with nodes relaying information across the network to ensure data delivery to the gateway. Various parties can subscribe to the incoming data stream. As well as being stored in a Geographical Information System, the data are used to inform flood simulations which are used to make flood predictions.

The fundamental trade-off between the need for timely data and the need to conserve energy is the research focus of the project. The system is adaptive so that the sampling and reporting rates of the devices vary according to the need, conserving power and minimizing the data volume required. Some intelligence is required in this adaptation, because the importance of a device at a given moment will depend upon both its environmental context and its role in relaying data from other devices, each of which will vary dynamically according to circumstances.

The process of adaptive sampling is mediated by the use of a predictive model running on the Grid back-end, which allows for the real-time collection of data to update the flood predictions regularly with refreshed data. The predictive model is required to carry out extensive processing in a short period of time (currently 20 min to 1 h). Upon each model iteration the network changes its behavior, altering the reporting rates of each individual node according to the time variable demand placed upon it by the predictive model.

The adaptive sampling is achieved through a series of control loops. The outermost loop enables the flood predictions to influence the reporting rates of individual nodes, so that closer monitoring can be achieved in anticipation of a possible flooding event. The inner loop is the peer-to-peer behavior of a set of nodes which can communicate with each other but have no external coordination and are therefore described as self-managing or autonomic. Other possibilities include one node, such as the gateway, taking a coordinating role.

Although not focused on users, FloodNet demonstrates many aspects of an AmI infrastructure, including the use of Grid processing in the back end in order to support the intelligent behavior of the devices. It also raises the question as to what extent, in the fullness of time, the predictive modeling could occur on the devices themselves.

11.7 Challenges to the Vision

There are challenges in realizing this vision. For example, an infrastructure challenge exists in allowing devices to have a place on the Grid as they often lack the power required to support the Web Services styles required by many infrastructures. Furthermore, the communication patterns differ from a traditional file-compute Grid: devices may interact using asynchronous event notification, or they may provide continuous data. Bridging between the Grid and these devices raises fundamental issues. Several challenges are articulated in Atkinson et al. (2002), which identifies the need for infrastructure advances in Semantic Grid, trusted ubiquitous systems, rapid customized assembly of services and autonomic computing systems. Here we consider the following issues that arise from our seven points of symbiosis:

1. *Devices need the Grid for computation.* Grid applications are already being used to handle the data deluge from sensor networks, but the use of Grid services on-the-fly by AmI applications is not yet established, nor the negotiation mechanisms to select appropriate Grid services when many are available.
2. *Devices need the Grid for integration.* We see examples of this, as in the Grid Based Medical Devices project, but on a project-specific basis. A framework is needed to bring devices into the Grid in a, flexible, open, and interoperable way. This challenge is attracting attention in the Global Grid Forum through the Appliance Aggregation Architecture Research Group (Bhatia et al. 2003).
3. *The Grid needs devices to interface with the physical world.* The e-Science program has been application-led and hence usability is a key issue. While many projects interface through a graphical user interface or portal, there are examples of AmI approaches, as in the CombeChem project. These ideas will mature with further case studies and emerging methodologies. The Semantic Grid aspect is important here, and suggests a new research area in "semantic interactive systems.".
4. *Virtualization.* Pervasive computing and the Grid are both distributed systems that, at the appropriate level of abstraction, pose similar challenges in terms of resource description, discovery, and composition, in a world where multiple applications compete for resources that may only

be intermittently available. The description mechanisms have yet to be agreed, and the techniques for service discovery and composition are not very sophisticated at this time. Semantic Web Services proposals exist but are still some time away from standardization.

5. *The information systems perspective.* The principles of information systems design for Grid and AmI applications are emerging as new systems are designed and deployed, but this perspective does not generally attract much attention in the Pervasive, AmI or Grid communities. Semantic Web technologies should be part of this.
6. *Grid computation on networks of devices.* This has attracted very little work so far. Given the inevitable deployment of larger numbers of devices, and the increasing computational power of some devices, the opportunity to create local Grids is set to increase.
7. *Self-Organization.* Autonomic computing is beginning to attract attention and there are some examples of systems that have some self-configuration, self-optimization and self-healing properties. These are not the norm, and realizing the autonomic vision perhaps requires a paradigm shift in approaches to system design, which will not happen overnight.

Addressing these challenges often requires collaboration across disciplines. The need to build bridges between Grid and AI is discussed in Goble and De Roure (2004) and between Grid and ubiquitous computing in Davies et al. (2004). The drivers are in place: we see devices needing Grid (sensor networks), Grid needing devices (instruments and interaction), Grid needing Semantic Web (Semantic Grid), pervasive computing needing Semantic Web (e.g., task computing) and Semantic Web benefiting from both Grid and pervasive to realize its own vision. All together, we have the basis for realizing Ambient Intelligence.

11.8 Conclusions

e-Science applications and Grid technologies have developed dramatically in the last 5 years converging with commercial approaches to distributed computing, namely Web Services. The middleware developments are maturing allowing a broad range of application development and deployment involving sensor networks, data federation, system virtualization, and remote access to resources. The development of automatic, autonomic, dependable, and robust e-Infrastructures still requires further research and development.

In this chapter, we have tried to illustrate the relationship of developments in Grid technologies and pervasive computing and the mutually beneficial bond that exists between the two areas. With advances in semantic capabilities we are able to realize a vision of an intelligent connected world with pervasive computing systems providing personalized access to content, applications, and services. The possible applications are numerous and the case studies provided although illustrative, by no means span the spectrum. They do, however, give an insight into what is likely to become the 'normal' mode of operation. Clearly there are many opportunities for social and environmental benefits from these technologies as well as new methodologies for scientific research.

As we have noted, however, there are still many challenges to be met and amongst those not discussed above are security and trust, where there is a substantial amount of effort in both pervasive computing and Grid technologies (PerSec, online; SPC-Conf, online; SPPC, online; Grid Security Papers, online; UK e-Science, online). Together with the challenges in our seven areas of symbiosis there are clearly many interesting research problems to be solved. It is clear that there is still quite a way to go down the road to Ambient Intelligent Grid systems, but the first steps have been taken and the fundamental building blocks are in place.

12

Context Aware Services

J.L. Crowley, P. Reignier, and J. Coutaz

"Awareness of human context is the key to limiting disruption."

12.1 Introduction

We are currently witnessing the integration of information technology into practically all aspects of the human environment. Current trends indicate that we will soon see technologies that allow extremely large numbers of extremely low cost (micro-euro) microscopic computing elements to be directly incorporated into common artificial materials such as plastics, fabrics, or paper. Wireless network technologies would allow such elements to coalesce into local ad-hoc networks, enabling ordinary objects such as tables, chairs, and walls to provide information technology functions such as sensing, display, communications, and computing. With such technology, ordinary objects can be made animate, with autonomous abilities to sense and communicate.

Endowing ordinary objects with animate abilities raises a serious threat to the ability of humans to control and focus attention. As can readily be seen from Internet-based services, when communication and information become free, attention becomes the limiting resource. Ambient Intelligence has the potential to greatly amplify the risk to disruption to human attention, as animated objects and autonomous migratory services vie for human attention. Unless such disruption is minimized and controlled, society will quickly rebel, rejecting the intrusion of animate machines in the human environment.

Within the current state of the art in informatics, one of the greatest sources of disruption is the autistic nature of software and services. Information technology, as practiced in the late twentieth century, is totally insensitive to humans. Software systems are devoid of any ability to sense and recognize the goals, the activities or much less the emotional state that defines the context within which humans interact with machines. Without awareness of the human context, any attempt at proactive assistance simply generates an unwanted distraction, as a certain software manufacturer discovered with an animated Paperclip.

In this chapter, we argue that awareness of human context is the key to limiting disruption of human attention by Ambient Intelligence. We present a conceptual framework and software architecture to allow systems and services to sense and model the context of users. This conceptual framework has been developed and validated in the construction of several context aware systems (Crowley 2003) developed in a series European IST Projects. Constructing these systems has allowed us to refine the framework and to create a software architectural model as well as a design method for context aware systems.

In the following section, we first review some of the various forms in which the awareness of context has been posed in different scientific communities. We then present a conceptual framework that allows the design of systems that sense and recognize human situations. This framework leads to a proposal for a component-based layered architecture for building context aware services. We illustrate this architecture by describing a working system for context aware acquisition of a synchronized audio-visual stream from multiple cameras and microphones. In the final section, we identify what could be the most difficult hard challenge of context modelling: continuous automatic development of context models without disruption.

12.2 A Brief History of Context

Since the early 1960s, the notion of context has been debated, modelled, and exploited in many areas of informatics. Winograd (2000) points out that the word "Context" has been adapted from linguistics. Composed of "con" (with) and "text", context refers to the meaning that must be inferred from the adjacent text. Such meaning ranges from the references intended for indefinite articles such as "it" and "that" to the shared reference frame of ideas and objects that are suggested by a text. Context goes beyond immediate binding of articles to the establishment of a framework for communication based on shared experience. Such a shared framework provides a collection of roles and relations with which to organize meaning for a phrase.

12.2.1 Context in Artificial Intelligence

Early researchers in both artificial intelligence and computer vision recognized the importance of a symbolic structure for understanding. The "Scripts" representation (Schank and Abelson 1977) sought to provide just such information for understanding stories. Minsky's Frames (Minsky 1975) sought to provide the default information for transforming an image of a scene into a linguistic description. Semantic Networks (Quillian 1968) sought to provide a similar foundation for natural language understanding. All of these were examples of what might be called "schema" (Bobrow 1977). Schema provided context for understanding, whether from images, sound, speech, or written text. Recognizing such context was referred to as the "Frame Problem" and became known as one of the hard unsolved problems in AI.

12.2.2 Context in Computer Vision

In computer vision, the tradition of using context to provide a framework for meaning paralleled and drew from theories in artificial intelligence. The "Visions System" (Hanson and Riseman 1978) expressed and synthesized the ideas that were common among leading researchers in computer vision in the early 1970s. A central component of the Visions System was the notion of a hierarchical pyramid structure for providing context for image understanding. Such pyramids were designed to transform successively abstract descriptions of global context into successively finer and more local context terminating in local image neighbourhood descriptions that labelled uniform regions. Reasoning in this system worked by integrating top–down hypotheses with bottom–up recognition. Building a general computing structure for such a system became a grand challenge for computer vision.

12.2.3 Context in Mobile Computing

Schilit and Theimer (1994) were among the first to introduce the term "Context Aware" to the mobile computing community. In their definition, context is defined as "the location and identities of nearby people and objects and changes to those objects". While this definition is useful for mobile computing, it defines context by example, and thus is difficult to generalize and apply to other domains. Other authors, such as Brooks (1986), Rodden et al. (1998), and Ward (1997), have defined context in terms of the environment or situation. Such definitions are essentially synonyms for context, and are also difficult to apply operationally. Cheverest et al. (2001) describe context in anecdotal form using scenarios from a context aware tourist guide. His system is considered one of the early models for a context aware application.

12.2.4 Defining Context

Dey (2001) reviews and provides definition of context as "any information that can be used to characterize situation". This is the sense in which we use the term context. Situation refers to the current state of the environment. Context specifies the elements that must be observed to model situation. However, to apply context in the composition of perceptual processes, we need to complete a clear and formal definition with an operational theory.

Context describes dynamic, structured, and shared, information spaces. Such spaces are designed to serve a particular purpose. In ambient informatics, the purpose is to amplify human activities with new forms of services that can adapt to the circumstances in which they are used.

12.3 A Conceptual Framework for Context Aware Systems

In the desktop computing environments, complete ignorance of the human task context or social context limits systems to direct reaction to individual human actions. A few limited attempts to endow systems with the ability to propose actions and services (such as the infamous Paperclip) have been widely rejected by users as annoying. Without proper understanding of the human task context, systems propose inappropriate services, thus distracting from the already overloaded attention of users. Generally most users quickly disable such features.

If proactive interruption by context ignorant services is annoying on a computer desktop, imagine the disruption that can be caused in an ambient informatics environment. Few people would want to live in an environment where nearly every object is demanding attention in a misguided effort to be useful. Yet there is clearly a role for services that respond and assist human activity without direct command. To avoid disruption, such services must operate with an understanding of human activity.

Human activity is complex, and thus building and maintaining such a model is a challenging task. We propose to describe human activity by building on the notion of script. A theatrical script provides more than dialogue for actors. A *script* establishes abstract characters that provide actors with a space of activity for expression of emotion. It establishes a scene within which directors can lay out a stage and place characters. Much of human social interaction follows a generalized form of stereotypical scripts. Our challenge is to encode such scripts in ambient informatics services.

A script describes an activity in terms of a scene occupied by a set of actors manipulating props. Each actor plays a role, thus defining a set of actions, including dialogue, movement, and emotional expressions. Ideally, the audience understands the theatrical piece by recognizing the roles played by characters and thus associating the social situations of the characters to personal experience.

The concept of role is an important (but subtle) tool for simplifying the network of situations. A *role* is an abstract agent or object that enables an action or activity. Roles act as bindings between entities that are observed by perceptual components and interpretations provided by a situation model. Entities are bound to roles based on an acceptance test. This acceptance test can be seen as a form of discriminative recognition. When an entity is assigned to a role it is said to play the role. Normally, roles are variables with a single value. Formally, assignment of an entity to a role can be modelled as a predicate. In most cases, roles have unique assignments, and are thus predicates of arity 1. However, in some cases, roles may be played by a group of entities, thus having arity N. An example of such an arity N role is "the audience" in a lecture. In a lecture situation, at any instant, one person plays the role of the "lecturer" while the other persons play the role of "audience".

Situations are defined in terms of relations between the entities that are playing each role. *Relations* include the individual properties of entities (arity 1 predicates) as well as relative properties (predicates of arity 2 or higher) of configurations of two or more entities. Situations are organized in network (modelled as a directed graph). Changes in the relations of entities playing roles generate events. Events determine changes in situation.

Roles and relations allow us to specify a situation model as a kind of "script" for activity in an environment. Theatrical scripts are organized as a sequential set of scenes, composed in turn of a sequence of actions and dialogue. Human activity generally involves alternatives. Rather than a simple linear sequence, a context model for human activity must express the alternatives. Thus, we propose to model human activity as a network of situations. Each situation corresponds to a set of relations between entities playing roles. A context model corresponds to a graph of possible situations. Each situation defines the possible set of appropriate actions that can be taken by a service.

Organizing situations into a network allows different interpretation for the same set of roles and relations, based on the temporal context as provided by the previous situation. In this sense, we say that a network of situations allows a context aware specification of system behaviour.

To summarize, a situation model is a composition of situations that concerns a set of roles and relations. A situation model determines the configuration of processes necessary to detect and observe the entities that can play the roles and the relations between roles that must be observed. The roles and relations should be limited to the minimal set necessary for recognizing the situations necessary for the environmental task.

The concepts of situation, role, relation, and entity provide a foundation for designing systems that sense and model human activity. However, applying such a model to the design of real systems requires a compatible software architectural model.

12.3.1 A Layered Model for Context Aware Services

This section proposes a layered architectural model for context aware services, as shown in Fig. 12.1. At the lowest level, the service's view of the world is provided by a collection of physical sensors and actuators. This corresponds to the *Sensor–actuator layer* in Fig. 12.1. This layer depends on the technology and encapsulates the diversity of sensors and actuators by which the system interacts with the world. Information at this layer is expressed in terms of sensor signals and device commands.

Hard-wiring the interconnection between sensor signals and actuators can provide simplistic services that are hardware dependent and have limited utility. Hardware independence and generality require abstraction perception and abstract task specification. Separating services from their underlying hardware requires that the sensor–actuator layer provides logical interfaces or standard APIs that are function cantered and device independent.

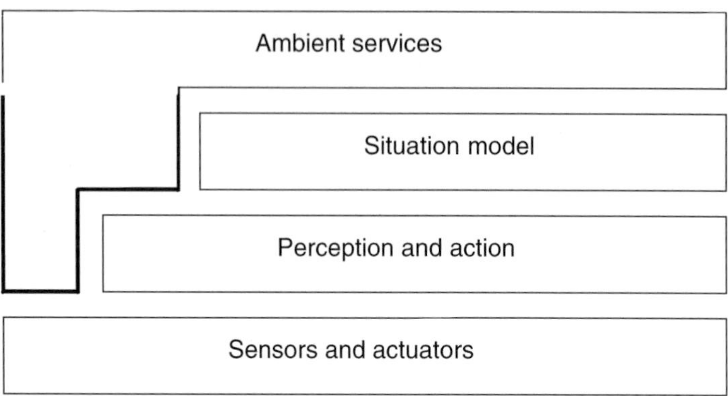

Fig. 12.1. Conceptual layers for context aware services

Perception and action operate at a higher level of abstraction than sensors and actuators. While sensors and actuators operate on device specific signals, perception and action operate in terms of environmental state. Perception interprets sensor signals by recognizing and observing entities. Abstract tasks are expressed in terms of a desired result rather than actions to be blindly executed.

For most human activities, there are a potentially infinite number of entities that could be detected and an infinite number of possible relations for any set of entities. The appropriate entities and relations must be determined with respect to a task or service to be provided. This is the role of the situation model. Situation models allow focusing attention and computing resources to determine the information required to provide services.

For a given state of a service, the situation model acts as a filter for events and streams from perceptual components. In certain well-defined cases, arrival of event or interpretation of a stream can result in an event being sent to the service components. In this sense, the situation model acts as a bottom–up filter for events from perceptual components.

Services specify a context. The context determines the appropriate entities, roles, and relations to observe with perceptual components. Information flow is inverted when a service changes state. In reaction to the user command, or to a change in situation, the service may send events to the situation model forcing a change in the situation graph, and possibly forcing a change in the configuration of perceptual components. In this case, the situation model acts in a top–down manner, elaborating and expanding the service event into commands to the situation model as well as to the perceptual component.

12.4 Defining Situation Models as Interaction Scripts

An ambient system exists in order to provide services. Providing services requires the system to perform actions, including modifying its internal configuration. We propose the following method for designing context aware ambient services.

One of the challenges of specifying a context model is avoiding the natural tendency towards complexity. Over a series of experiments we have evolved a method for defining situation models. Our method is based on two principles and leads to a design process composed of six phases.

12.4.1 Principle 1: Keep It Simple

In real examples, we have noticed that there is a natural tendency for designers to include entities and relations in the situation model that are not really relevant to the system task. It is important to define the situations in terms of a minimal set of relations to prevent an explosion in the complexity of the system. This is best obtained by first specifying the system behaviour, then for each action specifying the situations, and for each situation specifying the entities and relations. Finally for each entity and relation, we determine the configuration of perceptual processes that may be used.

The idea behind this principle is to start with the simplest possible network of situations, and then gradually add new situations. This leads to avoiding the definition of perceptual processes for unnecessary entities.

12.4.2 Principle 2: Behaviour Drives Design

The idea behind this principle is to drive the design process from a specification of the actions that the service is to take. The first step in building a situation model is to specify the desired service behaviour. For an ambient informatics, this corresponds to specifying the set of actions that can be taken, and formally describing the conditions under which such actions can or should be taken. For each action, the service designer lists a set of possible situations, where each situation is a configuration of entities and relations to be observed in the environment. Situations form a network, where the arcs correspond to changes in the roles or relations between the entities that define the situation. Arcs define the reaction to events.

These two principles are expressed in a design process composed of six phases.

Phase 1: Map Actions to Situations

The actions to be taken by the system provide the means to define a minimal set of situations to be recognized. The mapping from actions to situations need not be one-to-one. It is perfectly reasonable that several situations will lead to the same action. However, there can only be one action list for any situation.

Phase 2: Identify the Roles and Relations Required to Define Each Situation

A situation is defined by a set of roles and a set of relations between entities playing roles. Roles act as a kind of variable so that multiple versions of a situation played by different entities are equivalent. Determine a minimal set of roles and the required relations between entities for each situation.

Phase 3: Define Acceptance Tests for Roles

Define the predicates that must be true for an entity or agent to be assigned to a role. Currently these are logical predicates. We plan to move to probabilistic predicates, with the most likely entity being assigned to each role.

Phase 4: Define the Processes for Observation

Define a set of perceptual processes to observe the entities required for the roles and to measure the properties required for the relations. Define processes to assign entities to roles, and to measure the required properties.

Phase 5: Define the Events

Changes in situations generate events. Events may be results of changes in the assignment of entities to roles or changes in relations between the entities that play roles.

Phase 6: Implement Then Refine

Given a first definition, implement the system. Extend the system by seeking the minimal perceptual information required to appropriately perform new actions.

A situation graph implements a finite state machine. Human behaviour is, of course, drawn from an unbounded set of actions, and thus can never be entirely predicted by a finite state machine. Thus our model is most appropriate for tasks in which human behaviour is regulated by a well-defined, commonly followed, script. The lecture scenario is such an activity.

12.5 Example: Context Aware Automatic Video Acquisition

Over the last six years, our group has constructed a number of demonstration systems to explore concepts and techniques for context aware environments. These include the following elements:

- A system automatically determines the availability of an office worker for interruption.
- A system to support collaborative composition of a presentation by a group of people.

- A system to make visual and audio records of a meeting composed of several people.
- An automatic camera control system to record lectures.

Of these, the automatic camera controller is the most refined, and thus will be described below to illustrate the design and use of a context aware system. Version 1.0 of this system was used to record a series of eight lectures, or 3 h at the Forum of Cultures in Barcelona in July 2004. Successive versions have been demonstrated at the IST conference in Den Hague in October 2004, and the Festival of Science in Grenoble in 2004. In May 2005, version 2.0 has been used to record the In-Tech lectures on multi-media technology at INRIA Rhone-Alpes.

The intelligent camera controller controls a network of cameras and microphones to record a synchronized audio–video stream to capture the important events at a meeting or lecture. To accomplish this task, the camera controller must select and configure the most appropriate camera and microphone at each instant based on events in the lecture as interpreted using a situation model for a the current activity. A basic requirement for such a controller is to know where individuals are within the room, as well as what role each person is playing within the current activity. Such information can typically be obtained from a single static wide-angle camera using a system for robust detection and tracking of people. For robustness, such tracking can be driven by multiple detection modes.

The selection of the most appropriate camera and microphone depends on the current activities of individuals in the scene. Scene activity is described by a network of situations. Situations are defined by the relative spatial positions and activities of individuals in the scene. Activities include such things as talking, pointing with hands or faces, changing presentation slides, or arriving at the entrance to the room. Each situation specifies a microphone and camera configuration for use in composing the synchronized video recording. Each situation also specifies a set of possible next situations. Organizing situations into a network provides a more flexible configuration of system behaviour than a simple table of associations of situations with cameras and microphones. A graph of situations allows different interpretations to be made for the same configuration of people, based on the prior configuration (the situation context).

Roles are defined for the lecturer, the audience members, speaking agents, and new agents arriving in the scene (arriver). Perceptual processes determine if an object or agent can play a role by applying the appropriate acceptance test. When an entity or agent passes the acceptance test for a role, it is said to be able to "play" the role. For example, when a human "actor" walks to the front and addresses a group, he can be said to play the role of "lecturer". The acceptance test is defined as a unique person in specified region of the scene. A set of relational predicates is used to recognize the current situation.

Version 2.0 of the video acquisition is composed of the following seven processes running on five computers.

- Speech activity detector
- Movie encoder
- New-comer detector
- Eyes detector
- Pointing detector
- Supervisor
- Lecturer tracker

The supervisor process is implemented using a CLIPS rule interpreter in a JAVA environment. This process connects to an event bus. CLIPS rules in the supervisor react to events generated by the processes in order to select the current system actions, in this case, selecting the video stream that serves as the system output. The event bus provides a publish-and-subscribe mechanism for events. Each process connects to the event bus and specifies the classes of events that it can publish. Processes also subscribe to classes of events that they can receive.

The first task in defining a context model is to specify the available system actions. The available actions are as follows:

A_0	initialize processes and create a recording
A_1	record a global view of the scene from the panoramic camera and environmental microphone
A_2	record from lecturer's microphone and the lecturer's face using 3D tracking
A_3	record the audience with a dedicated camera and microphone
A_4	record the current slide and the lecturer's microphone
A_5	record a view of the screen area including the projected slide
A_6	stop recording

The available roles for this context are as follows:

director	the event coming from the human machine indicating that film recording should start
lecturer	the lecturer is detected using the microphone channel
audience	every person wearing a lapel microphone and not using the lecturer channel
slide	the image coming from the camera filming the screen region

speaker	any person wearing a lapel microphone producing voice activity on the associated channel
pointer	a hand within a predefined region nears the upper part of the body
new person	any person entering the room and detected using the tracker on the wide-angle camera

The situation graph for the lecture scenario is shown in Fig. 12.2.The situations in this graph are described as follows:

situation	name	description
S_0	initialisation	action: record view of screen, ambient microphone
S_1	start recording	role: director action: record room view and lecturer's microphone
S_2	lecturer speaks	role: lecturer, speaker relation: lecturer is same as speaker action: record lecturer and lecturer's microphone
S_3	question	roles: audience, person speaking relation: person in audience is person speaking action: record audience and audience microphone
S_4	pointing	roles: lecturer, hand, pointing region relation: lecturer's hand is in pointing region action: record view of screen and lecturer's microphone
S_5	new slide	action: record slide and lecturer's microphone
S_6	face to screen	roles: lecturer, face, screen relation: lecturer's face looking at screen action: record slide and lecturer's microphone
S_7	face to audience	roles: lecturer, face, audience relation: lecturer's face looking at audience action: record lecturer and lecturer's microphone
S_8	new person	role: person relation: new person enters the room action: record room
S_9	terminate	role: speaker relation: speaker exits lecture area action: terminate recording

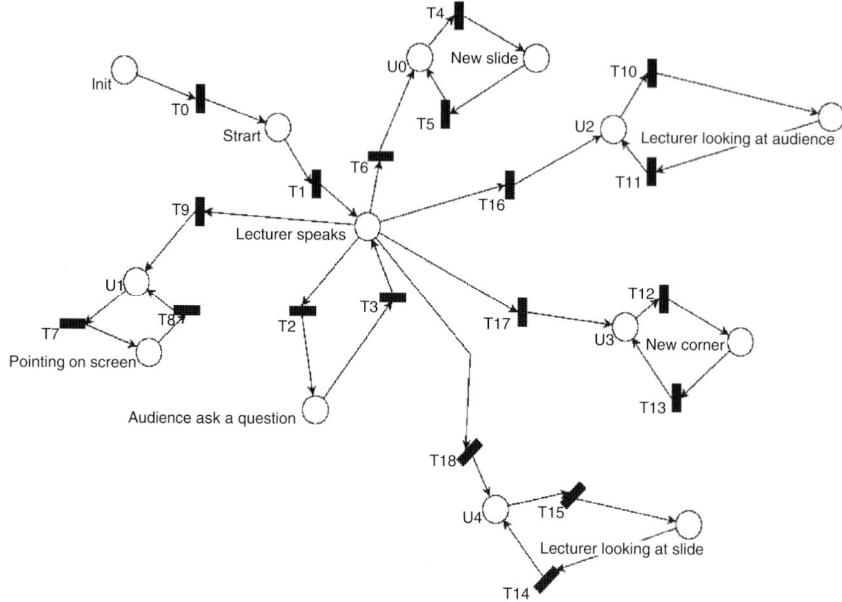

Fig. 12.2. The situation graph for the context aware video acquisition system

The final step is to define the temporal constraints between those situations, i.e., the context graph. Those constraints are expressed using the Petri Net formalism. This context graph is automatically compiled in a corresponding set of Jess rules, using the roles, relations, and situations definition.

12.5.1 Sample Recording from Context Aware Automatic Video Acquisition System

A film has been prepared in the FAME Augmented Meeting Environment at INRIA Rhone-Alpes to demonstrate version 2.0 of the context aware camera control system. The FAME room is equipped with six steerable Sony cameras, a fixed wide-angle camera, an array of four microphones, a set of wireless lavaliere microphones, and three video interaction devices constructed form the association of a camera with a video projector.

Figure 12.3. shows eight selected shots from a video sequence automatically generated by the camera controller. These shots correspond to situations S_0, S_1, S_2, and S_6, S_3, S_2, S_8, S_9, resulting in actions A_0, A_1, A_2, A_5, A_3, A_1, and A_6, respectively.

12.6 Learning Context Models: Adaptation and Development

It is not sufficient for a context aware system to behave correctly in any given context: such systems must also dynamically determine correct behaviour.

Fig. 12.3. Screen shots from the demonstration of version 2.0 of the context aware video acquisition system

Adaptation allows a system to maintain consistent behaviour across variations in operating environments. The environment denotes the physical world (e.g., in the street, lighting conditions), the user (identification, location, goals, and activities), social settings, and computational, communicational, and interactive resources. *Development* refers to the automatic acquisition of situation and context, and ultimately the acquisition of the entities, roles, relations from which situations and contexts emerge.

Context aware system must adapt and develop while retaining continuity and stability for users. Adaptation is necessary to maintain consistent behaviour while accommodating changes in the operating environment, task, user population, preferences, or some other factors. At the same time, context is too complex to be pre-programmed. A context model must develop through observation and interaction with users. A fundamental challenge is to provide both automatic adaptation and automatic development without disruption.

Adaptation and development are fundamental to providing useful and usable services to a variety of users in the presence of large variations in resources and activities. Context is too complex to be pre-programmed as a fixed set of stable variables: worse, the contract itself, which defines "correct behaviour", is not always precisely specifiable in advance. Thus, the context model, contract and adaptation process must develop through observation and interaction with the environment. At the same time, context models must not be disruptive. This creates a dilemma: how can context models evolve and develop without introducing disruption? The challenge is to find the appropriate balance between implicit and explicit interaction for providing the feedback required for development. We must determine the appropriate degree of autonomy, and this problem can impact every level of abstraction.

Current learning technologies, such as hidden Markov models and neural networks, require large sets of training data – something that is difficult to obtain for an extensible environment. Non-disruptive development of context models will require new ways of looking at learning, and may ultimately require a new class of minimally supervised learning algorithms. This requires that learning be studied as part of a semi-autonomous system. It requires that systems have properties of self-description, self-evaluation, and auto-regulation, and may well lead to new classes of learning algorithms specifically suitable to developing and evolving context models in a non-disruptive manner.

12.7 Conclusions

Context is key in the development of new services that will impact social inclusion for the emerging information society. For this to come true, we need to find the right balance between contradictory features. If context is redefined continually and ubiquitously, then how users can form an accurate model of a constantly evolving digital world. If system adaptation is negotiated, the question is how to avoid disruption in human activities. We believe that clear architecture and a well-founded explicit relationship between environment and adaptation are the critical factors – the key that will unlock context aware computing at a global scale.

13

Computational Intelligence

E. Aarts, H. ter Horst, J. Korst, and W. Verhaegh

"That's something I could not allow to happen"

HAL in Space Odyssey 2001

13.1 Introduction

Ambient intelligence is aimed at the realization of electronic environments that are sensitive and responsive to the presence of people (Aarts et al. 2002). The word "ambient" in ambient intelligence refers to our physical surrounding and reflects typical systems' requirements such as distribution, ubiquity, and transparency. The word "intelligence" reflects that the surroundings exhibit specific forms of social interaction, i.e., the ability to recognize the people who live and work in it, to grasp context, to learn and adapt to the users' behavior, and to show emotion.

As ambient intelligence is aimed at opening a world of unprecedented experiences, the interaction of people with ambient environments needs to become intelligent. Notions as media at your fingertips, enhanced-media experiences, and ambient atmospheres refer to novel and inspiring concepts that are aimed at realizing specific user needs and benefits such as *personal expression, social presence*, and *well-being*. These benefits seem quite obvious from a human perspective, but are quite hard to realize because of their intrinsic complexity and ambiguity. Obviously, the intelligence experienced from the interaction with ambient intelligent environments will be determined to a large extent by the computational intelligence exhibited by the computing platforms embedded in the environment, and consequently, by the algorithms that are executed by the platforms (Verhaegh et al. 2004).

The algorithmic techniques and methods that apply to *design for intelligence* in ambient intelligent environments are combined in the scientific and technological pursuit known as *computational intelligence*, which is aimed at designing and analyzing algorithms that upon execution give electronic systems intelligent behavior. For an introduction to the basic concepts in this field we refer to Engelbrecht (2002).

In this chapter we address the subject of computational intelligence in relation with ambient intelligence. The chapter is organized as follows. Since computational intelligence is rooted in the classical field of machine intelligence, we start our exposition with a brief outline of the developments in that domain. Next, we explain what the major behavioral characteristics are that need to be addressed and implemented by intelligent algorithms. The main part of the chapter is devoted to a discussion of a number of computational paradigms that serve as a basis for the realization of the intelligent and social user interaction within ambient intelligence. Next we discuss some elements related to the computational complexity related to these computational paradigms. We conclude the chapter with a short discussion of some challenges related to the future development of computational intelligence within ambient intelligence and as a preliminary conclusion we argue that new computing methods are needed to bridge the gap between the class of existing algorithms and the ones that are needed to realize ambient intelligence.

13.2 Machine Intelligence

According to Merriam-Webster's Collegiate Dictionary, "intelligence" refers to "the ability to learn or understand or to deal with new trying situations." Other definitions are the ability to reason, apply knowledge, manipulate one's environment, or think abstractly. Over the years mankind has proved indefatigable in its attempts to build electromechanical machinery that exhibits some intelligent behavior. "Counting" is clearly one of the most important of these intelligent activities that inspired engineers and scientist to functionally incorporate into automatic machinery. In this respect, the calculating device constructed by the famous mathematician Blaise Pascal in 1642 is widely recognized as the first "digital" computing device, where digital refers to making use of a finite set of internal states. Also the automatic table calculator called the *Differencing Machine*, which was constructed by the eccentric but brilliant mathematician Charles Babbage in 1822, can be seen as a landmark development in machine intelligence.

13.2.1 Artificial Intelligence

Since the invention of the computer in the mid-1940s machine intelligence has become a significant scientific sub-domain of computer science, which is often denoted as artificial intelligence. According to Minsky (1986), artificial intelligence is the science of making machines to do things that require intelligence if done by man. Classical subjects of investigation are vision, natural language and speech processing, robotics, knowledge representation and reasoning, problem solving, machine learning, expert systems, man–machine interaction, and artificial life (Rich and Knight 1991; Winston 1992; Nilsson 1998; Brooks 2002; Russell and Norvig 2003). McCorduck (1979) provides a good account of the early developments in artificial intelligence including some of the controversies that arose among scientists about the limitations of computers in comparison with human beings.

Over the years the discussions about artificial intelligence and the question whether mankind would be able eventually to build intelligent machines have been plentiful. Alan Turing, an early pioneer in computing science, developed the so-called Turing test to discriminate between computation and mind (Turing 1950). The Turing test entails that a keen interrogator questions a machine and a human volunteer at the same time. The interrogator cannot see either of the two, and he can pose the questions to both of them by making use of a keyboard only. The machine is said to pass the test if the interrogator cannot tell the difference between the machine and the volunteer from the answers given by the two.

Turing's proposition became controversial and was questioned by scientists working in the field of natural languages and learning psychology. The philosopher John Searle (1980) criticized the Turing test with his well-known Chinese Room argument, which demonstrated that a machine could reply to the posed questions in a way that was indistinguishable from the volunteer, but that it developed no other attributes that are generally considered as expressions of human intelligence such as understanding and self-consciousness. This controversy gave rise to what is known as the *mind–body problem*, which addresses the question whether a perfectly reconstructed body would have a mind by itself.

Hawkins and Blakeslee (2004) go in their recent book beyond the classical approaches to machine intelligence and develop a general theory of the human brain. They argue that the brain is not a computer, but a memory system that stores experiences reflecting the structure of the world. The memory system can generate predictions based on the nested relations among the memories, thus giving rise to intelligence, perception, creativity, and consciousness. Furthermore, the authors argue how eventually intelligent machines can be built based on their theory of intelligence. Although Hawkins and Blakeslee provide new and inspiring ideas, their theory cannot provide solutions to some of the open problems related to intelligence in the human brain. For instance, the above mentioned mind–body problem remains unresolved to a large extent.

13.2.2 Movie Script Scenarios for Ambient Intelligence

One of the salient features of ambient intelligence is the massive integration of intelligent features into the electronic background of our environments. Clearly, these features should enhance the interaction of users with their environments, thus enabling a means for nonobtrusive and social interaction that may increase productivity and creativity. Evidently, these statements are just abstract wordings and to make them more concrete one often resorts to the description of use-cases or scenarios. Examples can be found in the ISTAG report published in 2002, which provides four scenarios that cover different aspects in the daily life of ordinary people (ISTAG 2001). A good scenario should position the different use-cases within a realistic context that is kept consistent and that is maintained all over the story covered by the scenario.

The ISTAG scenarios, for instance, use different persons for different settings: Maria in *Road Warrior*, Dimitrios in *Digital Me*, Carmen in *Trafic, Sustainability, and Commerce*, and Anette and Solomon in *Social Learning*.

As another approach to the issue of scenario building one may resort to the many movie scripts that have been developed in the science fiction movie genre. A classical example is *2001: A Space Odyssey* (1968, MGM) directed by Stanley Kubrick. In this movie script Kubrick features an intelligent computer named HAL that serves three cosmonauts in a journey through outerspace. During the journey HAL develops certain elements of cognitive and social intelligence, such as the ability to have natural conversation, to plan complex tasks, and to show emotion. Kubrick gives with HAL a convincing and realistic rendition of an intelligent machine that is capable of developing an affective relation with its user. More recently, Stork (1997) edited a collection of book chapters in which renowned specialists cover specific aspects of HAL's behavior in relation with machine intelligence, including supercomputer design, reliable computing and fault tolerance, gaming, speech processing, vision, man–machine interaction, planning, affective computing, and computer ethics. From Stork's collection of contributed chapters in machine intelligence, one may conclude that HAL could have been built by 2001 and that it would pass the Turing test. Remarkably, Turing himself predicted that it would be possible to build a machine that would pass the Turing test somewhere between 2000 and 2010.

2001: A Space Odyssey describes a scenario of a computer that interacts with users in a natural and intuitive way, but HAL is still a computer, a machine that acts as a stand-alone object. In ambient intelligence the aim is on integration of such intelligent functionality into distributed environments that surround human beings. To obtain scenarios that describe how this could become effective one again may resort to the wonderful world of movie scripts. A compelling definition of ambient intelligence is visualized in *Mathilda* (1996, Tristar Pictures after a book with the same title by Roald Dahl) directed by Danny DeVito in a scene where the little girl Mathilda discovers her ability to control objects in her physical environment by speech and gesture. The scene also illustrates profoundly the ultimate aim of ambient intelligence to put the user in the center of the ambient environment and to give him full control of it.

In *Total Recall* (1990, Tristar Pictures), Arnold Schwarzenegger spends a virtual vacation on Mars, but afterwards events force him to go there for real. In one of the many science fiction scenes Schwarzenegger communicates with an interactive display wall that can switch to a peace giving scenery of the Colorado mountains, which helps him unwind after he has received bad news from Mars.

In *Minority Report* (2002, 20th Century Fox), directed by Steven Spielberg, Tom Cruise is positioned in a digital world in which he is surrounded by interactive screens and displays everywhere, allowing direct interaction with the environment. The concept of a ubiquitous digital smart interactive environment is consistently maintained throughout the movie. Newspapers have become interactive display foils that can play real-time video messages. Billboards have become public displays that provide urban annotation and support personal information access in the public domain. Even packaging material consists of interactive displays showing video commercials, and they can be controlled in a natural way, for instance by smashing it against the wall when it should stop advertising. Television viewing has become truly three-dimensional through the use of holographic display technologies that enable the viewer to actively participate in the scenes that are displayed.

The Harry Potter movies after the books by Joan Rowling provide other interesting examples of ambient intelligence scenarios in movie scripts. Especially, the third movie called *The Prisoner of Azkeban* (2004, Warner Bros. Pictures), directed by Alfonso Cuarón, depicts a world imbued with ambient intelligence features. The castle Hogwarts, school of witchcrafts, is flooded with pictures that "contain" living creatures and can act as interactive displays. They can even control access to rooms through the use of spoken passwords. The ceiling of the dining room is a huge display that can change the ambience of the room to match the occasion or the time of the day or season. Objects are smart and can be used to control the environment. They also can be used to access ambient information. For example, the *Rememberal* is an object that indicates through its color when the person who holds it has forgotten something. The ultimate example perhaps is the *Marauders Map*, which is an interactive two-dimensional map that reveals the geographical position of any person that is within range of the area displayed by the map.

All these scenes show examples of scenarios of distributed intelligent environments that support the people who interact with them, enabling them to perform daily tasks, and to become more productive, creative, and expressive. So, the notion of a computer as a physical device has disappeared. The computer has been integrated into the environment moving it to the background and bringing functionality to the foreground, leaving the user in total control. Another interesting observation is that technology is not going to make the difference in the end, but that the way technology is being applied and used will be the crucial factor. For instance, in both movies *Minority Report* and *The Prisoner of Azkeban* a world is depicted in which people are surrounded with interactive displays. Yet, the world in *Minority Report* is much more threatening than that in *The Prisoner of Azkeban* where the use of ubiquitous displays adds to the well-being of its inhabitants rather than providing a feeling of "big brother is watching you" as in *Minority Report.*

13.2.3 Social Versus Cognitive Intelligence

An important conclusion that can be drawn from the discussion presented in the previous sections is that intelligence in ambient intelligence is aimed at social intelligence rather than at cognitive intelligence. From the viewpoint of ambient intelligence, electronic doors that open automatically if a person is approaching them are more socially intelligent than Deep Blue, IBM's powerful chess computer that beats Gary Kasparov, the world champion of chess, in a direct confrontation in May 1997. This means that ambient intelligence provides a new kind of challenge for artificial intelligence. In ambient intelligence the level of intelligence of an electronic system is judged by the way users perceive it from a social interaction viewpoint. Evidently, the types of tasks, as well as the context in which they are carried out, play an important role in the user perception.

Csikszentmihalyi (1990) has introduced the concept of *flow* to discuss the observation of intelligent behavior of human beings when performing tasks. Flow is an experience concept in psychology that is universally perceived in all cultures and ways of living. Flow refers to the feeling people experience when in contact with the world around them, providing a sense of reason and purpose of their activities. Flow can be very helpful in the description of social intelligence in everyday life. Dunne and Raby (2001) present in their compelling book a number of interesting design studies in which they place functionally modified electronic daily life objects, such as their *parasitic light, GSM table*, and *compass table*, in the homes of people and describe how they develop patterns of social interaction with these objects over time. It is astonishing to see how people develop affective relationships with these objects that are meaningless at first glance.

Reeves and Nass (1996) present another interesting study on social intelligence in the interaction of people with electronic equipment. They introduce the so-called *Media Equation*, which states that people should be able to interact with media in the same way as people interact with each other. This is an easy statement to express but hard to accomplish. Nevertheless, it directly implies that natural human interaction paradigms such as multi-modality, personalization, expression, and emotion should be key elements in social machine intelligence, and this implication has become generally accepted over the years.

Picard (1997) addresses the compelling issue of emotions in computing in a more scientific way by developing a theory on affectiveness in man–machine interaction. She not only discusses how computers might develop emotions but also why they must do so, and this may become quite useful in the design of ambient intelligent systems.

13.3 AmI Elements of Social Intelligence

From the discussion presented above one might conclude that the development of ambient intelligence is still in a conceptual phase and that we have to resort to a model of applications or scenarios to illustrate what we mean by the concept when we want to make it more concrete. Evidently, scenarios as presented above can be very helpful in the discussion on the realization of ambient intelligence, but the proof of the concept evidently is in the realization of commercial products and services that are attractive to people. Over the years, a large variety of successful and less successful attempts have been made to develop such commercial products and services, and looking at the market there is already a lot that is available. Below, we present some of the most profound examples for the purpose of our discussion on machine intelligence and the corresponding computational methods.

To put the commercial products and services into perspective we distinguish between the following five classes of AmI (ambient intelligence) elements in social intelligence:

- See, hear, feel
- Understand, interpret, relate
- Look, find, remember
- Act, learn, adapt
- Create, express, emerge

This class division is rather arbitrary, and there is no theory that supports it. However, it combines the main social activities that human beings undertake in their daily lives into categories of well-defined and interrelated tasks that relate to the concept of flow as defined by Csikzentmihalyi (1990). Moreover, it provides a simple framework for the discussion of computational intelligence paradigms in ambient intelligence. Below, we present for each of these classes a number of examples of prototypes or commercially available products and services. Also this selection is chosen rather arbitrarily, but it may serve the purpose of providing an impression of what is already on the market, thus indicating where we stand in the development of ambient intelligence. Finally, we briefly sketch for each of the classes some of the basic computational paradigms that can be used to realize the socially intelligent interaction defined by the class. In the next section, we then discuss some of these paradigms in more detail with the purpose of presenting some of the mathematical models and computational details that are applied.

13.3.1 See, Hear, Feel

This class of AmI elements is concerned with the general aspects of visual, auditory, and other sensorial information processing. It refers to actions such as viewing screens, listening to spoken text or music, and sensing location and context. Examples of products and services in this domain are given in Table 13.1.

The computational paradigms that have been established in this domain are based on vision, speech processing, and sensor data fusion, which all heavily rely on signal processing techniques, often applied in the digital domain after an analog-to-digital conversion of the analog data. Most of the approaches follow three steps. Firstly, the environment is sensed through the

Table 13.1. See, hear, feel

OpticalSensors' "Talking Cane": warns blind or visually impaired persons through spoken language for obstacles in the vicinity.
Vivometrics's "LifeShirt": a lightweight, washable, sleeveless vest embedded with sensors for continuously collecting and classifying cardiopulmonary patient data.
Vos Systems's "IntelaVoice Voice Operated Dimmer": allows to dim your lamps with simple voice commands.
Logitech's "QuickCam Orbit" Desktop Video-Camera: automatically follows you when you move around.
EyeOn Trust's "Golfmate": allows to locate easily and immediately a golf ball on a golf course.
Singapore Technologies Electronics' Fever Screening System: shows whether passers-by are running a temperature.
Mitsubishi's ITS-ASV2 Car: provides audible and visual warnings when it detects that the driver is not alert enough.
Mercedes Benz' "Talking Alarm Kit": monitors the driving lane and alerts if pedestrians appear in front of the car.

collection of raw data from sensor networks, consisting of cameras, microphones, position detectors, motion trackers, chemical sensors, and others. Secondly, the data are processed and combined, and thirdly the data are classified.

Well-known techniques that are used in this domain are face and body recognition, geometric modeling of two- and three-dimensional objects, image segmentation, biometrical data processing, speech recognition, and synthesis. For the classification, one often resorts to *artificial neural networks* or *hidden Markov models*. The artificial neural networks are used as generalized class separators. They can deal with conflicting or incomplete data sets. The hidden Markov models typically use extremely large probabilistic networks of nodes that represent partial solutions. Final solutions are then obtained by computing the most likely feasible combination of partial solutions. This computation is often carried out using a mathematical programming approach called dynamic programming. This approach has regained much attention over the past years as a result of the huge increase in the capacity of present-day computing devices and the availability of large data sets.

13.3.2 Understand, Interpret, Relate

This class of AmI elements is concerned with aspects of giving meaning to certain events or activities that are sensed or recorded. In general terms this is the domain of reasoning, induction, and deduction. Again, we start the presentation with a short overview of some commercially available products and services as listed in Table 13.2.

Table 13.2. Understand, interpret, relate

Blissful Babies' "Why Cry": a calculator-sized battery-powered device that can analyze a baby's cry and give an indication of the cause.
Wow Wee Toys Ltd's "Speak2Click": a conversation robot that allows you to easily communicate with your PC.
Philips' ICat: user-interface robot that supports natural conversation and can be used for control and support tasks in the home.
Elite Care's "Smart House": assists occupants with diminished mental capability, by detecting their physical motions, movement of objects, and operation of appliances and lights.
K Laboratory's "FeliPo" Service: allows a mobile phone to detect information on a poster about advertisements that are customized for gender, location, time/date, and the weather conditions.
NewNow InetShop's "Sensor-Bin": a dustbin that will open its lid when you move directly in front of it.
Saab's "Alcokey": prevents a person from driving the car if not sober.

Most of the computational techniques that are applied in this domain are known as rule-based or expert systems. Upon input these systems argue and reason using data and relations among data. They typically apply heuristic rules to combine and transform data in order to reach valid conclusions or develop new pieces of information.

Expert systems constitute a classical field of research in machine intelligence, and substantial progress has been made over the years. Generally speaking, there are two elements that can be found back in most of the existing expert systems. Firstly, there is the *knowledge representation*, which determines not only the data structures that are used to specify the data items of the information that needs to be processed, but it also captures the relations among the data items in a data model that allows to manipulate the data items. Secondly, there is the element of *reasoning*, which refers to manipulating and transforming data items in order to reach conclusions that make sense within the context of use. The reasoning may also lead to a sequence of actions that need to be taken in order to accomplish a given goal and which is often referred to as a plan.

More recently, probabilistic methods have been introduced to reason about information. These methods apply the working hypothesis that different possible outcomes should be evaluated simultaneously to select eventually the most likely one based on the probability that certain initial conditions are met. *Bayesian methods* are a class of computational techniques that apply this approach with considerable success.

13.3.3 Look, Find, Remember

This is the classical domain of search and retrieval of information, browsing databases, and remembering specific pieces of information or events. This class of AmI elements has become particularly interesting as a result of the ubiquitous availability of information throughout society. Having unlimited access to any kind of information is not a privilege if one is not supported by intelligent means to handle the information overload. So, this domain is central to information handling in ambient intelligence. Table 13.3 gives some examples of commercially available products and services in this domain.

There are many search paradigms available in this domain. A major class of approaches uses heuristic rules that construct a solution to the search problem. Other techniques use iterative methods that continually try to improve a given solution using certain criteria. *Local search* is a well-known example of a search paradigm that applies this approach. *Constraint satisfaction* is yet another method that uses a tree-search approach to find a solution to a search problem that satisfies a well-defined set of constraints. Associative memories based on artificial neural network models are used to complete partial information in retrieval problems or to find approximate matches. Other methods that allow for data mining with uncertainty apply approximate pattern matching techniques, which are quite similar to the computational approaches used in stochastic Markov models. More recently, one has developed the so-called Bayesian classifiers, which are statistical methods applying conditional and prior probability distributions to evaluate classification hypotheses.

Evidently, we need to mention many search techniques that have been developed to browse the Internet. Most of these methods apply heuristic search and pattern matching rules, some of which are based on graph theoretical models. The recent introduction of the Semantic Web (Berners-Lee et al. 2001), which enables not only the search for data items, but can also account for relations between data items, has largely stimulated the development of intentional search techniques.

Table 13.3. Look, find, remember

Siemens "Digital Graffiti": appears as an SMS message when a user with a phone walks by a given physical location.
Philips' "Easy Access": allows a user to find back a song in a music database by humming a tune of the song.
Hutchison's "3FriendFinder": allows location of other users of this service in a map on the mobile phone's display.
GPSTracks' "GlobalPetFinder": allows building a fence of any size, and will alert you if your pet wanders outside it.
Cirrus Healthcare Products' "Angel Alert": tracks children and warns when they stray too far from adult supervision.
Shazam Mobile-Phone Service: identifies a piece of music with a mobile phone.
SmartHome's "Memory Key": shows the status of the door (locked or unlocked) and how much time has passed since you locked it with that key.

13.3.4 Act, Adapt, Learn

This class deals with characteristics of personalization and adaptation over time, which can be obtained from the processing of time sequences of data that contain behavioral patterns. Table 13.4 gives some examples of products that exhibit certain of these characteristics.

Most computational methods in this domain apply some form of *machine learning* techniques. Many of these techniques build an internal model that tries to capture the input–output relation of a set of learning examples, which is often referred to as supervised learning. If such a set of learning examples is not available, similar techniques can be applied to the relation between observed data items to group them into classes, which can be used later to classify new data items. These implicit models can also be used to interpolate or extrapolate from existing data upon input of new data items. Examples of machine learning techniques that are frequently applied are stochastic learning models, such as *reinforcement learning* and *artificial neural networks.* In learning, the artificial neural networks are used as generalized function classifiers where the learning examples can be viewed as the data items in a multidimensional learning space. The computational models mentioned above can also be used to build models of user behavior, and they can be determined for different types of contexts, and combined within the same actor model. An example of such an approach can be found in the development of recommender systems that support in the selection of video items. Here user preferences are captured in multiple models that reflect different contexts of use.

Table 13.4. Act, adapt, learn

Noxa Med's Anti-Snore Pillow: eliminates snoring of a specific person by producing vibrations if it detects the onset of snoring.
Adidas' Intelligent Shoes: adapt automatically their characteristics to the surface on which the person who is wearing them runs.
OSIM's "iMedic 500 Advanced Massage Chair": measures automatically the length of a person's back and the shoulder position, to determine where it should apply pressure.
Toyota's "Prius" Car-Parking Assistance: allows a car to park itself without the driver having to touch the steering wheel.
Emfitech Oy's "SafeSeat" and "Safefloor": chairs and floors that can monitor the health and well-being of patients, and send warnings if needed.
Omron's "OKAO Vision": a user interface for an ATM or a ticketing machine that automatically adapts after observing the user's gender, age, and other attributes.
Tomy's Sleep Watch Doll: observes the user's sleeping habits, and bothers the owner when he or she is not going to bed or getting up at the usual time.

13.3.5 Create, Express, Emerge

This class of AmI elements is probably the most interesting of all since it deals with creativity in the most absolute form, i.e., creating something meaningful that is new. It has to do with new elements of forms and shapes that enhance the interaction of people with media in the broadest sense. Here, affectiveness and emotion come into play, and in the end, this class of AmI elements may drive very well the most challenging innovations introduced by the concept of ambient intelligence. First, let us have a look at some of the examples, shown in Table 13.5.

The domain "create, express, emerge" deals with revolutionary concepts and is aimed at creating an enhanced experience for its users. One speaks in this respect of poetic interfaces, intimate media, and affective computing, which are all different concepts but with the very same purpose of creating immersive sensorial experiences resulting from revolutionary shifts in interaction paradigms. Examples are the automatic generation of multimedia narratives or the creation of ambiences through automatic generation of lighting and sound effects. Also the stimulation of sensorial arousals through physical motion and vibration is part of the enhanced experience.

There are only a few general computational methods in this domain. Most approaches applied in the examples presented in the table use heuristic rules to create the experience, and there are no general guidelines on how to apply these rules successfully. *Evolutionary computing* techniques however constitute a class of generally applicable methods that may become of great

Table 13.5. Create, express, emerge

Philips' Ambilight: creates a halo around the television with dynamically changing colored light that enhances your viewing experience.
Digital Fashion's 3D Simulation System for Virtual Modeling: allows virtual modeling and coordination of clothes, cosmetics, and accessories in real time.
Citroen's "C-Airlounge" Concept Car: is equipped with projectors in the armrests and floor that create particular light effects and moods.
Andersen Windows' "Concept House": Has windows that can also function as Loudspeakers and displays.
Jabberwock's Chat Software: simulates an automated chat person on the Internet.
Sharp's LN-H1W "Lumiwall" Window Panel: provides light during night and day, by combining daylight transmission, solar power generation, and illumination.
D-BOX' "Odyssey": brings a new dimension to entertainment by moving the viewer's seat in synchronism with the action on screen.
Lofty's "Hotaru" Smart Pillow: helps the user fall asleep easily and comfortably by emitting light that varies in response to the breathing pattern of the user.

significance in this domain, because of their natural approach to the problem of creating new meaningful results from existing ones. These methods follow the basic laws of Darwinian evolution, which create new elements of a population through mutation and recombination of existing elements within the population. Fitness rules are applied to select the best ones and in this way, surprising new results can be generated.

Another approach is that of using agent technology. *Intelligent agents* are small software programs that act autonomously within a distributed environment. They can create new agents and modify existing ones based on the use of external information. They are responsive to their environments and can contribute to the purpose of evolution.

13.4 Computational Paradigms

In this Section, we address four paradigms for computational intelligence: *Search*,*Reasoning*, *Learning*, and *Evolution*. For each of the paradigms we present a number of basic algorithmic approaches that can be applied in the design of AmI applications.

13.4.1 Search

Search algorithms are applied in situations where a solution has to be found in a large set of alternatives, subject to a number of constraints, and often with an objective that needs to be minimized or maximized (Papadimitriou and Steiglitz 1982). For instance, speech recognition requires finding a text for a given utterance, such that the probability that the utterance corresponds to the text is maximized. Music playlist generation is about finding a sequence of songs that meets the constraints set by the user on, e.g., songs and transitions.

Dynamic Programming

The first method we elaborate on is called dynamic programming, and is used, for example, in approximate pattern matching, in solving hidden Markov models, and in planning problems.

The key idea of dynamic programming is that decisions are taken one by one, and that optimal sequences of decisions are built in a "recursive" way. More precisely, the problem at hand is modeled as that of searching an optimal sequence of decisions $d_1, d_2, \ldots, d_n$. For instance, in a planning problem where a choice has to be made which of a given list of items to include, the ith decision could be whether or not to include the ith item. By successively taking the decisions $d_1, d_2, \ldots, d_n$, a complete solution to the problem is constructed. In this planning example, the sequence of decisions determines the set of selected items.

For determining an optimal sequence of decisions $d_1, d_2, \ldots, d_n$, the *principle of optimality of successive decisions* is used. This principle states that an optimal sequence $d_1, d_2, \ldots, d_n$ consists of a first decision, d_1, followed by an optimal sequence of decisions $d_2, \ldots, d_n$ for the remainder of the problem. The idea is now that for every possible situation after the first decision, first an optimal sequence of decisions $d_2, \ldots, d_n$ is computed. Then, the first decision d_1 is chosen such that the combination of its direct effect and the effect of the optimal finish $d_2, \ldots, d_n$ is optimized. Note that in this way we have reduced the problem from n degrees of freedom to $n-1$ degrees of freedom.

The above step can be applied repeatedly, leading to the following dynamic programming algorithm. First, the optimal decision d_n is determined for every possible situation after the first $n-1$ decisions. Next, for every possible situation after the first $n-2$ decisions, the optimal decision d_{n-1} is determined, given the optimal finishing decision d_n for the situation after choosing d_{n-1}. This is repeated until we have finally determined the optimal decision d_1.

Key in the above recursive procedure is the way "every possible situation after k decisions" is represented. In dynamic programming, this is modeled by a set of *states*, where a state describes the effect of the previous decisions in such a way that the result of the remaining decisions can be determined. In the worst case, the set of states after k decisions is exponentially large in k. For many problems, however, the set may be significantly smaller. For instance, in the planning example, the state may be given by the total duration of the already selected items, assuming that the problem is about selecting items with a total duration not exceeding some limit D. Knowing the total duration of the previously selected items is all one needs to know about the previous decisions in order to determine the remaining decisions optimally. If all items have a duration, i.e., a multiple of 5 min, then the set of states is given by all multiples of 5 min between 0 and D, which is independent of k.

The final element in dynamic programming is a recurrence relation to determine the combined effect of a decision d_i and the optimal finish $d_{i+1}, \ldots, d_n$ after it. Let for each possible state t after decision d_i the value of the optimal finish be given by $f_{i+1}(t)$. Then, for each possible decision d_i taken in each possible state s, the recurrence relation expresses the value $v(s, d_i)$ of taking that decision in that state as function of the direct effect of decision d_i and the value $f_{i+1}(t)$ of the state t that is reached after this step. Then, the optimal decision in state s is the one that maximizes $v(s, d_i)$, yielding the optimal value $f_i(s)$ for state s. In the planning example, the recurrence relation may be that the value of taking a decision d_i in state s is the value of the ith item if it is selected and zero otherwise (i.e., the direct effect), plus the optimal value of future selections given the resulting state t, where the resulting state t is given by s plus the duration of the ith item if it is selected, and s otherwise. More formally,

$$v(s, d_i) = \begin{cases} v_i + f_{i+1}(s + l_i) & \text{if } d_i = \text{"select item } i\text{" and } s + l_i \leq D, \\ f_{i+1}(s) & \text{otherwise,} \end{cases}$$

where v_i is the value of item i, and l_i its duration.

As we can see above, optimal sequences of decisions are calculated "backwards," i.e., first optimal last decisions are computed, then optimal one-but-last decisions are computed, etc. Dynamic programming variants exist in which the decisions are calculated "forwards." In contrast to the above approach, where one typically uses the same state set for each decision, a forward approach typically only maintains the set of states that are actually reached after the first k decisions. If two decision sequences $d_1, d_2, \ldots, d_k$ end up in the same state, the one with the best value is kept, and the other one is discarded.

A final remark that we make about dynamic programming is that, whereas the above presented algorithm gives a guaranteed optimal solution, it may be adapted to save computation time, at the cost of losing the guarantee of optimality. Many examples are known in the literature to turn a dynamic programming algorithm into an approximation algorithm, by pruning the state set. For instance, in the planning example, one may round the states to multiples of 15 min, even though the items' durations are multiples of 5 min. As this reduces the state set, it hence reduces the computation time. Other known approaches in, e.g., speech recognition is a so-called beam search, where the state set per decision is restricted by discarding states that are not likely to give an optimal solution.

Heuristic Algorithms

Many problems cannot be solved exactly in a reasonable time, for instance because of their intrinsic complexity. In this situation one may resort to the use of heuristics, which drop the requirement of finding an exact solution at the benefit of substantially shorter running times (Osman and Kelly 1996). In this section, we discuss two such approaches, a constructive one and an iterative one.

Constraint satisfaction. The first approach we mention, which is a common method for feasibility problems, is given by constraint satisfaction (Tsang 1993). Constraint satisfaction can be seen as a constructive approach in the sense that it gradually builds up a solution.

Key in constraint satisfaction is the definition of *domains*, which give for each of the involved decision variables a set of values from which it can be chosen. These domains, which initially may be quite large, are reduced during the course of the algorithm, until they each contain only one value, and hence a solution has been found, or until one of them becomes empty, meaning that no solution exists.

Domain reduction techniques in constraint satisfaction build upon combining constraints with other variables' domains, as well as on combining several constraints. This process is called *constraint propagation.* For instance, if a decision variable x has domain [2,10] and variable y has domain [5,7], and there is a constraint implying that x has to be at least equal to y, then the domain of x can be reduced to [5,10], as other values cannot lead to a feasible solution. An example of combining constraints is that, in a planning problem, if a task a is supposed to be executed on the same processor as another task b, and there is not enough time for a to execute before b has to be started (due to the domains), then one can draw the conclusion that a has to start after b has finished. The strength of constraint satisfaction is determined by the domain reduction power of combining constraints, and quite some research has been spent on the types of constraints that lend themselves best for constraint propagation.

If, at a certain moment, domains cannot further be reduced, while some domains contain multiple values, then a variable is chosen, and it is assigned a value from its remaining domain. Then again constraint propagation is applied as much as possible, to reduce the other variables' domains. If necessary, a new variable is assigned a value, etc. Obviously, the order in which the variables are considered for being assigned a value has a strong impact on the end result. Furthermore, value choices may turn out to lead to a situation where some domains reduce to empty sets. In that case, one may apply *backtracking* to undo (some of) the last decision(s) and to make other choices. Doing so, however, may increase the running time considerably, as one effectively is solving the problem exactly.

The field of constraint satisfaction can be positioned at the intersection of mathematical programming and artificial intelligence, and the fact that it contributed to the merger of these two fields is probably one of its major contributions in addition to the fact that it is quite a powerful method that can be applied to a large range of problems.

Local search. The second approach we mention is called local search (Aarts and Lenstra 1997), which works by starting with a rather arbitrary, yet complete solution, and iteratively making small changes to it. By defining the kind of alterations that can be made to a solution, e.g., replacing a song in a music playlist by another song, a so-called *neighborhood structure* is defined, which, for each solution, gives a set of solutions that can be obtained from it in one iteration. So, iteratively, a random solution is picked from the neighborhood of the current solution, and the objective function is evaluated for this new solution. If the effect is favorable, the new solution is directly accepted, and used for the next iteration. If the solution deteriorates, then in the simplest form of local search, called *iterative improvement*, the new solution is rejected. Doing so, local search may be trapped in a *local optimum*, meaning that no neighbor of the current solution improves the objective function. However, this local optimum may not be the overall (global) optimum.

To escape from a local optimum, *simulated annealing* uses a different way of treating deteriorating solutions, in the sense that they are accepted with a certain probability, which decreases with the amount of deterioration and over the course of the algorithm. The effect is that in the beginning, large deteriorations may be accepted, thereby ensuring a proper exploration of the solution space. As the algorithm continues, the chance of accepting deteriorations decreases, until in the end (almost) only improvements are accepted. Although simulated annealing uses a simple acceptance mechanism, it has the nice theoretic property that it converges to globally optimal solutions with a high probability.

Although simulated annealing does accept deteriorating solutions with a certain probability, it may run the risk of jumping back directly in the next iteration, and it may be very difficult to reach a certain good solution if that takes several deteriorating steps to get there. To overcome the former problem, *tabu search* maintains a list of decisions that are not allowed to be undone in the next couple of iterations. The latter problem is tackled by *variable depth search*, in which a given number of steps are performed in a sequence, deteriorating or not, after which the best intermediate solution is taken for the next iteration.

Local search is easily applicable in the sense that it requires (nearly) no specific problem knowledge, and it has given good results for various well-known problems. Furthermore, if problem-specific knowledge is present, this may be used to further improve the performance. For instance, the neighborhood structure may be reduced such that only promising solutions are kept. Although this may affect the theoretical underpinning of the approach, it may speed up the algorithm drastically, by which it can obtain better solutions in practical running times.

13.4.2 Reasoning

Reasoning algorithms use knowledge for drawing conclusions. Motivation for using a separate knowledge component was originally derived from the "combinatorial explosion" encountered with many early artificial intelligence programs (Russell and Norvig 2003). An expert system, or knowledge-based system, consists of a reasoning program and a knowledge base (Stefik 1995). Knowledge bases can take the form of rule bases or ontologies, which provide machine-understandable definitions of terminology. Bayesian methods are widely used to handle uncertainty in reasoning algorithms.

Expert Systems

One of the earliest expert systems is Dendral, which inferred molecular structures from measurements done with a mass spectrometer. The system used rules which connect peaks in mass spectrometer graphs to specified subgroups of molecules. Another famous early expert system is MYCIN, which used rules to diagnose blood infections. MYCIN's rules connected symptoms to possible causes, while the likelihood of different causes was distinguished by means of numbers called certainty factors. In addition to the term expert system, the term decision support system has also come into use. Many decision support systems contain a diagnostic reasoning component and advise people to perform a certain new observation or to execute a certain corrective action, for example. As an example, in the area of medicine, clinical decision support systems are attracting attention (Sim et al. 2001).

The area of expert systems and the related area of knowledge representation and reasoning face two central issues: knowledge acquisition and the tradeoff between expressive power and reasoning complexity. With respect to knowledge acquisition, the issue is to obtain declarative knowledge in such a way that it can be used by a system. The standard approach is to interview domain experts in order to obtain the required knowledge bases. However, in many cases, it has been hard or impossible to obtain and maintain suitable knowledge bases in this way; the problem to obtain suitable knowledge is also called the Feigenbaum bottleneck, after one of the originators of the field. Representatives from the application domain may develop the required knowledge bases, when a suitable metamodel of knowledge for the application domain is available. As an alternative to the involvement of domain experts, learning techniques may be useful to discover knowledge on the basis of evidence provided by data contained in databases or on the web, for example.

The second central issue that was mentioned, the tradeoff between expressive power and reasoning complexity, concerns the fact that if much freedom is allowed to express knowledge, then reasoning may become intractable. For the classical formal paradigm of reasoning, first-order logic, the problem to determine whether a conclusion is valid is undecidable in general; this was proved in the paper that introduced Turing machines as a general model of computation (Turing 1936). In order to enable reasoning, practical systems need to use restricted formalisms for expressing knowledge. The two central issues together lead to the challenge to develop a knowledge model for an application domain, enabling knowledge bases to be recorded and maintained in practice, in combination with a corresponding reasoning procedure realized by tractable algorithms.

Traditionally, a knowledge base typically consists of rules. In addition to rule bases, knowledge bases can also be *ontologies*, which provide machine-understandable definitions of the meaning of concepts. The combination of rule bases and ontologies leads to appealing possibilities for intelligent algorithms. For example, a system may get as input data from certain sensors, describing the context of use of the system in a low-level fashion. The sensor data are used in combination with an ontology to perform "context determination," making sense of the context of use by describing it in terms of high-level concepts defined in the ontology. Subsequently, a rule base, entirely phrased by means of high-level concepts on the level of the user, can be used to trigger actions performed by the system, for example, to realize preferences which a user has stated to apply to the current context of use.

The W3C is developing standard languages to support machine reasoning by means of knowledge on the web, to realize the vision of the semantic web (Berners-Lee et al. 2001). Two semantic web languages are already available in standard form: RDF (Resource Description Framework) and OWL (Web Ontology Language). RDF plays a basic role by allowing the expression of statements. OWL allows the expression of ontologies, which define meaning of terms used in RDF statements. Simple ontologies can already be expressed with the RDF Schema (RDFS) vocabulary. The W3C is standardizing a rule language in the future. Even without rules, the languages RDF and OWL lead to the possibility of automatically drawing conclusions from information on the web. For RDF and OWL, the valid conclusions, commonly called entailments, are determined by a logic and its semantics. Although the standard syntax for RDF and OWL uses XML, the meaning of RDF and OWL knowledge bases is independent of XML, and abstracts from the XML serialization used. Here the notion of RDF graph plays a role. An RDF or OWL knowledge base is formalized as an RDF graph, which is a set of RDF statements, i.e., subject–predicate–object triples; subjects and objects of triples are viewed as nodes, linked by predicates. Predicates are usually called properties. RDF includes variables, which are called blank nodes, and which are, implicitly, existentially quantified. An RDFS or OWL ontology describes concepts (i.e., classes) and relationships (i.e., properties). The RDFS vocabulary enables in particular the definition of classes and subclass relationships, and the specification of domain classes and range classes for properties. OWL extends RDFS in various ways. For example, properties (i.e., binary relations) can be stated to be functional or transitive or symmetric, or to be each other's inverse; classes can be stated to be disjoint or to be the union or intersection of other classes; it is possible to use constraints to define classes, for example, to define the class of persons all of whose parents are American, or the class of persons with two children. There are two variants of OWL with different semantics: OWL Full and OWL DL. OWL Full entailment is undecidable. OWL DL imposes restrictions on the use of the language to ensure decidability. For example, classes cannot be used as instances in OWL DL. OWL DL is supported by techniques developed in the area of description logics (Baader et al. 2003). Although description logics form decidable fragments of first-order logic, for which optimized reasoners exist, verification of OWL DL entailment requires nondeterministic exponential time (Horrocks and Patel-Schneider 2003). In analogy to RDFS, a weakened variant of OWL has been described which does not impose restrictions on the use of the language and for which entailment is NP-complete, and in P when the target RDF graph does not contain blank nodes (ter Horst 2004).

Bayesian Methods

It has been common practice to handle uncertainty in expert systems by means of numbers. For example, we mentioned already the uncertainty factors of MYCIN. Bayesian methods use probabilities, and seem to form the most popular way of handling uncertainty in expert systems (Pearl 1988; Jensen 2001). A central role is played by Bayes' rule. For random variables V and W with values v and w, respectively, Bayes' rule connects the conditional probability that $V = v$ given that $W = w$ to the converse conditional probability and the probabilities that $V = v$ and $W = w$:

$$P(V = v|W = w) = \frac{P(W = w|V = v)P(V = v)}{P(W = w)}.$$

This rule follows from the relationship between conditional probabilities and joint probabilities:

$$P(V = v|W = w) = \frac{P(V = v,\ W = w)}{P(W = w)}.$$

A *Bayesian network* is an acyclic, directed graph for which the nodes are labeled with random variables $V_1, \ldots, V_n$. A Bayesian network is essentially a compact representation of certain conditional independent relations: it is assumed that each variable V_i is conditionally independent of each set of variables $\boldsymbol{A}_i$ that are not descendants of V_i, given the set of parent nodes $\pi(V_i)$ of V_i: $P(V_i|\boldsymbol{A}_i, \pi(V_i)) = P(V_i|\pi(V_i))$. More precisely, this equation is assumed to hold for each value of each variable included. The conditional probabilities $P(V_i|\pi(V_i))$ are stored for each variable V_i, and include the probabilities $P(V_i)$ at the root nodes. Bayesian inference allows the computation of many probabilities and conditional probabilities by means of these probabilities $P(V_i|\pi(V_i))$. For example, the full joint probability distribution $P(V_1, \ldots, V_n)$ is the product of the conditional probabilities $P(V_i|\pi(V_i))$ for all $i = 1, \ldots, n$. This shows that if each variable has two values and if each node has at most k parents, then all the 2^n joint probabilities can be computed with at most $n2^k$ conditional probabilities. For values $w_1, \ldots, w_k$ of variables $W_1, \ldots, W_k$ and values $e_1, \ldots, e_m$ of evidence variables $E_1, \ldots, E_m$ appearing as descendants of the variables $W_1, \ldots, W_k$ in the network, there are standard, recursive procedures to compute the conditional probabilities $P(E_1 = e_1, \ldots, E_m = e_m|W_1 = w_1, \ldots, W_k = w_k)$ by means of the given conditional probabilities $P(V_i|\pi(V_i))$. Bayes' rule can be used in combination with such a procedure to go in the other direction, i.e., from "effect" to "cause":

$$P(\text{cause}|\text{effect}) = \frac{P(\text{effect}|\text{cause})P(\text{cause})}{P(\text{effect})}.$$

The computation of conditional probabilities given a Bayesian network is NP-hard in general. When the Bayesian network has a relatively simple structure, for example, when there is at most one undirected path between any pair of nodes, there exist polynomial time algorithms. The most efficient general exact algorithms for probabilistic inference use the Bayesian network to form a kind of parallel computer which exchanges messages in both directions between neighboring nodes, for example, by performing lazy propagation in junction trees (Jensen 2001). There exist stochastic approximation techniques, which can handle much larger Bayesian networks than the exact algorithms.

Bayesian methods have been extended to decision graphs and dynamic Bayesian networks (Jensen 2001; Russell and Norvig 2003) using sensor data values, utilities, and actions. Using probabilistic reasoning, actions are selected that maximize expectation values of utilities. This forms a widely used way to realize the "utility-based agents" that will be mentioned in a later section. Robotic perception systems use variants of these methods that involve continuous random variables rather than discrete random variables (Russell and Norvig 2003).

Bayesian reasoning methods are powerful, even though knowledge acquisition often remains an issue. The structure of Bayesian networks can in many cases be obtained by means of causal knowledge that is available from domain experts, but the conditional probabilities $P(V_i|\pi(V_i))$ are typically more difficult to obtain; the question "where do the numbers come from?" continues to raise discussion. Learning methods for this problem have been widely investigated. We discuss two methods for obtaining the conditional probabilities when the structure of the network is known, and a database of cases (training data) is available, possibly with missing values of certain random variables. In the gradient ascent learning method (Russell and Norvig 2003), a maximum likelihood hypothesis for the conditional probabilities is found. Hypothesis h represents values of the conditional probabilities $P(V_i|\pi(V_i))$. The objective function to be maximized by a gradient descent procedure is the probability $P(D|h)$ of the observed training data D given a hypothesis h. This method leads to a local optimum. Another method that is widely used for obtaining conditional probabilities from a database of cases with missing data is the expectation maximization (EM) method, which can be described as follows (Nilsson 1998). The starting point is a database of cases which do not all have values for all the random variables $V_1, \ldots, V_n$. For example, in a situation where the variables are Boolean, there may be ten cases where the variables $V_1, \ldots, V_{n-1}$ are known to be one but where the value of V_n is unknown. These ten cases are then handled in terms of "weighted cases," with value $V_n = 1$ with weight $P(V_n = 1|V_1 = 1, \ldots, V_{n-1} = 1)$ and value $V_n = 0$ with weight $P(V_n = 0|V_1 = 1, \ldots, V_{n-1} = 1)$. In the following step the conditional probabilities $P(V_i|\pi(V_i))$ are given random values, and Bayesian inference is used to compute the conditional probabilities used as weights in the cases with missing values. The database of cases, with missing values thus interpreted in terms of weights, is then used to estimate the conditional probabilities $P(V_i|\pi(V_i))$ as fractions of frequencies $N(V_i, \pi(V_i))/N(\pi(V_i))$. The conditional probabilities $P(V_i|\pi(V_i))$ thus obtained are used to obtain new values for the conditional probabilities used as weights in the cases with missing values. This procedure is iterated, and again converges to a local optimum.

13.4.3 Learning

Learning is the ability to improve one's performance through experiences in the past. Learning algorithms are typically applied in situations where initially only partial information is available to solve a given problem and gradually over time additional information becomes available and in situations where adaptation to changes in the environment is essential to obtain high-quality solutions. Examples of these types of applications are learning the preferences of a user for TV-programs on the basis of his/her viewing history, and adaptation to changes in voice and background noise in speech recognition.

Learning can range from simply memorizing past experiences to the creation of scientific theories (Russell and Norvig 2003). We focus on two specific computational paradigms, namely neural networks and reinforcement learning.

Neural Networks

An artificial neural network is a computational model that tries to follow the analogy with the human brain. It is built from artificial neurons or nodes, based on a simplified mathematical model of the biological neurons in the brain. A given node is connected to multiple input nodes, of which the states are given by $x_1, x_2, \ldots, x_n$. These connections have weights $w_1, w_2, \ldots, w_n$. The output or state y of the node may be discrete, say $y \in \{0, 1\}$, or real valued. For deterministic neurons, the output y is a function of the inputs

$$y = f(\Sigma_{i=1,\ldots,n} w_i x_i - b), \tag{13.1}$$

where b represents a threshold and f is some nonlinear function. If the neuron has a discrete output, then f will be a step function. For stochastic neurons with discrete output, the right-hand side of (13.1) gives the probability that the neuron has output 1.

Neural networks usually consist of many nodes that are connected in some way. A well-known example is a layered feed-forward network, where the nodes are arranged in multiple layers, from a first layer of input nodes via possibly multiple layers of intermediate nodes to a last layer of output nodes. In such a network, the nodes on layer i can only be input to the nodes in layer $i + 1$.

Such a feed-forward network can be used for classification purposes, where the states of the nodes in the input layer are directly determined by external input. They represent the features of the object that is to be classified. In successive steps, (13.1) is used to determine the states of the nodes in the successive intermediate layers, until in the last step the states of the output nodes are determined. These give an encoding of the class to which the object is supposed to belong. A simple encoding is given by using as many output nodes as we have classes that we want to distinguish, where a classification requires exactly one of the output nodes to receive state 1 and the others state 0. To realize a correct classification of objects, the connection weights w_i and the threshold b must be set appropriately for each of the neurons. If the relation between the states of input nodes and the required state of the output nodes is sufficiently understood, then these could be hard-coded. Usually, however, this is not the case.

Neural networks are especially interesting for applications where the relation between the states of input nodes and output nodes is unknown. By repeatedly providing it with examples of correct input–output combinations, a neural network will ideally be able to learn the underlying input–output relation by adapting the connection weights w_i and thresholds b of its nodes.

Research on feed-forward networks dates back to the 1950s and 1960s, when networks consisting only of input nodes and output nodes were extensively studied. These so-called perceptrons (Rosenblatt 1957; Minsky and Papert 1969) can represent only linearly separable concepts. The field has seen a strong revival in the 1980s and 1990s, when learning feed-forward networks with one or more intermediate layers using the back-propagation learning algorithm (Rumelhart et al. 1986) became common practice. The algorithm works in small iterative steps. In each step, observed errors at the output nodes are propagated back to the intermediate nodes to make small adjustments to the weights and thresholds. The success of learning algorithms in neural networks can be hindered by the possibility of getting stuck in local optima, by the slow speed of learning, and by the uncertainty on choosing an appropriate structure of the neural network. The structure of a neural network determines the number of layers and the number of nodes per layer.

Reinforcement Learning

Reinforcement learning is an example of unsupervised learning. Instead of learning from given input–output examples, reinforcement learning can best be characterized as learning from interaction. We next explain it in the context of sequential decision problems.

A sequential decision problem can be modeled as a Markov Decision Process (MDP) as follows. It models an agent whose environment can be in one of a finite set of states $s_1, s_2, \ldots, s_n$. In successive iterations, the agent has to choose one of a finite set of actions $a_1, a_2, \ldots, a_m$. The next state is determined by a probability distribution that depends on both the current state and the agent's action. Hence, the MDP is characterized by transition probabilities $P(s, a, s')$, which gives the probability of going from state s to state s' when action a is chosen. In addition, a reward or reinforcement function $r(s, a, s')$ gives after each action a reward that depends on the current state, the action, and the next state. Both the transition probabilities and the reward functions are assumed to be stationary, i.e., they do not change over time, and memoryless, i.e., they depend on the current state but not on previous states. The goal is to learn a policy that specifies (probabilistically) which action to choose in a given state such that the expected overall reward is maximized.

An optimal policy for a finite number of states and a finite number of iterations can be computed off-line using, for example, dynamic programming if transition probabilities and reward functions are given explicitly. In many applications, this is not the case. Instead, the agent starts with zero knowledge and it has to learn from interacting with the environment. A well-known on-line reinforcement learning algorithm is Q-learning (Watkins 1989). The algorithm maintains state-action values $v(s, a)$ for each state s and action a, representing the expected benefit of choosing action a in state s. Initially, they are chosen randomly. Based on these estimates, the agent chooses a next action. Based on the resulting state and reward, the estimates are updated. A parameter balances between exploiting the current estimates and exploring new possibilities to improve the estimates. Many alternative learning algorithms have been proposed. For further details, we refer to Sutton and Barto (1998). Reinforcement learning has been applied in various settings, such as in dynamically adapting the quality of service in video processing in set-top boxes and digital TV sets, to adapt the highly fluctuating processing requirements to the available processing resources in programable hardware (Wüst and Verhaegh 2004).

13.4.4 Evolution

Evolutionary computing is inspired by the way species are thought to evolve over time in nature. Evolution can be viewed at the level of species and at the level of individuals. Intelligent agents are autonomous software systems that sense their environment and that can achieve a form of evolution by being responsive, goal-directed, and socially able. As a computational paradigm, evolution can be viewed as extending and including the previous paradigms (search, reasoning, learning).

Evolutionary Computing

Evolutionary computing covers various computational paradigms that draw their inspiration from nature, such as genetic algorithms, genetic programming, evolutionary programming, and genetic local search. The two key mechanisms in evolution are *selection* and *variation.*

In nature, selection is achieved by a process called survival of the fittest, meaning that individuals of a population that are better adjusted to their habitat have a higher chance of surviving. These individuals also have a higher chance of mating, which implies that their genotype is more probable to remain in successive generations of the population. As a result, the population as a whole is better fit to its habitat. Variation is established by *recombination* and *mutation.* Recombination is obtained by sexual reproduction, where the genotype of two parents is combined in a new genotype of their child. Mutation corresponds to random perturbations of the genotype of an individual. Variation creates individuals that are potentially better suited to their habitat.

The application of this computational paradigm to combinatorial optimization problems or, more general, to search problems is quite straightforward. If we want to find a high-quality solution from a large set of solutions, then this problem can be formulated in evolutionary terms as follows. A solution can be considered as an individual. Starting with a set of initial solutions as starting population, the quality of the solutions in successive generations can be improved by preferably combining high-quality solutions to generate new solutions (recombination) and by realizing occasional random changes (mutation).

Genetic algorithms (Holland 1992) aim at faithfully mimicking this evolutionary process by representing each solution as a DNA-like string of characters and by realizing recombination through crossover operations. Other variants such as genetic local search aim at following the same basic principles more loosely, for example, by improving the individual solutions through iterative improvement until they are transformed into local optima, before the next recombination step is carried out. Genetic programming is closely related to genetic algorithms. Instead of recombining and mutating strings that represent solutions, genetic programming operates on computer programs (Koza 1994).

Intelligent Agents

Agents are computer systems that observe their environment by means of sensors and autonomously act in their environment by means of actuators; intelligent agents respond to changes in their environment and take initiatives in order to satisfy their objectives, and are moreover capable of interaction with other agents in order to satisfy their objectives (Wooldridge 1999). In other words, an intelligent agent is reactive, goal-directed, and socially able. Social ability implies negotiation and cooperation with other agents. Although there has been much work on reactivity and goal-directedness, there does not yet exist a widely accepted method of integrating goal-directed and reactive behavior. Investigations on intelligent agents often make significant use of other fields, such as game theory or modal logic. The semantic web languages already mentioned enable agents to understand the meaning of and to reason with information on the web.

The notion of intelligent agent has been used for the global organization of major recent textbooks on artificial intelligence (Nilsson 1998; Russell and Norvig 2003). Several kinds of agents are considered, with increasing capabilities, and with increasing use of search, reasoning, planning, and learning. Simple reflex agents work with simple condition–action rules, triggered by sensor observations. Model-based reflex agents use a model of the world and have an internal state, and react to new sensor input by updating the state of the world and by choosing an action that takes this state of the world into account. Goal-based systems use their continuously updated model of the world in combination with information about goals to determine their actions. Utility-based agents use numerical utility functions distinguishing the desirability of states, enabling more refined strategies for action than when only Boolean information about goals is available, for example, by using Bayesian methods as described above. Finally, learning agents are able to evaluate their actions and improve their performance.

We briefly discuss several approaches that have been investigated for developing intelligent agents, reflecting the diversity of the field (Wooldridge 1999). Layered architectures seem to have formed the most popular approach. There does not exist a widely accepted way of designing layered agent systems. There are systems where each layer connects to sensor data and action output, while other systems work with a kind of pipeline of layers from sensor input to output actions.

As another way of developing intelligent agents, the belief–desire–intention (BDI) approach (Bratman et al. 1988) has been widely used. The BDI model presents a kind of cognitive structure for agents, consisting of several parts. The information that an agent has about its environment is represented as a set of beliefs. The intentions of an agent represent the state that the agent is committed to reach. The desires of an agent consist of options available to the agent. A BDI agent continuously cycles through a sequence of steps. First, the sensor input is used to update the beliefs. Then, given the current beliefs and intentions, the desires (options) are updated, for example, in view of a plan to reach the intentions. Then, the set of intentions is updated, for example, to include new intentions to realize existing intentions. Finally, an action is chosen on the basis of the current intentions. The BDI architecture does not determine how to achieve a balance between two central issues already mentioned: reactivity and goal-directedness.

In another approach to develop intelligent agents, called reactive architectures, the idea is that complex intelligent behavior can emerge from combinations of simple behaviors, executed by simple reflex agents. One of the principal approaches in this direction is the subsumption architecture (Brooks 1986), which organizes situation–action rules in a subsumption hierarchy. An action is inhibited when there is an action lower in the hierarchy. In this way, an "avoid obstacle" action can be given the highest priority, for example. With this approach, agents can only react to local information. However, a form of cooperation between agents can be realized when agents perform actions such as "dropping" evidence leading to actions of other agents present at close distance. By using such mechanisms, it is envisaged that intelligent behavior can indeed "emerge" from the actions of a collection of simple agents. It is not clear how this approach, based on local information, can be combined with the use of relevant global knowledge.

We conclude this discussion of intelligent agents by briefly turning to an approach, which so far has remained primarily theoretical. If agents intend to coordinate their actions to achieve common goals, they might make use of models of the knowledge of other agents. In this connection, much work has been done using modal logics of knowledge and belief (Fagin et al. 1995). These logics are decidable extensions of propositional logic. We briefly discuss a logic that has been relatively popular, $S5_m$, which extends propositional logic with modal operators $K_1, \dots, K_m$ for m agents. If P is a statement, then the statement K_iP indicates that agent i knows P. The logic $S5_m$ has several axioms. First, the operators K_i distribute over implication:

$(K_i(P:Q) \wedge K_iP) \Rightarrow K_iQ$. Another axiom states that what each agent knows is true: $K_iP \Rightarrow P$. The last two axioms state that each agent knows that it knows something: $K_iP \Rightarrow K_iK_iP$, and also knows that it does not know something: $\neg K_iP \Rightarrow K_i\neg K_iP$. There are two inference rules. In addition to the familiar modus ponens rule, the logic $S5_m$ also has the so-called necessitation rule: given P, infer K_iP. Just as for propositional logic, various reasoning questions for these modal logics can be reduced to the satisfiability problem; for propositional logic, state-of-the-art SAT solvers can handle relatively large cases (Lynce and Marques-Silva 2002). The satisfiability problem for the logic S5 (i.e., only one agent) is NP-complete, just like the satisfiability problem for propositional logic; for $m > 1$, the satisfiability problem for $S5_m$ is PSPACE-complete (Halpern and Moses 1992). The logic $S5_m$ has been useful for reasoning about distributed systems, and has also been popular in connection with agents. However, for agents the logic is considered to be too strong, as it implies that an agent knows all the consequences of what it knows (this is called "logical omniscience"): if a statement P entails a statement Q, then in $S5_m$ we also have that K_iP entails K_iQ. It is believed that a realistic approach would involve weaker logical capabilities on the part of agents.

13.5 Intrinsic Limitations

Since the introduction of electronic computing devices around 1950, scientists have been fascinated by the question whether there exists an intrinsic complexity of problems that makes them intractable. For a long time they have been speculating about the question of the intrinsic computational hardness of problems. In an early letter, Gödel (1906–1978) addresses questions to Von Neumann (1903–1957) about the existence of efficient primality tests and about the conjecture made by Gödel on the existence of a general procedure that reduces the N steps needed to exhaustively search all possible solutions to a problem – which Gödel calls *dem blossen Probieren* – to $O(\log N)$ or $O(\log N^2)$ steps. Unfortunately, Neumann's reply to this letter is not known, most probably because it was sent to him less than a year before he died of cancer at the age of 51. Gödels optimism about the complexity of combinatorial problems was not shared by a group of Russian scientists in cybernetics who were convinced that there existed a class of problems that could only be solved by complete enumeration, something they called *perebor*, and they were working on the proof of the conjecture that *perebor* could not be removed.

Theoretical computer science has provided a foundation for the analysis of the complexity of algorithms in computational intelligence, based on the original computational model of the Turing machine (Turing 1936). This has led to a distinction between *easy problems*, i.e., those that can be solved within polynomial running times, and *hard problems*, i.e., those for which it is generally believed that no algorithm exists that solves the problem within a time that can be bounded by a polynomial in the instance size. Consequently, instances of hard problems may require running times that grow exponentially in their size, which implies that eventually certain tasks cannot be accomplished successfully within reasonable time. For some problems, running times can easily extend beyond a man's lifetime if the instance size is sufficiently large. Garey and Johnson (1979) provided an extensive list of intractable problems in their landmark book. The so-called *intractability* calls for workarounds. One frequently resorts to methods that do not carry out the task optimally but rather approximate the final result. These so-called approximation algorithms indeed can reduce computational effort, but if one imposes the requirement that the quality of the final result is within certain limits then again for many well-known problems exponential running times are inevitable. Therefore, one often resorts to the use of *heuristics* without performance guarantees, which improve running times considerably, but at the cost of final solutions that may be arbitrarily bad in theory.

The computational complexity theory of learning studies a number of problems that are all related to the question how closely a learning hypothesis that is built by a learning system resembles the target function if this function is not known. *Identification in the limit* refers to the capability of a learning system to converge to the target function for a given problem representation. Early work in computer science, building on Popper's theory of falsification, implies that this may not be possible, which implies that there are certain learning tasks that cannot be achieved. A general principle in computational learning is *Ockham's razor*, which states that the most likely hypothesis is the simplest one that is consistent with all training examples. The notion of *Kolmogorov complexity* (Li and Vitanyi 1993) provides a quantitative measure for this, given by the length of the shortest program accepted by a universal Turing machine, implementing the hypothesis. It thus provides a theoretical basis for the intuitive simplicity expressed by Ockham's razor. The theory of *Probably Almost Correct* (PAC) learning addresses the question how many training examples must be applied before a learning hypothesis is correct within certain limits with a high probability (Kearns and Vazirani 1994). The theory reveals that for certain nontrivial learning tasks this number may be exponentially large in the size of the input of the learning system, thus requiring exponentially many training examples and exponential running times.

13.6 Concluding Challenges

From the discussion of the computational paradigms presented above one can draw the conclusion that ambient intelligence may be very well served from a computational point of view by many paradigms that exist already. This indeed is correct, but it does not mean that there are no challenges in the design of algorithms for computational intelligence from an ambient intelligence point of view. Recently we have presented an in-depth discussion of this topic

(Aarts 2005) and below we summarize the most important conclusions from that discussion. In general terms there is a need for the following eight types of algorithms:

1. *Small-footprint algorithms* that allow for the execution of intelligent algorithms in low-power mobile systems such as personal digital assistants and personal communication devices.
2. *Real-time mathematical programming algorithms* that allow real-time execution in complex decision making processes based on mathematical programming techniques such as the dynamic programming algorithms mentioned in the previous section.
3. *Intentional search algorithms* that enable the effective and efficient handling of queries based on user intentions. This calls for approximate string search, pattern matching, and classification techniques that can deal with ultra-large databases.
4. *Feature extraction algorithms* that extract metadata from raw data, thus enhancing the classical functionality of storing and retrieving data items by adding dynamical data manipulation that allows for semantic information processing.
5. *Humanized algorithms* that account for human perceived values such as perceptive measures, inconsistency, and feedback. Consequently, rigid physical objective measures resulting from classical measures such as completion time, capacity, speed, and length need to be replaced by more psychological measures that reflect user experiences.
6. *Aware algorithms* that connect the physical world and the digital world and can process and fuse large amounts of physical data, thus providing awareness and contextual information.
7. *Life-long learning algorithms* that can develop dynamical user models that capture specific knowledge over time of users in relation to the tasks they want to perform.
8. *Collaborative algorithms* that support the distributed nature of ambient systems by allowing the use of autonomous components that run on local devices to perform collaborative tasks and generate immersive experiences.

These algorithmic challenges provide some main directions for future research into computational intelligence in ambient intelligence. Their accomplishment does not necessarily require the development of new computational paradigms. They require the development of ultra-fast implementations and the development of techniques that can handle huge amounts of information. Furthermore, the incorporation of human-centered evaluation techniques that can quantify human perceived values and can capture user behavior is a subject that requires further investigation. Evidently, it should be clear from the many new ideas emerging from the detailing and realization of the ambient intelligence vision that this new field poses great challenges on the design of intelligent algorithms.

14

Social User Interfaces

A. Nijholt, D. Heylen, B. de Ruyter, and P. Saini

"Knowing others leads to wisdom, knowing the self leads to enlightment."

Lao Tzu

14.1 Introduction

Current technological and research developments pertaining to Ambient Intelligence, Ubiquitous Computing or Pervasive Computing share an impetus towards embedding computation in our social and physical environments making it an inseparable part of our daily lives. One consequence of embedding technology in this way in our everyday life is that today's user–system interaction paradigm will change substantially. Interaction is expected to be continuous, prolonged, and tied in with the physical spaces that surround us. It will involve a disparate range of interaction devices, affect social interactions and will often bridge physical and virtual worlds. Additionally, technology is expected to become more intelligent and to adapt itself to our needs and the dynamics of the environments we live in. To make this possible, AmI systems will need to be equipped with sensorial and reasoning capabilities. One consequence of confronting users with, for example, home dialog systems that have perceptive and reasoning capabilities is that additional expectations are created. Since people already have a tendency to attribute human like properties to interactive systems (Reeves and Nass 1996), it is expected that implementing human like properties in such home dialog systems will have an important impact on the user–system interaction. Similar observations can be made for other environments, e.g., smart office environments, (virtual) collaborative work or meeting environments and smart educational environments.

In this chapter we look at user–system interaction from this particular viewpoint. How can we include aspects of human social interaction in the interface and what difference will it make? We will investigate this in the paradigm of Ambient Intelligence. That is, we assume that there are sensors that perceive the user or inhabitant of a smart environment and that the information that is obtained from the sensors is being processed, fused, and interpreted in order to give useful, acceptable, and effective responses, taking into account characteristics of the user (interests, personality, moods, emotions, background, culture). Hence, in this chapter we look at smart environments to see how social interaction can be improved rather than simply the efficiency of the interaction.

Obviously, in order to do experiments, and in order to be able to build prototypes showing our ideas, we have to confine ourselves to rather stripped-down versions of smart environments, but not in a way that we loose the principles that we want to research. Rather than looking at traditional human–computer interaction we want to look at environments in which smart and mobile objects, human and virtual inhabitants share the environment and have sensorial capabilities that allow them to communicate with each other and to provide support to each other. Another principle is that we want to explore the role of verbal and nonverbal cues in social interaction in smart environments. That is, we need to take into account multimodal interactions between users and the smart environment. On the one hand, the user may be allowed and may decide to interact with the environment using a diversity of modalities and combinations of modalities in order to get his or her intentions clear; on the other hand, the environment can be designed in such a way that it understands and anticipates what the user wants, by tracking and interpreting not only those things the user decides to inform the system about, but also by interpreting the user's actions and whatever else it can determine from sensors that allow their input, e.g., facial expressions, body posture, biometrical information, etc.) to be interpreted from social and emotional points of view. Even when the user does not make an attempt to interact with the environment, the environment can gather useful information about the user.

In this chapter we introduce case studies where users interact with so-called *dialogical robots* (physical robots, embodied agents, smart objects) and where some level of affect and social intelligence in the interaction was introduced. The first case study concerns a home dialog system that has been extended with verbal and nonverbal interaction cues in a Wizard of Oz experiment in order to measure the impact of social intelligence characteristics. The second case study describes the design and the development of a prototype of a state-of-the-art multimodal dialog system with underlying models of affect. Both systems have also in common that the interaction with the dialogical robot is meant to provide real-time support to other actions of the user in the environment.

The remainder of this chapter is organized as follows. In the next section we present a short overview of state-of-the-art human–agent interaction from a multimodal, social, and emotional interaction point of view. Agents may be human, embodied virtual agents, fully virtual with no visual representation or physical robots. In Sect. 14.3 we introduce the iCat robot system and discuss how, in a Wizard of Oz experiment, it has been used to show how adding social intelligence to a human–robot interaction makes a difference in the appreciation of the human partner of the topics that are discussed. Sect. 14.4 is about the Intelligent Nursing Education Environment System (INES) and our attempts to make the system aware of what a student experiences during the interaction with the system. These experiences need to be translated to an internal state of the system from which actions can be planned in order to adapt the system and its inhabitants to a new situation. Finally, in Sect. 14.5 we draw some conclusions and discuss future research.

14.2 Social Interfaces and Multimodal Interaction

In our investigations we have looked at the design of a social interface between humans and dialogical robots in a smart environment. The users must be able to address the environment and this environment needs to be aware of the interaction that is going on. In this context, we investigate how verbal and nonverbal interaction with embodied agents and situated devices can take place in smart environments. We consider especially interactions that are also guided by affect and maybe even by an implicit desire to have some development of a social relationship between user and system. This will require us to build systems that have social intelligence.

The notion of social intelligence is not a well-defined one. It is useful to take as a starting point the ability of an agent to relate to other actors or agents in a society, understand them, and interact effectively with them. This ability can be contrasted with other kinds of intelligence, for example, the ability to solve complex logical problems and the ability to monitor one's own and others' emotions and to use that information to guide one's thinking and actions (Nishida et al. 2005). Being able to relate with other agents in a society also means being able to develop personal relationships. Social intelligence in this sense means knowing how to judge and behave in everyday social situations. However, as the same authors point out, social intelligence can also deal with social structures and social conventions. The questions in this case are how do these structures and how do these conventions constrain, guide or stimulate the way individual agents interact with each other. When designing environments that can be shared among participants this latter point of view should put emphasis on facilitating common ground building and community development. Although we certainly do not exclude these latter points of view in the design of our systems, i.e., we do not exclude the building of community knowledge in the home and educational dialog environments that we consider here, the main emphasis of this chapter concerns the first notion of social intelligence, i.e., the ability to relate not only in an intellectual way but also on an affective and interpersonal level to the agents we interact with.

Social intelligence requires an understanding of social situations in which people meet each other. When people meet face to face, they receive input from all the sensory modalities. They may see, feel, hear, smell, and touch each other. They assess social situations by the input to their sensors that comes from facial expressions, body posture, gestures, paralinguistic features of the speech such as prosodic signals, and from bodily contact or the distance that people keep (proxemics). They are aware of many of the cues that they provide to the other in response. Their actions are not just motivated as a response but they are chosen to make a specific impression on the other. Being able to perceive and interpret these cues and respond to them appropriately is an important part of showing social intelligence. In short, getting along with people requires intelligence, and applying this intelligence requires understanding of all the cues that are signaled not only verbally but also nonverbally.

When a computer system, such as a communication robot or an embodied conversational agent, lacks social intelligence, there will typically be situations where it will not understand its human conversational partner and will make wrong assumptions about how to proceed in an interaction. Moreover, it will display nonverbal behaviors that do not fit the social situation. It may show a lack of social awareness by displaying emotions, body postures, facial expressions, and gaze behavior that disrupts the interaction rather than taking care of a smooth, socially appropriate communication between system and conversational participants. Among the basic behaviors that have clear implications for the social, interpersonal relation – particularly when they are not timed and executed properly – are gaze behavior and keeping the appropriate distance. These cues can be very subtle. Establishing eye contact, or avoiding it, is determined by several factors. For instance, when the speaking process demands a high cognitive load, people tend to look away from the listener to avoid distraction from the feedback provided by the listener through the visual backchannel. However, by looking at a speaker, a listener not only may show that he is attending to what is being said, but also mark his interest. Conversely, when not looking at the speaker, this might be interpreted in all kinds of ways: the listener is distracted, is not showing any interest, and is starting to think about a reply. The way the eye contact is broken and where the listener is looking at alternatively, may distinguish between these interpretations. In the interactions with dialogical robots, and other forms of Ambient Intelligence, people will expect similar cues that tell a user when and how to engage with the system, whether or not the system is attending to the user.

Another way in which social interaction between humans is regulated is stated in the form of equilibrium theory. *Equilibrium* theory, proposed by Argyle and Argyle and Dean (1965), is a hypothesis about the way different nonverbal behaviors, such as gaze, distance, and touch, compensate for one another in order to signal the "social distance" between people. For instance, if the social distance between people is high (for instance people who do not know each other waiting for a bus) but they are forced to stand close together, they will try not to look at each other as both of these behaviors mark close relationships. In an interesting experiment, Bailenson et al. (2001) have investigated the equilibrium theory in a virtual setting. The question here is how a system can mediate the intimacy level with users in an ambient environment.

In the following sections we discuss experiments with systems that we have designed to show socially intelligent behavior.

The case studies that we will introduce here deal with the role of affect and social intelligence in a home environment and in an educational environment. Although the studies were not conducted from this point of view from the start, it turned out to be useful to look at them this way, allowing observations on modeling, developing, and maintaining social relations with artifacts and the environment. Maybe more importantly, it proved interesting to investigate how the user in his or her interaction behavior is influenced by the introduction of affect and social intelligence in the interface.

One case study is about human–robot interaction and the other about human–virtual human interaction. Obviously, for both human–human interaction studies are relevant. Also, the two studies certainly have to take into account the context in which the interactions are taking place. In fact, the possibility to do this and to learn from this has been the main reason to discuss them in this chapter.

In both cases we are designing and modeling interactions that take place in an environment where many different sensors gather information. Multimodal models of interaction are able to fuse such information and a multimodal dialog manager is able to provide an interpretation – on a semantic and a pragmatic level. From that interpretation the system should be able to decide about an effective next action that aims at supporting its human partners in the environment.

The two systems we will look at in this chapter have been built and investigated in our research environments. They are the iCat system on human–robot interaction, and the INES on human–embodied agent interaction. In both cases there is face-to-face interaction between a human conversational partner and a synthetic partner (robot or embodied agent). However, more importantly, the interaction between human and dialogical robot is guided by what is happening in the environment which itself can provide information about the interaction.

Our first case study is devoted to the so-called iCat system (de Ruyter et al. 2005). This study is an experiment on simulated human–robot interaction in a home environment. In a Wizard of Oz simulation it is shown how positive feelings about the interaction and the topics that are discussed can be induced in the human partner by making the iCat robot react in a socially intelligent way.

The second case study is devoted to the role of affect in a smart educational environment. We discuss a specific environment called the INES where multimodal interaction between a student, a patient and a tutor is modeled (Heylen et al. 2005). Again, we have a situation where there is direct interaction between a user (a student) and a virtual human (the patient), where there is an environment, embodied by a virtual tutor, that knows what is going on and tries to understand the multimodal interactions between patient and student.

14.3 The Impact of Affect and Social Intelligence: the iCat Case

In the continuous strive for natural interaction, research into computational and robotic characters is demonstrating an increasing interest into the topic of social intelligence. Nevertheless, the benefits that it might offer to users are to this point only hypothesized. Rather than examining the effect of singular factors conducive to social intelligence, this case study examines what broader benefits could be brought upon the interaction experience by a more socially complex and coherent home dialog system that is perceived as more socially intelligent. Such a holistic examination that would show the relevance and importance of social intelligence in the domain of human–computer interaction has not been attempted before.

In this case study we report on a controlled experiment into the effects of perceived social intelligence in a home dialog system addressing the following research questions:

- Will test participants be able to perceive the level of social intelligence implemented in the home dialog system?
- What is the effect of bringing the concept of social intelligence into a home dialog system on the perception of quality of the interactive systems (other than the home dialog system) in the environment?
- Will the participant's acceptance for home dialog systems increase if the concept of social intelligence is implemented into these systems?

In the following sections, we present, a home dialog system using a robotic interface, the iCat. We describe an experiment addressing the research questions raised above. The results of this experiment are discussed leading to some conclusions in the final section.

The home dialog system used in our study takes the form of an "interactive Cat," or just iCat. The iCat is a research platform for studying social robotic user-interfaces. It is a 38 cm tall user-interface robot and is implemented as a desktop robot since it lacks mobility facilities; see Fig. 14.1. The robot's head is equipped with 13 standard R/C servos that control different parts of its face, such as the eyebrows, eyes, eyelids, mouth, and head position. With this setup we are able to generate many different facial expressions that are needed to create an emotionally expressive character.

A camera installed in the iCat's head is used for different computer vision capabilities, such as recognizing objects and faces. The iCat's foot contains two microphones to record the sounds it hears, perform speech recognition, and to determine the direction of the sound source. By determining the direction of a sound source, the iCat can exhibit turn-to-speaker behavior. Also, a loudspeaker is installed to play sounds and to generate speech. Furthermore, iCat is connected to a home network to control in-home devices, e.g., light, VCR, TV, radio, and to access the Internet. Finally, touch sensors and multicolor LEDs are installed in the feet and ears to sense whether the user touches

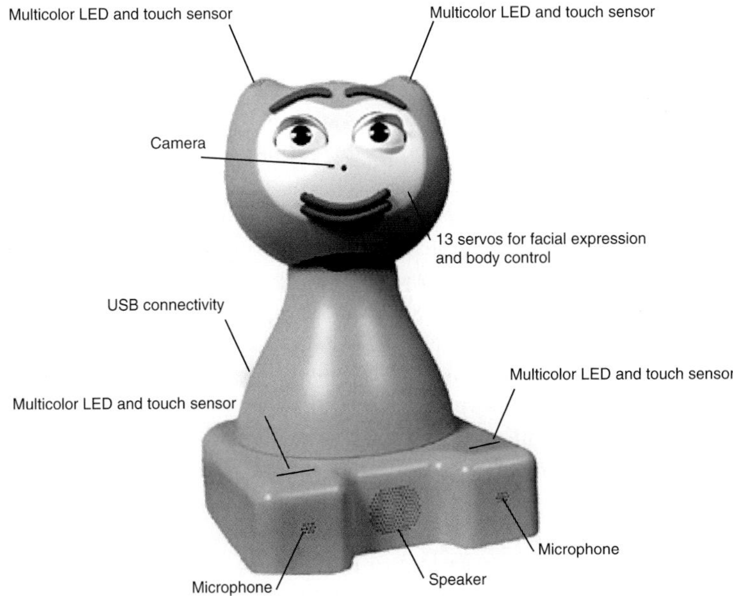

Fig. 14.1. Hardware setup of the home dialog system "iCat"

the robot and to communicate further information encoded by colored light. For instance, the operation mode of the iCat such as sleeping, awake, busy, and listening, is encoded by the color of the LEDs in the ears.

14.3.1 Experiment

A total of 37 paid subjects participated in the experiment (15 women and 22 men). These participants were selected to have at least some basic experience with e-mail and Internet and were externally recruited and randomly assigned to one of two experimental conditions.

The experiment took place at the HomeLab (de Ruyter and Aarts 2004). Participants would be left with the iCat in the living room of the HomeLab, while the experimenter would observe and control the experiment from the observation station of the HomeLab. We adopted a one-factor between-subjects design in which social intelligence was manipulated. There were two conditions.

Condition 1: Social Intelligence

During this condition the robot would talk (using synthesized speech) with lip synchronization, blink its eyes throughout the session, and display facial expressions while exhibiting the following selected social intelligence aspects:

- *Listening attentively*: by looking at the participant when she or he is talking and occasional nodding of the head.
- *Being able to use nonverbal cues the other displays*: responding verbally to repeated wrong actions of the participant by offering help.
- *Assessing well the relevance of information to a problem at hand*: by stating what is going wrong, before offering the correct procedure.
- *Being nice and pleasant to interact with*: by staying polite, mimicking facial expressions (smile when participant smiles for example), being helpful.
- *Not ignoring affective signals from the user*: by responding verbally or by displaying appropriate facial expression to obvious frustration, confusion, or contentment.
- *Displaying interest in the immediate environment*: the immediate environment being the participant and the equipment used in tasks, by carefully monitoring the person and the progress of the tasks.
- *Knowing the rules of etiquette*: by not interrupting the participant when she or he is talking.
- *Remembering little personal details about people*: addressing the participant by name, remembering login information, and passwords if asked.
- *Admitting mistakes*: by apologizing when something has gone wrong, but also when no help can be provided upon participant's request.
- *Being expressive*: by showing facial expressions while talking, if appropriate.
- *Thinking before speaking and doing*: by showing signs of thinking (with facial expression) before answering questions or fulfilling the participant's request.

The behaviors for the social intelligence condition were available as pre-programmed blocks for the experimenter who would observe (the socially intelligent condition) and listen to the participant and would type in responses for the iCat to utter. Further, the experimenter would initiate these pre-programmed social behaviors at appropriate moments in a Wizard of Oz fashion.

Condition 2: Social Neutrality

The iCat did not display any facial expressions and did not blink its eyes. It talked and used lip-synchronization, but the aspects of social intelligence listed above did not drive the talking. It responded verbally only to explicit questions from the participant. The only self-initiated help was when the participants really got stuck and could not continue without help.

We underline that contrary to studies listed in the previous section, we did not seek to assess the impact of each of the low level behaviors listed here. Rather, we hoped that their combination would lead the iCat to be perceived as socially intelligent and it is the impact of this perception that we aimed to assess, see Fig. 14.2.

Fig. 14.2. Snapshot from the test sessions showing respectively: the participant (LIV1), an overview of the living room (LIV2 and LIV3), and a close-up of the iCat (LIV4)

14.3.2 Tasks

Participants were asked to perform two tasks: (a) program a DVD recorder to record three broadcast shows for the upcoming week and (b) complete an online auction. The auction task involved registering for the service and buying several items on a list. For registration as a new user, the site required a valid web-accessible e-mail. Participants could give iCat their e-mail details (login and password) if they wanted iCat to monitor their selected items. In asking the help of the iCat, participants would also need to entrust it with their password.

14.3.3 Measures

A multiple set of measures was designed to test both the direct effects of iCat's behaviors and the potential implicit spillover effects like satisfaction with the DVD recorder.

Social Behaviors Questionnaire (SBQ). In the absence of existing validated instruments to assess social intelligence in interactive systems, a questionnaire was developed for the purpose of this study. Its purpose was to verify whether we succeeded in creating two separate conditions that the participants would rate differently in terms of social intelligence. The questionnaire (described in a separate publication) was built up of five-point scales rating the agreement of subjects to statements such as the following:

- The robotic cat takes others' interests into account.
- The robotic cat does not see the consequences of things.
- The robotic cat says inappropriate things.
- The robotic cat is not interested in others' problems.
- The robotic cat tells the truth.

User Satisfaction Questionnaire (USQ). The USQ is an instrument developed previously in-house for assessing user satisfaction with consumer products (de Ruyter and Hollemans 1887). The USQ was used to assess the satisfaction with a DVD recorder that participants had to operate during the experiment.

The Unified Theory of Acceptance and the Use of Technology (UTAUT). This questionnaire measures technology acceptance (Venkatesh et al. 2003). We used the UTAUT with some adaptations for the home domain, to measure the extent to which participants would use iCat at home after the experiment.

In a post-experimental questionnaire participants could indicate on a five-point scale what they thought about their own performance during the experiment.

Finally we noted the number of times participants asked the robot general questions and the number of times they asked questions about the experimental tasks. We also noted the number of times that participants looked at the robot during the entire session.

14.3.4 Results

The results from the SBQ verify the distinctness of the experimental conditions that we wanted to create: participants rated the socially intelligent iCat as more social than the neutral one, which validates the design of behaviors exhibiting social intelligence.

The USQ also had a differential effect between the two conditions. Since the USQ was developed to test satisfaction with a consumer product after thorough interaction with that product and the DVD recorder task only consisted of exploring one function in a time frame of 10 min, the significant difference found between the two experimental conditions is remarkable.

The UTAUT was applied to the iCat and, as such, it shows the explicit positive effect of the social intelligence manipulation.

There was no significant effect regarding perceived auction performance; most participants thought they did pretty well in both conditions. The task of buying items was for most of them not a hard one. As such, many of them felt they did very well. Participants would have liked to delegate more chores to iCat and to have asked it more questions. Most participants asked iCat to monitor their items for bids from others (83% of participants).

The only participants not very satisfied with how well they had performed were those who in their daily lives did not spend much time on the web or on the computer.

Overall the impressions were that participants were more "social" with the socially intelligent iCat: they were much more inclined to laugh, ask questions, and ask for elaborations, than with the neutral iCat. They were more curious about the reasons the social robot said the things it said than when they were interacting with the neutral robot. For example, when asked which LDC monitor was a good one (to buy in the auction task) they were happy that iCat could help by naming a product. But they were curious how it knew this and why it was the best. They were also more inclined to ask about iCat's opinion on the other LCD monitors. They asked these questions politely and using full sentences. In the case of the neutral iCat, they were more inclined to take the suggestion of the best LCD monitor for what it was and not continue asking further. In cases that they did ask more, it was usually in shorter and to-the-point command like sentences than in the social condition.

Participants in the socially intelligent condition liked the fact that the robot was expressive in terms of facial expressions, that it nodded and shook its head in response to their talking. Overall they agreed that it was only natural for the iCat to use its potential this way. However, participants in the neutral condition also liked iCat with its more neutral behavior. After all, it is a robot and it should not try or pretend to be anything other than that. Moving and facial expressions would only look like a poor attempt to seem alive and it would likely annoy and distract from whatever you are doing. This finding shows how hard it can be to imagine something you have not experienced. Neither group of participants could imagine iCat being the opposite of what they experienced.

14.4 Modeling Affect in a Dialogical Robot

INES is an application designed at the University of Twente that allows students to learn procedural medical tasks. The procedures of subcutaneous injection have been implemented in our system as a first example. This task requires the execution of several subtasks, for example, taking care that the instruments are sterilized, that there is communication with the patient, and that the injection is done in a correct way. Besides as a research project on multimodal interaction, the INES application serves as a basis for research on affective computing.

14.4.1 Multimodal Interaction

In Fig. 14.3 we show the current INES configuration, consisting of a student interacting multimodally with the system using speech, keyboard, and a haptic device. It also shows a virtual tutor interacting with the student, a virtual

Fig. 14.3. Student interacting with the Intelligent Tutoring System

patient that can be addressed by the student and objects in the virtual world, e.g., the patient and the needle that can be manipulated using haptics and speech. The virtual tutor has been implemented as an embodied agent (a talking face). When it is considered useful it displays its (dis)approval by its facial expressions. It interacts with the student mainly by using speech recognition and speech synthesis.

The student can also communicate with the patient. For example, for this particular application, asking him or her to move his or her arm or to roll up a sleeve. Communication with patient or tutor is mostly related to the handling of the haptic device. This device is, for this particular application, represented as an injection needle in the virtual world that is displayed on the screen.

The input for the virtual tutor in the current INES environment consists of keystrokes, mouse movements and clicks, movements and force using the haptic device, and speech. There is a limited build-up of the interaction history. A more complete interaction history will be obtained by embedding the characteristics of a generic multimodal interaction architecture in the INES environment; see also Hofs et al. (2003). This earlier work contained models of multimodal interaction – in that particular case tuned to speech and pointing – that can be used for haptic input combined with speech, mouse, and keyboard input, and that allows the embedding of multimodal input in a discourse model that also keeps track of the history of the interaction. However, even now, the multimodal input, embedded in its situational and dynamic context, allows the tutor agent to make assumptions and from that compute possibly emerging emotions of the student that is performing a (sub-) task in the INES environment. Being able to respond in an appropriate way corresponding to the student's emotion (sympathizing) will make the tutoring process more effective (Kort et al. 2001).

This nursing education environment has been built using our own multiagent platform. The virtual tutor receives input from different agents, for example, from a collection of error agents that keep track of what the student is doing with the haptic device. The agents that currently have been implemented in the environment track the activities of the student, notice the errors that are made, interact with the student, and change the teaching environment. In particular the so-called ErrorAgents know about the direct performance of the student: does the student use the right angle of the needle when trying to give an injection, what is the speed and what is the force that is used, has the needle been sterilized, does the student take too much time, does he or she have many questions or is asking too often about explanations, etc.

14.4.2 Affective Interactions

In our work on INES, we have also started to build a tutor agent that tries to be sensitive to the mental state of the student that interacts with it. Tutoring situations are essentially a social encounter, the goal of which is for a student to learn some task or acquire knowledge with the tutor acting in all kinds of ways to assist the student with this goal. The actions of a tutor are also not just restricted to pure instructions but they should also create the right emotional conditions for a student to act. The fact that the tutoring situation is a social encounter means that influencing the emotional state proceeds through social acts with emotion changing potential. For instance, the tutor has the status to judge (criticize or praise) the student for his actions.

The emotional state of the student contributes a lot to whether a student is motivated or challenged, which are key conditions for certain actions. Curiosity and puzzlement may lead to investigate problems. But also frustration may lead to action, even though it is a more negative affect. The tutor can choose to consider taking certain actions to bring about a change in the emotional state. Learners can be motivated by challenging them, giving them confidence, raising their curiosity, and making them feel in control. These goals can be achieved by means of various tactics. The student can be challenged by selecting appropriately difficult tasks, or by having the difficulty emphasized or by having some kind of competition setup. Confidence can be boosted by maximizing success directly (praising) or indirectly ("it was quite a difficult task, you managed to do"). Curiosity is typically raised in Socratic methods when the student is asked to ponder many questions. The tutor can decide to leave the initiative to the student or offer options that suggest the student can make choices and thereby influence the student's feeling of being in control.

In our prototype system we are experimenting with different kinds of pedagogical strategies: Socratic methods, active student learning, using deep explanatory reasoning, etc. Each of these requires different kinds of interaction (verbal and nonverbal) with the virtual tutor. The tutor in our system is aware of the history of the interactions, in particular the activity level of the student, the number and kinds of errors made by the student, and it uses this information to introduce affect in the interaction. The system not only figures out what teaching strategy fits the current context best, but also what kind of speech act is appropriate, which wording should be chosen – choosing between a reprimanding or more encouraging turn of phrase – and which facial expression should be displayed. For instance, the tutor may ask the student a series of questions first, before moving on to having the student make the injection. Or it might decide to give a demonstration first.

Presently, we are not really keeping track of the student's emotional state. Instead, the tutor tries to make reasonable assumptions about the student's mental state based on information obtained from ErrorAgents and InputAgents. For the tutor we distinguish emotions that allow the tutor to feel content or discontent (i.e., joy and distress) and that allow the tutor to feel sympathy for the student (i.e., happy-for and sorry-for). Three agents have been designed that take care of emotions: the EmotionTutor, containing the tutor's emotions, the EmotionStudent, containing the student's emotions, and the EmotionalResponse, containing the algorithms that determine in what emotional way will be responded. Experiments to evaluate the affective behavior of the INES TutorAgent have been reported by Heylen et al. (2005) and Nijholt (2003).

As a next step in the INES research, we have looked at possibilities to measure more accurately what the student is actually experiencing during the exercise that has to be followed and to feed the tutor with this information. Lisetti and Schiano (2000) have shown that facial expressions can reveal something about the cognitive and emotional states of students interacting with a tutoring system. Still, it is not at all clear how a facial expression displayed during an interaction should be associated with a mental state of the student, let alone how to use them to optimize the tutoring process. Nevertheless, facial expressions appear and we can make an attempt to understand what triggered the expression in a particular context, and from that, hopefully, get some information about aspects of the mental state of the student that can be used to provide useful feedback and to adapt the teaching.

In a pilot experiment we collected video material from students interacting with the system; see Fig. 14.4.

We looked at Scherer's component process approach to find a way to come to grips with the relation between facial expressions, the situation they occur in and the mental state of the student (Scherer 1987; Wehrle et al. 2000).

Fig. 14.4. A student working with the INES during the experiment

Scherer tries to present a coherent picture about what elements of a situation trigger what kinds of emotions. His model explains how an organism evaluates stimuli in a series of appraisal checks. The general idea is that the outcome of these checks results in specific facial expressions. This can be used to relate stimulus (situation), facial expression, and appraisal (mental state). This includes dimensions such as novelty (suddenness, familiarity, and predictability), pleasantness, goal significance (relevance, expectation of outcome, etc.), coping potential, and compatibility standards.

In the data we have looked at what facial expressions occur and we analyzed the situations in which they occurred in terms of the stimulus evaluation checks. In this way we can build a database that gives a crude indication of the relation between facial expressions and appraisal. Hence, when our system captures a facial expression in a particular situation it allows us to infer some aspects of the mental state of the user. In our corpus we found the following facial expressions: smile, raise eyebrows, pull down mouth corners, and frowns. Smiles often occurred when the students were manipulating the haptic device, as this seemed pleasurable. This is work in progress, but the aim is to derive a table associating elements of our particular tutoring situation, the facial expressions that occur in that situation and the mental state one might assume to hold that is consistent with the data and that might be of use for the tutoring system (Heylen et al. 2005).

14.5 Conclusions and Future Research

As technologies in the area of connectivity and computational platforms are moving from ubiquitous, i.e., available throughout the environment, to pervasive, i.e., embedded in out daily lives, and to ambient, i.e., creating intelligent environments, we observe that such systems will have perceptive and reasoning capabilities. This will in its turn lead to raised expectancies by end users.

In addition, the vision of Ambient Intelligence is advocating that technologies should be hidden into the background and that only the system's functionality should be available to the user. This introduces the need for advanced home dialog systems since users will still have to interact with the functionality embedded into the environment.

Combining this ambient and intelligent characteristic of future systems, we see an opportunity for the concept of social intelligence to facilitate user–system interaction. More specifically, we have investigated the effects of applying this concept to interactive systems in both the physical and virtual world. This research has led us to the conclusion that of social intelligence is an essential higher-level context for implementing intelligent and affective systems. Our research has shown that by the application of this concept of social intelligence we cannot only manipulate the user's perception of quality in an interactive system, but we can also significantly increase the user's acceptance of system intelligence embedded into the environment.

15

Multi-modal Human–Environment Interaction

R. Wasinger and W. Wahlster

"The environment is everything that isn't me."

Albert Einstein

15.1 Introduction

AmI environments require robust and intuitive interfaces for accessing their embodied functionality. This chapter describes a new paradigm for tangible multi-modal interfaces, in which humans can manipulate, and converse with physical objects in their surrounding environment via coordinated speech, handwriting, and gesture. We describe the symmetric nature of human–environment communication, and extend the scenario by providing our objects with human-like characteristics. This is followed by the results of a usability field study on user acceptance for anthropomorphized objects, conducted within a shopping context.

The talking toothbrush holder is an example of a consumer product with an embedded voice chip. If activated by a motion sensor in the bathroom it says "Hey don't forget to brush your teeth!". Talking calculators, watches, alarm clocks, thermometers, bathroom scales, greeting cards, and the pen that says "You are fired" when one presses its button are all products that are often mass-marketed as gimmicks and gadgets (Talkingpresents, online; Jeremijenko 2001). However, such voice labels can also offer useful help for people who are visually impaired, since they can be used to identify different objects of similar shape or to supply critical information to help orientate the users to their surroundings (Talkingproducts, online). All these voice-enabled objects of daily life are based on very simple sensor–actuator loops, in which a recognized event triggers speech replay or simple speech synthesis.

This chapter presents a new interaction paradigm for Ambient Intelligence, in which humans can conduct multi-modal dialogs with objects in a networked shopping environment. In contrast to the first generation of voice-enabled artifacts described above, the communicating objects in our framework provide a combined conversational and tangible user interface that exploits situational context such as whether a product is in or out of a shelf, to compute its meaning.

15.2 Tangible Multi-modal Dialog Scenario

Our experimental scenario attempts to combine the benefits of both physical and digital worlds in a mixed-reality setting by targeting an in-store scene, but augmented by instrumented devices like a Personal Digital Assistant (PDA) and a shopping trolley with a mounted display. Whereas the PDA is used as a communication channel through which users can associate directly with the products rather than through a sales assistant, the shopping trolley is capable of offering shopping advice based on its current contents. An in-store setting encompasses the down-to-earth basics that only a traditional store in a real world and with real physical products can provide such as the sense of touch. When instrumented, it further provides the convenience inherent in digital worlds such as ubiquitous information access. The unification of these two worlds is achieved through a Tangible Multi-Modal interface (TMM) that is seamlessly integrated into existing shopping practices. TMMs are now being incorporated into a wide range of fields, up to and including safety-critical applications (Cohen and McGee 2004).

Tangible User Interfaces (TUIs) (Ullmer and Ishii 2001) couple physical representations (e.g., spatially manipulable physical objects) with digital representations, e.g., graphics and audio, yielding interactive systems that are computationally mediated. In our scenario, we use an intuitive "one-to-one" mapping between physical shopping items on the shelf and elements of digital information. The spatial relation of a physical token partially embodies the dialog state, which can be seen in our example in that a product can be either on a shelf, in the shopping trolley, or outside of these containers. The position of the product is mapped to a physical action of the user, where the physical movements of the artifacts serve as a means to controlling the dialog state.

The Mobile ShopAssist (MSA) is a demonstrator that aids users in product queries and comparisons. The goal is to provide rich symmetric multi-modal interaction and the ability for users to converse directly with the products. Using the MSA, a shopper interested in buying a digital camera would, for example, walk up to a shelf and synchronize its contents with their PDA. After synchronization, they may ask a product about its attributes, e.g., "*What is your optical zoom?*" or even compare multiple products together, e.g., "*<gesture> Compare yourself with this camera <gesture>*." Comparisons may be made among products from the physical world, digital world, or a mixture of both, i.e., mixed-reality.

When interacting with the digital cameras, the user may decide to communicate indirectly with the object "*What is the price of this camera <gesture>*" or directly "*What is your price?*". The input modalities available to the user include speech, handwriting, gesture, and combinations thereof. It is direct interaction and the concept of *anthropomorphization*, i.e., assigning inanimate objects human-like characteristics that we focus on, see also Sect. 15.5.

Fig. 15.1. Anthropomorphized object initiating a dialog

Assuming the user has chosen to interact directly with the objects, the objects will in return communicate directly with the user and may also initiate mini-dialogs when picked up or put down on a shelf or in a shopping trolley, similar to (1) in Fig. 15.1. Once the users have finished conversing with the products, they may decide to buy the product, or to simply take the information that they have downloaded back home with them to think about later on. The objects, not limited to digital cameras, can then be placed into the shopping trolley and taken to the cashier. On request, the user's interactions are logged and summarized in a personal shopping diary (Kröner et al. 2004).

The MSA is a mixed-initiative dialog system, which means that both a product and the user can start a dialog or take the initiative in a sub-dialog. For instance, when the product is picked up – and no accompanying user query is issued – the product will introduce itself. Another system-initiated dialog phase is that of cross-selling, which occurs when a product is placed into the shopping trolley. Such a dialog might give advice on accessories available for the product, for example: "*You may also find the NB-2LH batteries in the accessories shelf to be useful.*"

Instrumented environments containing RFID tagged products have till now primarily benefited the retailer through improved inventory management and tracking. Our scenario also highlights user benefits in the form of comparative shopping, cross-selling recommendations, and product information retrieval based on real physical indexes.

15.3 Instrumented Environment Infrastructure

The main infrastructure components that exist in our shopping environment include the mobile device, which is used as a communication channel, the containers, e.g., shelves, trolley, shopping products and the belonging to a shop; see Fig. 15.3. Each shelf is identified by an infrared beacon that is required when a user synchronizes the shelf's product data. The products are identified through the use of passive RFID tags, which allow a product to be classified as being either in or out of a container. Each container has an RFID antenna and a reader connected to it, and this allows the shelves and shopping trolley to recognize when products are put in or taken out of them.

The instrumented shelves may be scattered over several rooms, and communicate via a WLAN connection with the AmI server, as shown in Fig. 15.2 (similar instrumented shopping environments without a tangible multi-modal interface are the Metro Future Store and MyGrocer (Kourouthanasis et al., 2002)). It is the AMI server that maintains the product database, and the event heap (Fox et al. 2000), which is used for recording extra-gesture events. As described in Butz et al. (2004), a searchlight in the form of a steerable projector further allows the system to find and highlight products based on optical markers. This is important in establishing a link between the physical products and their digital counterparts (and vice-versa), which do not need to be sorted in the same way. Such a situation could for example arise when digital objects are re-sorted based on specific product features such as price or manufacturer, instead of their physical location.

After a client device such as a PDA has been synchronized with a shelf, it will maintain its own blackboard of events, on which it stores not only the extra-gesture interactions broadcast by the server, but also speech, handwriting, and intra-gesture interactions that the PDA is capable of recognizing and

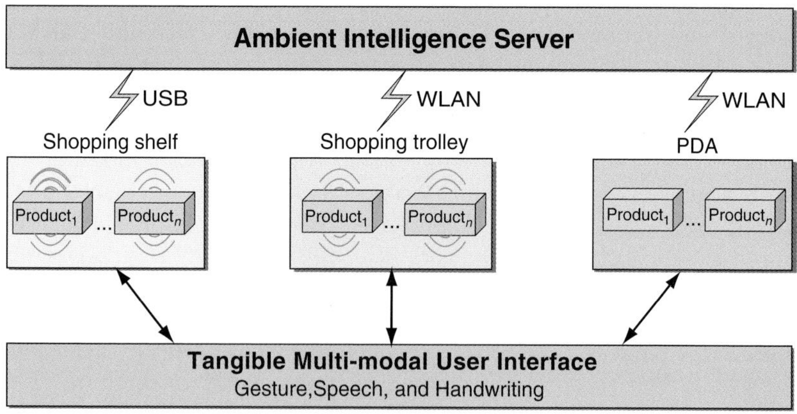

Fig. 15.2. Distributed architecture of the MSA

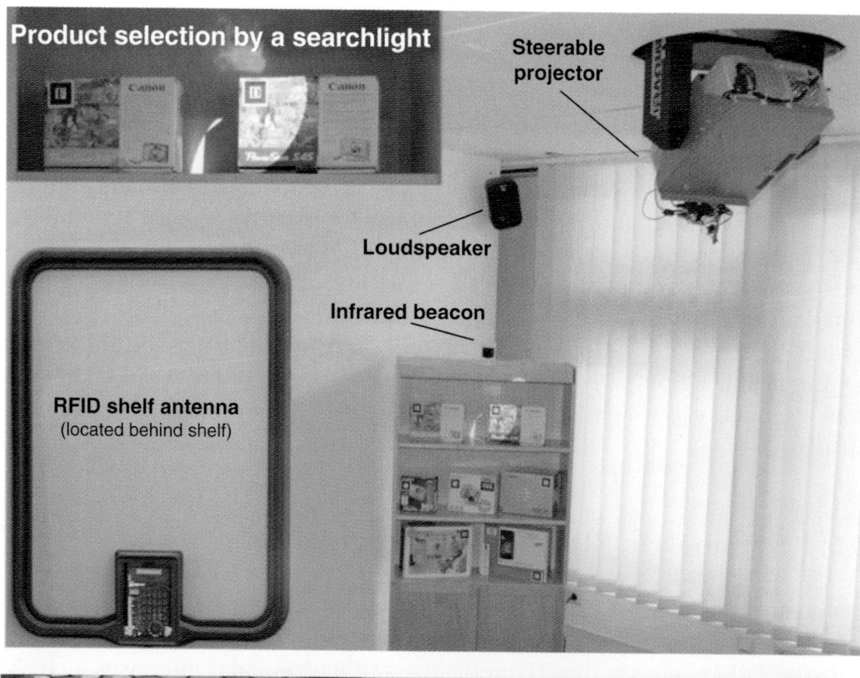

Fig. 15.3. Instrumented shopping environment

interpreting locally. When a shopping trolley is added to the scenario (Schneider 2004), the contained products are listed on a trolley-mounted display, and the trolley will offer advice on additional products that may be relevant to the user (see Fig. 15.3).

The data downloaded upon shelf synchronization is contained within the product database. This is located on the AmI server, and contains product feature–value lists for attributes like "price," "optical zoom," and "mega pixels." The database also contains images, links to URL manufacturer sites, RFID and optical marker values for the products, and a reference to the associated grammar file used for input recognition. This data is retrieved by SQL queries and transferred from the server to the mobile device in XML format.

The input grammar files contain a similar feature–value list, in which grammar entries for each feature are defined for the different modalities like speech and handwriting. The input grammars are assigned to a group of products based on their product type, which allows multiple products to share a single grammar file, as is the case for the product type "digital camera."

After a user has synchronized with a shelf and starts to browse through the products, internal data representations are created for both the objects, and feature keywords displayed on the PDA's PocketPC display. This representation is for example used by the modality of intra-gesture, first during input interaction as screen coordinates are mapped to underlying graphical objects and visual What-Can-I-Say (WCIS) keywords currently on the screen, and later during output presentation through the use of object and keyword lookup functions, which locate a particular reference and highlight it on the display.

15.4 Symmetric Multi-modal Interaction

As defined in Wahlster (2003), symmetric multi-modality refers to the ability of a system to use all input modes as output modes, and vice-versa. Empirical studies have shown that the robustness of multi-modal interfaces increase substantially as the number and heterogeneity of modalities expand (Oviatt 2002). Information provided by one or more sources can be used to resolve ambiguities or manage recognition and sensor uncertainties in another modality, thereby reducing errors both in the system's interpretation of the user's input, and the user's understanding of the system's output. Whereas modality fusion maps multi-modal input to a semantic representation language, the modality fission component provides the inverse functionality of the modality fusion component, since it maps a communicative intention of the system onto a coordinated multi-modal presentation.

Most of the previous multi-modal interfaces such as the ones presented by Oviatt and Wahlster (1997) and Maybury and Wahlster (1998) do not support symmetric multi-modality, since they focus either on multi-modal fusion, e.g., QuickSet (Cohen et al. 1997) and MATCH (Johnston et al. 2002), or multi-modal fission, e.g. WIP (Wahlster et al. 1993). Symmetric multi-modal dialog systems like SmartKom and the MSA create a natural experience for the user in the form of daily human-to-human communication, by allowing both the user and the system to combine the same spectrum of modalities for the input as for the output. The MSA represents a new generation of multi-modal dialog systems that deal not only with simple modality integration and synchronization, but cover the full spectrum of dialog phenomena that are associated with symmetric multi-modality. Symmetric multi-modality supports the mutual disambiguation of modalities, as well as multi-modal or cross-modal deixis and anaphora resolution.

15.4.1 Base Modalities

Multi-modal interaction in the MSA is based on the modalities: speech, handwriting, and gesture, whereby gesture can be further grouped into the types intra and extra. Intra-gestures refer to product and feature selections on the display of the PDA (intra_point), while extra-gestures refer to actions in the physical real world such as picking an object up from a shelf (extra_pick_up), or putting an object back onto a shelf (extra_put_down).

From this limited number of base modalities, a wide range of mixed and overlapped input combinations can be formed. Wasinger and Krüger (2004) outline a total of 23 input modality combinations that were tested within a laboratory setting for use with the system. The modalities included both unimodal (e.g., speech-only) and multi-modal (e.g., speech–gesture) combinations, as well as overlapped and non-overlapped modality combinations. Overlapped modality combinations are ones in which (possibly conflicting) information is provided multiple times in potentially different modalities, as seen in the following non-conflicting speech–gesture overlapped feature interaction: "*What is the price <intra_gesture = price> of the EOS10D?*" Such redundant information is useful for reference resolution.

All of these input modality combinations are however only one side of the interaction equation. The flipside encompasses the output modalities used by the anthropomorphized objects when replying to the user. Speech output for example is presented to the user via an embedded synthesizer. We currently use two synthesizers, one is a formant synthesizer which requires a small memory footprint (around 2 MB per language), while the other is a high quality concatenative synthesizer that has a much larger footprint, between (7 and 15 MB) per language for a single voice. Although the formant synthesizer sounds robotic, it provides far greater flexibility in manipulating voice characteristics such as age and gender, which is important in providing the anthropomorphized objects with their own personality (see Sect. 15.5.2). The output equivalent to handwriting is the use of system fonts that are displayed in a predefined location on the PDA's display. Intra-gesture output for object selection is achieved by drawing a border around the selected object, while intra-gesture output for feature selection is achieved by highlighting the active keyword within the visual WCIS text bar, which scrolls across the bottom of the PDA's display. Extra-gesture output is made possible through the use of a steerable projector, which provides for real-world product selection by placing the product under a spotlight. Figure 15.4 shows the use of the primary modalities within our system, for both input interaction and output presentation. This figure also shows how objects and features can be referred to within the modality types. The output for intra-gesture for example shows a selected feature and below it a selected object.

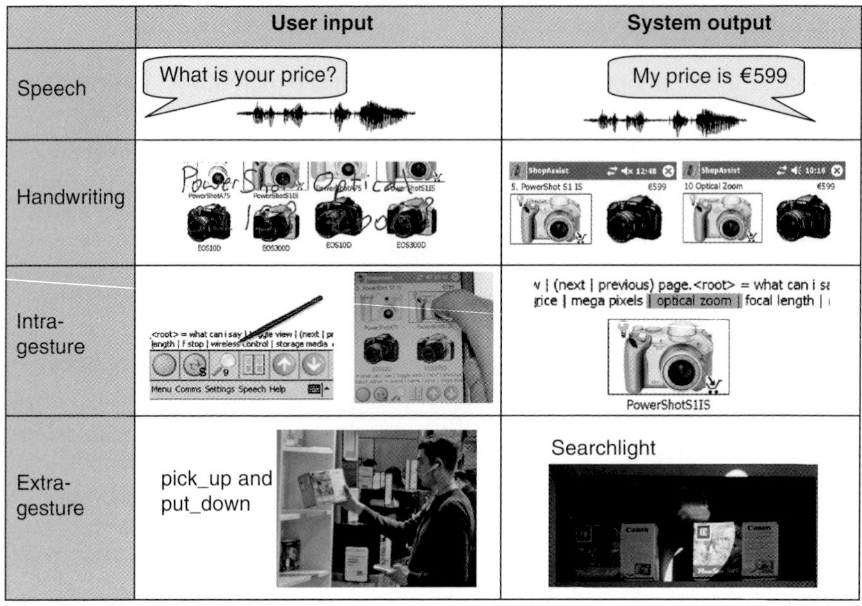

Fig. 15.4. The modalities used for both input and output

15.4.2 Symmetric Modality Combinations

Systems that support multi-modal interaction such as speech, handwriting, and gesture, require an efficient means of fusing the interactions together to form a single unambiguous dialog result, which can then be passed onto subsequent modules in the system such as a retrieval component. Multi-modal user input interaction within our system generally consists of a single feature and one or more object references, for example: "*What is your price* *<gesture = PowerShot* S70>?". Valid values for the feature tag include (in reference to digital cameras) "price," "optical zoom," and "mega pixels," while valid values for the object tag include "PowerShot S70" and "CoolPix 4300." Before such interactions can be parsed however, they must first be converted into a modality-free language. This language is formatted in XML and closely resembles the W3C EMMA standard (see www.w3.org/TR/emma/) in that each tag (i.e., FEATURE and OBJECT) contains a number of attributes like the modality type, timestamp, confidence value, and N-best list values.

On the flipside, multi-modal output from our anthropomorphized products must provide the resulting value information alongside reproducing the feature and object information, and be flexible enough to cater for both direct interaction: "*My price is* €500," and indirect interaction: "*PowerShotS50, price, €500.*"

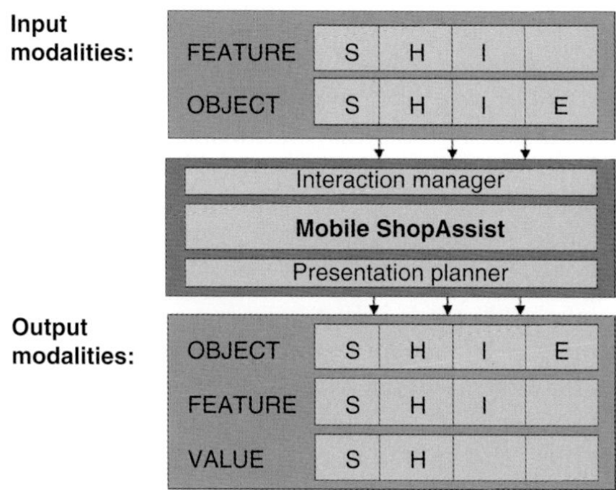

Fig. 15.5. Symmetric modality matching in the MSA

Figure 15.5 summarizes the potential range of modality combinations that exist for user input and anthropomorphized object output, when the modalities speech (S), handwriting (H), intra-gesture (I) and extra-gesture (E) are available. For input alone, possible multi-modal combinations can be seen to include: SS, SH, SI, SE, HS, HH, HI, HE, IS, IH, II, and IE. This figure does not consider multiple object referents, or overlapped input, which would create an even larger number of modality combinations to choose from. In this diagram, the interaction manager is responsible for recognizing and interpreting user interactions with the system, while the presentation planner is responsible for coordinating output for presentation back to the user. This output must be consistent not only in providing the correct information in response to user queries, but also in the choice of modality combinations that are used to present the information.

15.4.3 Output Modality Allocation Strategy

The output strategy within the MSA uses speech as a base modality to present object (O), feature (F), and value (V) information. Complementing speech is the modality of handwriting, which is used first to present the user with the transient information (O) and (F), and then a short time later with the non-transient information (V). Gesture is additionally used to show the selected (O) to the user as non-transient information, either solely via an intra-gesture on the PDA display or also as an extra-gesture via the searchlight, if it is available. Intra-gesture output for the feature (i.e., highlighting an active keyword from the scrolling WCIS text) only occurs if the scrolling text is currently visible. At the end of an interaction dialog, a user will have been presented with the same information in two complete modalities (speech and handwriting), and part of the modality gesture.

In comparison to recent usability tests in which our subjects stated that they preferred non-overlapped input modalities to overlapped ones ($\chi^2(2, N = 27) > 24.889, p < 0.000$), our users were keen to be provided with overlapped modalities for the output. Redundancy in the output modalities as used above compensated for the names of objects such as "PowerShot S1IS" and "EOS 10D" being pronounced incorrectly by the speech synthesizers, and for the transience required in presenting the written language as two separate events on the limited display space.

The current output strategy is just one of several possibilities. Other output allocation strategies include for example the exact replication of modalities used for input as for output (mimicking), user defined profiles, or profiles that limit the media to types that third person parties cannot observe (e.g., handwriting, intra-gesture, and speech output through a PDA-based headset), or that do not require a PDA (e.g., server-sided speech and extra-gestures).

15.5 Anthropomorphized Products

In this section, we outline the concept of anthropomorphization. We describe the difference between direct and indirect interaction, and also outline how we account for anthropomorphized objects in the MSA, with particular focus on the language grammars, the product personalities, and the state-based object models that define when our objects may initiate dialog interaction with the user.

15.5.1 The Role of Anthropomorphization

Anthropomorphism is the tendency for people to think of inanimate objects as having human-like characteristics. Many early cultures made no distinction between animate and inanimate objects (Todd 2002). Animism is looking at all Nature as if it were alive. It is one of the oldest ways of explaining how things work, when people have no good functional model. When users interact with AmI environments rather than with a desktop screen, there is a need for communication with a multitude of embedded computational devices in mass-marketed products. For human–environment interaction with thousands of networked smart objects, a limited animistic design metaphor seems to be appropriate (Nijholt et al. 2004); see also Chap. 14 of this book.

Although there are various product designs that use an anthropomorphic form (like the Gaultier perfume bottles that have the shape of a female torso), in the work presented here we stimulate anthropomorphization solely by the pretended conversational abilities of the products. Since the shopper's hands are often busy with picking up and comparing products, in many situations the most natural mode to ask for additional information about the product is the use of speech. When a product talks and answers the shopper's questions with its own voice, the product is being anthropomorphized.

There is a longstanding tradition among some HCI researchers against the use of anthropomorphism (Don 1992), because it may create wrong user expectations. This has lead to taboos like "Don't use the first person in error messages." People are however used to dealing with disembodied voices on the telephone, and our empirical user studies also provide evidence that most shoppers have little concern about speaking with shopping items such as digital cameras (see Sect. 15.6). In addition, through the world of TV commercials, shoppers are used to anthropomorphized products like "Mr. Proper," a liquid cleaning product that is morphed into an animated cleaning Superman, or the animated "M&M" round chocolates.

Of course, anthropomorphized interaction can be irritating or misleading, but our system is designed in such a way that it presents its limitations frankly. The WCIS mechanism in the MSA guides the users in their decision-oriented dialog and makes it clear that it has only restricted, but very useful communication capabilities. We contend that anthropomorphism can be a useful framework for interaction design in AmI environments, if its strengths and weaknesses are understood.

15.5.2 Adding Human-like Characteristics

Apart from the assortment of modality combinations available in the MSA, users may choose to interact either directly or indirectly with the shopping products. These products will in return also need to respond correspondingly. We derive the terms direct and indirect interaction from the mode of reference being made to the "person" segment of a dialog. In English for example, there exist the tenses: first person (the person speaking), second person (the person being spoken to), and third person (the person being spoken about). From an input perspective, direct interaction refers to the second person (e.g., "*What is your price?*"), while indirect interaction refers to the third person (e.g., "*What is the price of this/that camera?*"). From an output perspective, direct interaction (as used by the anthropomorphized objects) takes the first person (e.g., "*My price is* € 599"), while indirect interaction takes the third person (e.g., "*The price of this/that camera is* € 599").

Within the MSA, grammar files exist for each product type, such as "digital camera", and for both English and German. These grammar files define the recognizable input (e.g., product and feature information) for the modalities handwriting, gesture, and speech. Although the individual modalities may be used to communicate complete dialog acts (i.e., product and feature information), speech is the only modality in which complete sentences may be used. Three forms of speech input are accepted by the system, namely "keyword," i.e., speaking only the keyword, (e.g., "*price*"), "indirect", (e.g., "*What is the price of <product>?*"), and "direct," (e.g., "*What is your price?*"). The grammar files for each of the product types are downloaded onto the PDA together with the product information, each time the user synchronizes with a particular shelf container. These files are then parsed by the PDA to create the individual grammars required for each of the recognizers.

Objects within the MSA are further personalized by one of five different formant synthesizer voice profiles (three male, two female, and all adult), which are based on parameters such as gender, head size, pitch, roughness, breathiness, speed, and volume. A limitation of our approach is that five different voices cannot provide each product in a shelf, let alone an entire store, with a unique voice. An alternative would be to use pre-recorded audio samples for each product, but this would require different magnitudes of storage space. A different approach might be to allow the PDA to assign the voices to products, which would allow at least the first five products interacted with to have a unique voice. Such an approach would also allow the use of personality matching strategies to better market products to specific user groups. Dynamic voice assignment would however also create the need for storing voice to product mappings for future use, so that returning users are not faced with anthropomorphized objects with multiple personalities.

15.5.3 State-Based Object Model

A further feature of our anthropomorphized objects is their ability to initiate interaction with the user when in a particular state; see Fig. 15.6. These states are based on variables such as a product's location, a recent extra-gesture action, and an elapsed period of time. The location of a product may be either "in a shelf," "out of a shelf," or "in a shopping trolley," and extra-gesture events include: "pick_up" and "put_down." Thus, the physical acts of the user like "Pick_Up (product007, shelf02)" and "Put_Down (product007, trolley01)" are mapped onto dialog acts like "Activate_Dialog_With (product007)" and "Finish_Dialog_With (product007)," respectively. In this case, the Put_Down

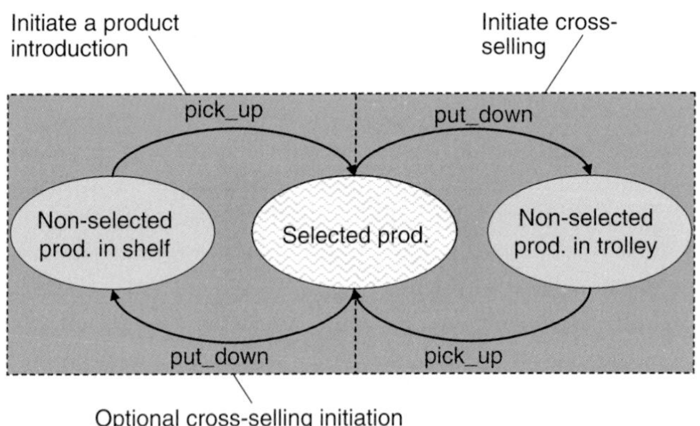

Fig. 15.6. Base product states used for object-initiated interaction

action reflects a positive buying decision as the product was placed inside the trolley, but the product could just as equally have been put down on the shelf instead, thus reflecting a negative buying decision. As an example, an object will initiate a dialog interaction if it is picked up from the shelf for the first time and no further user interaction is observed within a 5 s time frame. Silence as a powerful form of communication is well documented (Knapp 2000), and in our case such silence forces the product to introduce itself; see also (1) in Fig. 15.1. A product might also initiate an interaction when for example placed inside the shopping trolley, in order to alert the users of any further products (e.g., accessories) that they might be interested in purchasing, i.e., cross-selling.

15.6 Usability Study

Ben Shneiderman, as a prominent critic of anthropomorphized user interfaces, stated at a panel discussion documented by (Don 1992) "I call on those who believe in the anthropomorphic scenarios to build something useful and conduct usability studies and controlled experiments." That is exactly what we have done in the described research.

In this section, we describe an empirical field study on user interaction with anthropomorphized objects. The goal of the study - which was conducted at an electronics store of the "Conrad Electronic" - was to identify how accepting people would be to conversing with shopping products such as digital cameras. This study was part of a larger experiment designed to test modality preference and modality intuition. A total of 1,489 interactions were logged over the two-week test period, averaging 55 interactions per subject. Each test session generally took between 45 and 60 min to complete, during which time an average of 13.8 shoppers could be seen from the shelf's location.

15.6.1 Method

Our sample of test persons consisted of 27 people, 16 females, and 12 males, and ranging in age from 19 to 55 (mean: 28.3 years). We advertised the study by posting notices around the University of Saarland in Germany, and setting up a registration desk at the main cafeteria. Only two subjects were from the faculty of computer science. Our setup consisted of a shelf of digital cameras located in a prominent part of a local electronics store. Each participant was allocated a PDA and headset, and asked to stand in front of the shelf containing real-world camera boxes. The subjects were briefed on how to use the system and the individual modality combinations. They were then instructed to interact indirectly using the third person tense, e.g., "*What is the price of this camera?*" and then later on directly by using the second person tense, e.g., "*What is your price?*". In each case, the products responded in an aligned manner, i.e., third and first person tenses, respectively. To ensure that our subjects spent enough time interacting each mode, they were given a series of smaller sub-tasks to complete, such as to find the cheapest camera on the shelf, or to find the camera with the largest number of mega pixels. During the test, system output was limited to a single female concatenative synthesizer voice. This configuration was chosen to minimize the effect that voice quality and limited number of voice types might have on the study. After having completed the practical component, the participants were given a small questionnaire.

15.6.2 Results

The first question that we asked our subjects was which of the two interaction modes they preferred best. The proportion of subjects that preferred direct interaction over indirect interaction (18 from 27, 66%) signifies a distinct trend for anthropomorphization, $\chi^2(1, N = 27) = 3.00, p = 0.083$. This result is seen clearer in men than in women, in which 10 from 12 men (83%) stated that they preferred direct interaction: $\chi^2(1, N = 12) = 5.22, p = 0.021$, which is significant. An advantage seen by several subjects with direct interaction was that the dialog interactions were shorter and simpler, e.g., "*What is your price?*" compared to "*What is the price of the PowerShot S50?*".

Following this question, we asked our subjects if they would reciprocate with direct interaction if the objects only spoke directly to them. Twenty-two from 27 subjects (81%) stated that they would allow themselves to be coerced into communicating directly: $\chi^2(1, N = 27) = 10.70, p = 0.001$, which is significant. Courtesy and conformity were cited reasons for this allowed coercion. Note that a "no" response to this question would result in incoherent language similar to the following:

U: "*What is the price of this <gesture> camera?*"
O: "*My price is* €599."
U: "*How many mega pixels does this camera have? <gesture>.*"

We then asked our subjects whether they would interact directly with a given range of products (soap, digital camera, personal computer, and a car), first as a buyer (B), and then as the owner (O) of the product. For brevity, we report only the resulting significance values obtained from our non-parametric χ^2 tests, where df = 1, and $N = 27$. Whereas only around 30% of people would interact directly with a bar of soap (as B: $p = 0.034$, as O: $p = 0.201$), around 70% of people said that they would interact directly with digital cameras (as B: $p = 0.034$, as O: $p = 0.033$), personal computers (as B: $p = 0.012$, as O: $p = 0.003$), and cars (as B: $p = 0.336$, as O: $p = 0.003$). Our subjects were more inclined to interact directly with the products as the owner rather than as a buyer, and this difference is best seen for the product type "car," in which a Wilcoxon signed rank test bordered on statistical significance

($z = -1.890, p = 0.059$). As the owner of the products "personal computer" and "car," men were more inclined than women to talk directly with the objects, with a Mann–Whitney U-test showing this trend in gender difference to be: $U(16, 12) = 40.5$, equating to $p = 0.072$ for both product types. Other objects that our subjects said they would consider talking directly with included plants, soft toys, computer games, and a variety of electronic devices like TVs and refrigerators.

Finally, we tested which modalities people would be comfortable using in a public environment, e.g., when surrounded by other shoppers, compared to a private environment, e.g., when no shoppers are around. Given the choice of "comfortable," "hesitant," and "embarrassed," the results showed that our subjects would feel comfortable using all modalities except speech when in a public environment ($\chi^2(2, N = 27) > 12.667, p < 0.002$). Moreover, they feel comfortable when using all modalities in a private environment ($\chi^2(2, N = 27) > 10.889, p < 0.004$).

15.6.3 Lessons Learnt

From this empirical study, our hypothesis that subjects would not simply reject the concept of anthropomorphized objects was confirmed, and indeed many of the subjects actually enjoyed the concept. The study has also shown that product type, e.g., toiletries, electronics, automobile, relationship to a product, e.g., buyer, or owner, and gender, (male, female) all have an effect on a person's preference for direct interaction with anthropomorphized objects. Future tests on the benefits of anthropomorphization could focus on a broader set of product types, the acceptance of cross-selling, and richer product personalities including distinct voices.

15.7 Conclusions and Future Work

This chapter has described a new interaction paradigm for instrumented environments based on tangible multi-modal dialogs with anthropomorphized objects. For this purpose, we introduced the concept of symmetric multimodality and applied it to speech, handwriting, and gesture. Finally, we showed via a usability field study that direct interaction with anthropomorphized objects is accepted and indeed preferred by the majority of users. Such findings have already been exploited in two other projects of our research group in which interactive installations for museums and theme parks are being developed.

Future work will now focus on scalability aspects of our approach, which will be particularly important if the system is to provide a shop full of differing products with rich forms of communication and personalities. The underlying grammars of this mobile system have currently been handcrafted for each product type. This is acceptable when many products all have the same attributes, such as with digital cameras, but is less acceptable when many different product types exist, as would be the case when modeling the products of an entire store. We are currently developing a module to automatically generate the direct and indirect grammars based on keyword information available in the product database, and the type of question to be associated with the keyword, e.g., a wh-question (who, what, when, where, why, and how), or a yn-question (yes and no), and perhaps later also alternate and tag questions.

16

Intelligent Media

P. Treleaven and S. Emmott

"The media is the message."

Marshal McLuhan

16.1 Introduction

A development is underway which has the potential to reshape our culture, communication, commerce and our relationships. It is the emergence of Intelligent Media. Intelligent Media is physical or digital media, e.g., software, fabrics, music, moving images, materials, "content", which exhibits or into which is embedded some intelligence, e.g., ability to learn, adapt, communicate and/or interact with its environment. Intelligent Media is already emerging, and nowhere is it showing more importance than in its potential to underpin and drive the future of the creative sectors, from advertising to architecture, fashion to film, museums to music, the performing arts to publishing and television to tourism. In some sectors such as animation and architecture, this transform is well underway. In other sectors such as fashion design and music, it is just emerging. Intelligent Media is also transforming the creative process with "creatives" increasingly able to "mix" the physical with the digital and the "smart". In this chapter, we give a selected overview of the emergence of Intelligent Media.

A new phenomenon is emerging, fuelled by the convergence of computing, media and networks, the so-called "digital revolution", and the emergence of novel physical materials. It is the advent of *Intelligent Media.* Intelligent Media is physical or digital media, e.g., software, fabrics, music, moving images, materials or content, that exhibits or has embedded into it "intelligence" – an ability to adapt, communicate and interact with its environment, and with us, in novel ways.

Intelligent Media is already showing the potential to create an enormous impact on our relationship with the things around us that influence, underpin and define social relationships, culture and society – from television to toys, film to fashion, product design to packaging, architecture to advertising and cities to commerce. And as these elements form the basis of the "Creative Sectors", Intelligent Media therefore has the potential to create equally significant new innovation and market opportunities in the Creative Industries, and lead a fundamental paradigm shift in today's Creative Industries, creative practice and the creative process. It has the potential even to form the basis of entirely new markets, sectors and processes, which we cannot imagine today.

Just as the Information Technology "revolution" completely re-shaped commerce and business productivity, in this paper, we explore the early emergence of Intelligent Media, with a focus particularly on some examples of how it might have an impact in re-defining and re-shaping the creative sectors and creativity.

16.2 Emergence of Intelligent Media

When architects design a building today they will almost certainly do so using Intelligent Media. Firstly, they explore the design using a virtual building. This comprises a computer model of the structure and the application of "intelligent systems" to optimise such things as heat, light and sound, even the interaction of people; and virtual reality to visualise not only the exterior and interior of the building, but also the social interactions afforded by the design. Moreover, the use of Intelligent Media in architecture and building is now spreading into the building itself leading to the emergence of ' "*Ambient Intelligence*": invisible, ubiquitous computing and networks embedded throughout a building to monitor and interact with the building's environment and its occupants. And the use of Intelligent Media in just this one example does not stop. The actual building materials are now starting to be constructed from so-called "smart" materials (Addington and Schodek 2004); see also Chap. 4 of this volume.Or consider how Intelligent Media is already having an impact on how we acquire and listen to music. Shazam (online) is a service that enables anyone to find out about an artist, title and availability of music through their mobile phone, by thephone "listening" to and recognising music. It could be music in a bar, in the street, on the television or from someone singing. The music Shazam identified can then be purchased from the Internet or (more likely) downloaded directly from a complete stranger's computer via the Internet (e.g., Kazaa or Napster). And the listener is probably doing so on an "invisible", powerful computer that has become part of the "new everyday". It is called an iPod, from Apple.

The pattern recognition (or "perception") algorithm behind Shazam is a company secret (we might perhaps hazard a guess that it is broadly based on a very clever recognition and referencing of phase information), but there is no doubting that the resulting technology, service and experience of the incredible Shazam are excellent examples of a glimpse of what we can expect to see in the rise of Intelligent Media.

The third example is the inventions and innovation of scientist and artist Danny Rozin. While a student at the NYU Interactive Telecommunications Programme, Danny invented a beautiful form of Intelligent Media that allows the ability to paint on real canvas using real bristle paintbrushes using "virtual" paint. The artist simply puts the paintbrush into an empty paint pot of a range of colours and applies the (non-existent) paint to the actual canvas, and on it is painted in light of the colour of the paint applied.

These three examples of Intelligent Media illustrate how it is starting to emerge, and how it is starting to influence the creative process and creative sectors. The first is transforming the world of architecture and our relationship with our surroundings. The second has the potential to revolutionise the music industry, and the "creative" process of how music is marketed to potential listeners and buyers, through advertising. The third augments and transforms the creative process itself, and challenges our notion of what is "real" and what is "virtual", what is physical and what is digital, and is an excellent glimpse into what Intelligent Media may offer, and how it can challenge and change everything we know today about the relationship between media, technology, creativity and culture.

16.3 Economic Importance of the Creative Industries

For the past 5–10 years high-tech business and research has been increasingly dominated by the biosciences and information technology for "business productivity". Areas such as biotechnology, genetic engineering, bioinformatics and systems biology, etc., have attracted massive levels of research funding and private investment. Now the economic significance of the Creative Industries for employment and new businesses is being recognised; see Table 16.1. As in the late 1990s with the emergence of the Internet, we expect a fountain of exciting new high-tech start-ups to be spawned from the infusion of technology and creativity; a new generation of Pixars and Industrial Light and Magic.

Recently London government and the UK government highlighted the importance of the Creative Industries to the economic prosperity of London and the United Kingdom; adding £85 billion annually to the UK's output; and in London adding £25 billion to the economy and employing over 500,000 people (GLA Economics 2002).

16.4 Intelligent Media: Key Concepts and Technologies

The whole premise of Intelligent Media is the convergence of science and technology with physical and "traditional" digital forms of media, ranging from music, television and film to buildings, product packaging and textiles. In this section, we outline some of the key concepts and technologies underlying the evolution of Intelligent Media; see Table 16.2.

Table 16.1. Creative Industries

Entertainment
Film
Television
Music
Performing arts
Radio
Computer games
Cultural
Libraries
Museums
Design
Architecture
Product design
Fashion and "Beauty"
Packaging
Leisure
Sports
Tourism and travel
Advertising

Table 16.2. Intelligent Media: key concepts and technologies

Key enabling concepts
Virtual artefacts
Integrated (digital and physical) media
Analysis of media
Ambient Intelligence
Interaction, perception and experiences
Key enabling technologies
Virtual reality and virtual environments
Intelligent and evolutionary systems
Ubiquitous communications and ubiquitous computing
"Smart" materials

16.4.1 Key Concepts

In terms of the focus of this paper, that of the relationship between Intelligent Media and the creative sectors, this can be characterised into the following concepts:

- *Augmented Creativity.* The creative process is assisted and takes on new forms through enabling the exploration, imagination, expression and realisation of real or virtual artefacts and environments.
- *Smart Products.* Products communicate with, learn about and/or adapt to their environment. This is an example of how "Ambient Intelligence",

involving ubiquitous computing, ubiquitous networks and/or "smart" materials (examples include active sensors, intelligent plastics, adaptive and reactive building materials), is starting to emerge.
- *Intelligent Environments.* The seamless integration of media, people and technology in social, commercial, and culturally sensitive environments, covering interaction, perception and experiences.

Again, if we look at Intelligent Media in the context of the creative sectors, there are a number of emerging "creativity" paradigms, which can be formulated as follows:

- *Virtual artefacts.* "Creatives" are increasingly working with virtual artefacts, using computer models, as in animation and architecture.
- *Integrated or "mixed" media (digitaland physical).* Products will increasingly embrace "multimedia" in its broadest sense, integrating virtual and physical media and materials, and interacting with the five senses: sight, sound, smell, touch and taste.
- *Analysis of media.* As demonstrated by the mathematical and computer analysis and recognition of text and music, it will become increasingly possible to analyse media to identify, for example, cultural characteristics such as genre and optimise composition. What is currently being achieved in text and music will also be applied to 2D pictures, 3D graphics, film/video animation, and possibly even smell.
- *Ambient Intelligence.* The emergence of near-ubiquitous intelligence embedded into the things and the environments that surround us, which are connected to ubiquitous networks – the concept of "intelligence" as the new infrastructure for the twenty-first century (as electricity, water, road/rail and telecommunications emerged as key infrastructures in the eighteenth, nineteenth, and twentieth centuries).
- *Interaction, perception and experiences.* Artefacts will become increasingly sensitive and responsive to their environment, embracing interaction and in the future, perception and even experiences.

16.4.2 Key Technologies

At the moment, technologies underpinning the emergence of Intelligent Media include ubiquitous computing and networks, intelligent systems, virtual and "mixed reality" environments, intelligent materials and advanced machine learning techniques. But it is worth noting that even today's state-of-the-art will mature over time and be replaced by newer technologies that we cannot imagine today, in their form, function or significance.

- *Virtual Environments*: virtual environments technology supports creativity, design and visualisation in a virtual, digital world. This includes *Virtual Reality* (visualisation, haptics, perception), *3D scanning* (scanning of artefacts and humans to create digital clones) and *Digital Clones* (3D digital objects, avatars, intelligent behaviour); see also Chaps. 17 and 18 of this volume.

- *Intelligent systems*: to model, predict, adapt to, recognise and perceive elements of our world and augment the creative process and the creative product, Intelligent systems are largely based on the notion of perception, learning and generalisation of things through *Machine Learning*, which covers neural networks, evolutionary systems (genetic algorithms, genetic programming, classifiers, evolutionary programming and evolutionary strategies), Bayesian statistics and Inductive Logic Programming. In addition *Multi-level modelling* involves attempting to understand how complex systems function through hierarchical models, by studying the relationships and interactions between various levels of the system; see also Chap. 13 of this volume.
- *Ubiquitous computing*: increasingly sophisticated and increasingly miniaturised computing continues to rapidly become embedded into an increasingly expanding range of "things" other than what we think of as "computers", i.e., a PC, such as cars, taxis, telephones, toys, washing machines, televisions, cash-point machines, key fobs, whiteboards, trainers and packaging. At the same time, there is an increasing proliferation and expansion of a wide variety of networks that are connected to these things, from 802.11 wi-fi to ultra-wideband (UWB) and from Universal Mobile Telephony Service (UMTS) to bluethooth to various Personal Area Networks (PANs), and of course existing Internet and satellite technologies. The combination is creating a world where more and more people and everyday things are connected to each other through this network or networks infrastructure of networks creating a world of "ubiquitous computing"; see also Chaps. 9 and 10 of this volume.
- *Smart materials*: "Smart" materials technology will become increasingly important. A smart material is one that interacts with its environment, responding to changes in various ways. A simple example is photochromic glass, darkening on exposure to light. Other technologies include *Technical Textiles* (materials meeting high technical and quality requirements – mechanical, thermal, electrical, durability), and nanotechnologies, i.e., devices of nanometer scale: thin films, fine particles, chemical synthesis and advanced micro-lithography; see also Chaps. 4–6 of this volume.

16.5 Intelligent Media and the Creative Sectors

Unlikely though it may seem, analogy with the UK Financial Services industry highlights the potential importance of Intelligent Media. In the past ten years the convergence of information technology (IT) and mathematics has transformed finance in terms of products, services, (de)regulation and how "customers" buy, interact with and use finance and financial services.

The "disruptive" effect of Intelligent Media on the Creative Industries will be even more profound, because of the pervasive impact of Creative Industries' *product* on the fabric of society – our culture, communication and relationships, with each other and with others and other things.

16.5.1 Film and Television

One only has to look at the latest blockbuster film to see the pervasive influence of technology, most notably digital animation. Correspondingly, the "hot topic" in television is *"mixed reality intelligent environments"* is becoming increasingly important because they allow producers the flexibility to combine real and digital actors in real or imaginary worlds, or even the ability to bring real actors back to life (GTI online).

In entertainment, Pixar and Industrial Light and Magic (ILM) are legendary for creating, developing and producing computer-based effects and entire feature films with a new three-dimensional(3D) appearance, and memorable characters. In fact, the visual effects created by ILM have appeared in eight of the ten highest grossing movies ever. Now, new companies are starting to emerge that take computer-based effects and films to the next stage through more "intelligent" digital media. Mental Images (online) – a small Germany-based company founded by a handful of mathematicians – provide the Intelligent Media technology behind the special effects in The Matrix and several other Hollywood hits.

Intelligent Media

Mental images products are based on the application of "artificial intelligence" technology into existing computer graphics and effects technology. Similarly, Natural Motion (online) is the creator of Active Character Technology (ACT), a claimed break-through in 3D character animation based on Oxford University research on intelligent systems to model of the behavior or biological systems. The company has received funding from the games, film and finance sectors. Its development team comprises software engineers, zoologists, biologists, physicists and animators.

Other technologies such as 3D laser scanners (Fig. 16.1) and motion capture devices (Fig. 16.2) play an increasingly important role in animation. Currently actors are scanned in devices such as the highly accurate, but expensive Cyberware laser scanner and actors' motion is captured using devices such as the Vicon system. However, both processes require highly labour-intensive manual post-processing.

16.5.2 Music

Music is also being transformed by Intelligent Media. The impact ranges from exploitation of intelligent "packet switching" to enable peer–to–peer networking to distribute music (e.g., Napster), to science-based innovation that allows

Fig. 16.1. 3D body scanner

Fig. 16.2. Motion capturing device

the analysis of composition (e.g., Shazam, Polyphonic HMI), to the rise of smart consumer products such as the iPod for listening to music.

Intelligent Media

Polyphonic HMI has developed an intelligent system for music analysis and is currently utilising it in two key areas: individual music recommendation and as a tool for music label services for the record industry. For music label services,

Polyphonic's Hit Song Science (HSS) analyses the underlying mathematical patterns in unreleased music and compares them to the patterns in recent hit songs. The new technology can isolate individual patterns in key aspects of the music that humans detect and that help determine whether or not they like a given song.

This is an excellent example of Intelligent Media. The first step in the "analysis of composition" process is to use Intelligent Systems to analyse the composition of millions of songs and isolate patterns in many musical structures involving melody, harmony, tempo, pitch, octave, beat, rhythm, fullness of sound, noise, brilliance and chord progression. By doing that, Polyphonic and others found that hit songs share similar characteristics. It was rare to see a song that fell outside of the clusters and had become a hit.

The second step in the process takes the analysed data and overlays extra parameters relating to the commercial success of the music or a listener's personal preferences. For commercial success these parameters are data such as total sales, highest chart position, date of release and others.

16.5.3 Cultural Sectors

The Cultural Sectors cover libraries and museums, exhibitions and fairs. The major challenge for libraries, museums and exhibitions is to make their collections available to ever-larger groups, for example over the Internet, and to augment the experience through technology. For example, the "Intelligent Cultural Heritage" industries are starting using virtual reality to enhance the experience of visiting a museum, exhibition or tourist attraction.

An example of companies in this area is Google, essentially ("historically") a digital library company. The world's largest search engine is based on the Ph.D. work on online searching of Stanford students Larry Page and Sergey Brin. At the core of Google is its unique ability to analyse the "back links" pointing to a given web site.

Intelligent Media

The "digital library" heart of Google's software is PageRank a system for ranking web pages. PageRank relies on the uniquely democratic nature of the web by using its vast link structure as an indicator of an individual page's value. In essence, Google interprets a link from page A to page B as a vote, by page A, for page B. But, Google looks at more than the sheer volume of votes, or links a page receives; it also analyses the page that casts the vote. Votes cast by pages that are themselves "important" weigh more heavily and help to make other pages "important".Important, high-quality sites receive a higher PageRank, which Google remembers each time it conducts a search. Of course, important pages mean nothing to you if they do not match your query. So, Google combines PageRank with sophisticated text-matching techniques to find pages that are both important and relevant to your search. Google goes beyond the number of times a term appears on a page and examines all aspects of the page's content (and the content of the pages linking to it) to determine if it is a good match for your query.

16.5.4 Design

The design sectors cover architecture and product design, and we have also included packaging. Industries in this sector look set to undergo a major shift as a consequence of Intelligent Media that ranges from intelligent packaging to virtual environments to *Biomimetic design* – creating good design (of buildings, textiles, furniture, etc.) from lessons of nature.

Packaging is particularly active. There are numerous innovative companies in the packaging and product design area. Cypak is developing packaging technology for making objects smart, secure and connected. Their core technologies are disposable microelectronics and sensors, a new contact-less data transfer technology and new security solutions to guarantee authenticity and integrity. Commotion Printing Display's core technology is proprietary electro-active ink that changes colour when a low current is applied. ScentSational Technologies, as the name implies, is developing Olfaction packaging technology for food and beverage companies based on "enhanced aroma delivery" in consumer products.

Intelligent Media

Intelligent packaging developments include labels with embedded diagnostic capabilities that monitor temperature, humidity and time and communications based on RFID; see for example Fig. 16.3. These labels are typically used to protect the integrity, the freshness and the safety of foods. Intelligent packaging subdivides into: primary packaging of individual items and secondary packaging at the case and pallet level. At the primary level are smart active labels (SAL) that can monitor temperature. At the secondary level are time temperature indicators (TTI) that can use electronics and RFID to store temperature history data, and can transmit it to a reader.

Low-cost printed electronics displays, from companies such as COMMOTION, are printed conductives on a flexible substrate. These tags can be programmed to display "expired" either when a case or pallet has pasted its date, or been exposed to adverse conditions.

16.5.5 Leisure Sectors

The Leisure sectors cover the sports industry, and also tourism and travel. Technologies in the highly competitive sport sector are becoming ever more important. Most of the major sports companies such as Nike, Adidas and Speedo are seeking to increase the "intelligence" of their products either by embedding sensors or using smart materials.

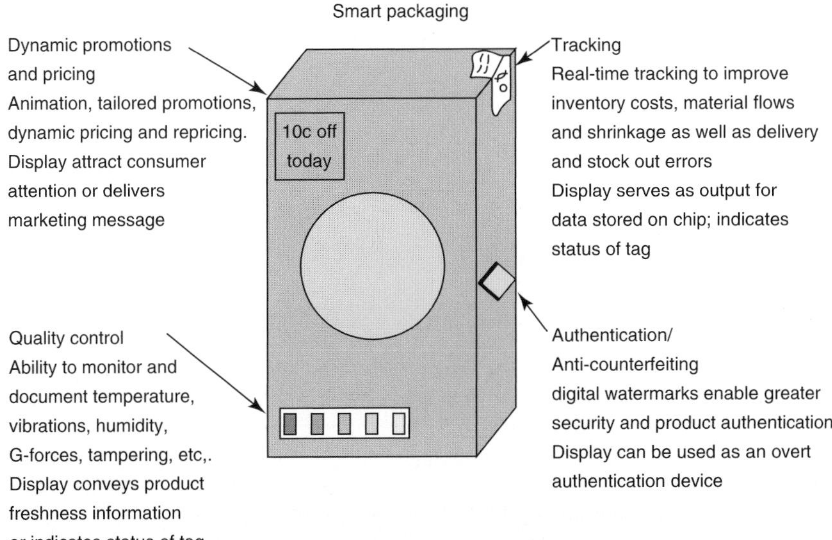

Fig. 16.3. Future Intelligent Packaging: display, freshness indicator, RFID tag and brand protection devices

Intelligent Media

Typical of materials technology is Speedo's "shark-skin" inspired swimsuit. The company claims that its Fastkin FSII swimsuit can reduce friction in water by up to 4% over its rivals by mimicking a shark's skin. Studies of sharks discovered that friction is different over different parts of its body, so the shark's skin changes in texture to better manage the flow of water. Speedo reportedly applied the same principles to the suit.

The suits took four years of top-secret testing to develop. Knitted fabric is constructed with tiny hydrofoils with V-shaped ridges that decrease drag and turbulence by directing water flow over the body and allowing surrounding water to flow more effectively. Muscle compression components reduce muscle vibration that is a major source of power loss and fatigue for swimmers.

Another example is Adidas's "intelligent shoe"; see Fig. 16.4. As the name suggests this is an attempt to make shoes "smart". The shoe contains a battery-powered sensor, microprocessor and electric motor, allowing the shoe to respond to changing conditions and the user's running style.

16.5.6 Advertising

This sector is spawning an increasing number of "Intelligent Media" companies, such as Vert offering geo-targeting technology. Vert is the developer of an innovative outdoor digital advertising network comprising Video Interactive Displays (VIDs) and geo-targeting of customers with adverts customised

Fig. 16.4. Adidas "smart" shoe

Fig. 16.5. IDEO's interactive dressing

to location. The Vert system utilises wireless networks using a web-based infrastructure linked to the Internet via a wireless cellular modem. Based on information from the Global Positioning System, VIDs use Internet access to feed live ads through a wireless connection to the screen.

Intelligent Media

An interesting use of technology was IDEO's interactive dressing room developed for PRADA; see Fig. 16.5. The enabling technology for the store is radio-frequency ID tagging (RFID). All merchandise has its own RFID tag. An RFID tag is also part of a PRADA customer card. Customer preferences are stored on the database, and only the customer card provides access. This information is used to customise the sales experience and further enhance the service provided to the card-holding customer.

Once inside a dressing room the customer can directly access information that relates to their particular garment selection. As garments are hung in the closet their tags are automatically scanned and detected via RF antennae embedded in the closet. Once registered, the information is automatically displayed on an interactive touch screen, enabling the customer to select alternative sizes, colours, fabrics and styles, or see the garment worn on the PRADA catwalk as slow-motion video clips.

16.6 Conclusion

We have provided here a brief review of just a small set of examples of initial instantiations of Intelligent Media as it is emerging today and their potential impact on the creative sectors. There seems little doubt that the continuing emergence of Intelligent Media is set to bring about, or force, a sea change in the creative sectors. The future of the creative sectors will absolutely depend upon, and be driven by, the convergence of computing, digital networks, science, culture and social artefacts, which define Intelligent Media. This is not some trite "futurology" type prediction; it is already happening and the building blocks of Intelligent Media are already in place and evolving and proliferating rapidly.

Importantly, the innovation required to create and deliver the types of Intelligent Media that people will want as part of the fabric of their lives will require a radical new approach in research and development that is at the interaction, intersection of science, technology and creativity. It is at this intersection where entirely new types of "Intelligent Media" will be created that will re-define and drive the future of the Creative Industries, and the future of the social, cultural, economic and commercial potential of these sectors.

17 Smart Environments

T. Kirste

"Bringing computers into the home won't change either one, but may revitalize the corner saloon"

Alan J. Perlis

17.1 Introduction

Catchwords like *Ubiquitous Computing*, *Pervasive Computing*, and, the youngest term, *Ambient Intelligence* (Shadbolt 2003; Aarts 2004), paraphrase the vision of a world, in which we are surrounded by smart, intuitively operated devices that help us to organize, structure, and master our everyday life. They share the notion of a smart, personal environment, which characterizes a new paradigm for the interaction between a person and his everyday surroundings: smart environments enable these surroundings to become aware of the human that interacts with it, his goals and needs. So it is possible to assist the human proactively in performing his activities and reaching his goals. If my car stereo tunes in to exactly the station I just listened to at the breakfast table, then this is a simple example for such an aware, proactive environment; just as the mobile phone that automatically redirects calls to my voice mail in case I am in a meeting, or the bathroom mirror that reminds me of taking my medications.

Hitherto, it is the user's responsibility to manage his personal environment, to operate and control the various appliances and devices available for his support. But, the more technology is available and the more options there are, the greater is the challenge to master your everyday environment, the challenge not to get lost in an abundance of possibilities. Failing to address this challenge adequately simply results in technology becoming inoperable, effectively useless. The goal of smart environments is to take over this mechanic and monotonous control task from the user and manage appliance activities on his behalf. Through this, the environment's full potential for assisting the user can be mobilized, thus allowing tailoring to the user's individual goals and needs.

Technical foundation of smart environments is ubiquitous computing technology: the diffusion of information technology into all appliances and objects of the everyday life, based on miniaturized and low cost hardware. In the near future, a multitude of such *information appliances* and *smart artifacts* will populate everyone's personal environment. In order to make the vision of smart environments come true, a coherent teamwork between the environment's appliances has to be established that enables a co-operative, proactive support of the user. Wireless ad-hoc networking and embedded sensors provide the basis for coherent and coordinated action of an appliance ensemble with respect to the current situation. By enabling multi-modal interaction, such as speech and gesture, an intuitive interaction becomes possible. On top of this, strategies, models, and technologies for the self-organization of appliance ensembles are required that allow an adaptation to the user's needs and desires.

In this chapter, we will outline some of the challenges that smart environments pose with respect to creating a coherent teamwork from an ad-hoc ensemble of information appliances.

17.2 Smart Environments

What are smart environments? In their recent book, Cook and Das (2005) define a smart environment as "one that is able to acquire and apply knowledge about an environment and also to adapt to its inhabitants in order to improve their experience in that environment."

This definition seems fair enough. However, it has the drawback of defining smart environments by a specific implementation strategy using explicit knowledge, and learning, rather than just by their user-visible behavior. For the purpose of this chapter, we will therefore revert to a less implementation-specific definition, which can be formulated as follows:

> Smart environments are physical spaces that are able to react to the activities of users, in a way that assists the users in achieving their objectives in this environment.

The central characteristic that justifies the adjective "smart" is the environment's capability to select its actions based on the user's objectives and not just on the current sensor data. The notion of objectives implies that the environment has to have a certain level of understanding of the *user's* view of the world.

The concept of *assistance* describes the benefit of smart environments for the user: they off-load work from the user. Today, the type of work off-loaded is primarily memory and control tasks. This results in a reduction of the characteristic cognitive "glitches," and their consequences, associated with these tasks: forgetting to turn the off the oven, forgetting one's point in a speech while fidgeting with the projector's remote control, wasting energy because of forgetting or being too occupied to turn off the bathroom heat, not doing video conference because of not wanting to take the pain of having to memorize a 200 page operation manual, etc.

The literature presents many typical examples of smart environments. Below we recall some of the scenarios considered in some of these examples.

- A quite early development, the *Reactive Video-Conferencing Room* (Cooperstock et al. 1997) is an example for streamlining the interaction with complex video conferencing systems based on predefined reaction patterns, which are triggered by sensors dispersed throughout the environment as described by the following scenario:

  > Just before noon, Nicole arrives at the university and enters the lab. The room lights turn on automatically and an audio message greets her. [...] An electronic calendar that has been awaiting her arrival then activates the presentation equipment and initiates a video connection with the conference room automatically. Nicole begins her presentation by placing a diagram under the document camera. The remote participants immediately receive a view of this diagram, along with a small 'picture-in-picture' of the presenter [...].

- The Intelligent Classroom (Franklin et al. 2002) aims at supporting a lecturer through anticipating his activities and adjusting the room's infrastructure in a suitable way, as outlined by the following scenario description:

  > The classroom observes the speaker walk away from the podium and over to the chalkboard [...], once he has reached the chalkboard and stopped, the Classroom adjusts the lights and sets the camera to show the portion of the chalkboard he is likely to write on.

- The objective of Microsoft's *EasyLiving* project (Brumitt et al. 2000) is to simplify the control of home infotainment infrastructures by automatically selecting devices based on the spatial relations between user and device location as described in the following scenario:

  > Tom is at home. He enters the living room sits down at a PC in the corner. He surfs through a selection of MP3's, and adds them to a playlist. He gets up and sits down on the couch. His session follows him to the large wall screen across from the couch. This screen is selected because it is available and in Tom's field of view.

- Michael Mozer's *Neural Network House* (Mozer 1998) tries to learn how to optimize both energy consumption and inhabitant satisfaction with respect to lighting and heating control as depicted by the following scenario:

  > We call the system ACHE, which stands for adaptive control of home environments. ACHE monitors the environment, observes the actions taken by occupants (e.g., adjusting the thermostat; turning on a particular con figuration of lights), and attempts to infer patterns in the environment that predict these actions. ACHE has two objectives. One is anticipation of inhabitants' needs. Lighting, air temperature, and ventilation should be maintained to the inhabitants' comfort; hot water should be available on demand. When

> inhabitants manually adjust environmental set points, it is an indication that their needs have not been satisfied and will serve as a training signal for ACHE. If ACHE can learn to anticipate needs, manual control of the environment will be avoided. The second objective of ACHE is energy conservation. Lights should be set to the minimum intensity required; hot water should be maintained at the minimum temperature needed to satisfy the demand; [...].

- Using a conceptually similar approach, Cook and Das (2005) aim in their *MavHome* project at additionally off-loading routine appliance operation tasks from the user as described in the following scenario:

  > At 6:45am, MavHome turns up the heat because it has learned that the home needs 15 minutes to warm to optimal waking temperature. The alarm sounds at 7:00, after which the bedroom light and kitchen coffee maker turn on. Bob steps into the bathroom and turns on the light. MavHome records this interaction, displays the morning news on the bathroom video screen, and turns on the shower. (Cook et al. 2003).

- In the spirit of earlier work at MIT on intelligent environments (Coen 1998), the *AIRE* sub-project of MIT's *Oxygen* initiative emphasizes multi-modal interaction with the environment, as outlined by the following scenarios:

  > [...] 'Alright then,' Alice asserts, 'let's get this going. Computer – —start the meeting.' Back in the conference room, the meeting agenda is projected onto a wall, and the first agenda item is highlighted: 'David's Presentation on ≪Adapting Traditional Games in Intelligent Environments: iBoggle.≫'
  >
  > [...] At this point, your computer beeps to remind you that you only have two minutes left in the presentation. You skip ahead to your last slide and summarize your major points. 'Thank you, David, for the presentation,' Alice remarks. 'Computer, move on to the next agenda item.' On the agenda display, which is back to the front of the navigation panel, 'David's Presentation' is checked off. The next agenda item, 'Set up the New Product Focus Group,' is now highlighted. [...] (Hanssens et al. 2002).

Below we list some interesting examples of projects that aim at investigating different aspects of smart environments. This list is by no means intended to be comprehensive. Links to additional literature can be found at the AAAI web page on smart rooms (AAAI, online).

- In the EMBASSI project Herfet et al. (2001) aimed at providing speech and gesture interaction with everyday appliances.

- The *Anthropomorphized Product Shelf* presented in Chap. 15 of this book enables objects in the environment to engage in multi-modal interaction with the user. The goal is to simplify getting background information on goods in shopping scenarios.
- The *iCat* home dialog system presented in Chap. 14 of this book looks at strategies for exploiting social intelligence as a means for manipulating the user's perception of system quality and the user's acceptance.
- Stanford's *iRoom* (Johanson et al. 2002) concentrates on providing *direct manipulative* interaction metaphors for smart environments. For instance, using a 2D floor plan of the iRoom, users can control devices by selecting them from the floor plan visualization on a PDA.

It is interesting to note that the projects outlined above can be grouped into the following three distinct classes, based on their preferred primary *interaction metaphor*:

- Implicit interaction: Reactive Video-Conference Room, Intelligent Classroom, EasyLiving, Neural Network House, and MavHome.
- Explicit interaction that addresses an environment proxy: AIRE (proxy: "Computer"), iCat (proxy: "dialogical robot"), and iRoom (proxy: PDA).
- Explicit interaction with individual appliances/objects: Anthropomorphized Product Shelf.

The environments are ordered based on an increasing visibility of the objects to be controlled by the user and on increasing interaction requirements on behalf of the user.

17.3 Building Smart Environments: the Ensemble Challenge

17.3.1 Another Scenario ...

As can be gleaned from the previous section, the *smart conference room*, or *smart living room* for consumer-oriented projects as shown in Fig. 17.1, which automatically adapts to the activities of its current occupants, is a typical example of smart environments. Similar to the *Intelligent Classroom* scenario, it might automatically switch the projector to the current speaker's presentation as she approaches the lectern – for the smart living room this reads: "switch the TV set to the user's favorite show, as he takes seat on the sofa." – and subdue the room lights, turning them up again for the discussion. Of course, we expect the environment to automatically fetch the presentation from the speaker's notebook. And the speaker should be able to use her own wireless presentation controller to move through her slides, although she might just as well choose to pick up the lectern's presentation controller.

Such a scenario does not sound too difficult. It can be readily constructed from common hardware available today, and, using pressure sensors and RFID

Fig. 17.1. Environments we would like to smart: conference rooms (left) and living rooms (right)

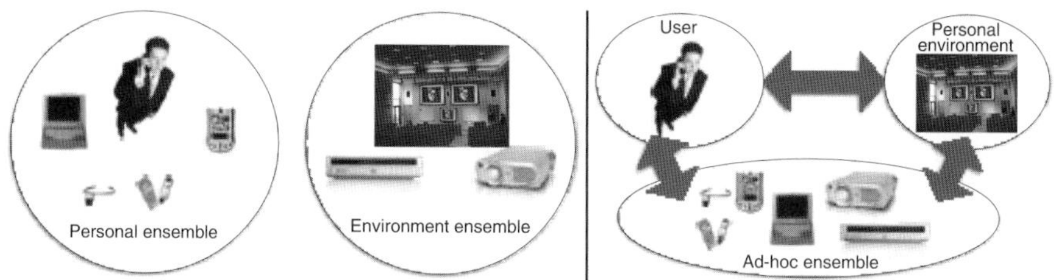

Fig. 17.2. Appliance ensembles: physical constituents (left) and compound ad-hoc ensemble (right)

tagging, does not even require expensive cameras and difficult image analysis to detect who is currently at the lectern. Setting up the application software for this scenario that drives the environment's devices in response to sensor signals does not present a major hurdle either. So it seems as if smart environments are rather well understood, as far as underlying information technology is concerned. Details like image and speech recognition, as well as natural dialogs, of course need further research, but building smart environments from components is technologically straightforward, once we understand what kind of proactive behavior users will expect and accept.

17.3.2 ... and Its Implications

The above assertion only holds true as long as the device ensembles, that make up the environment, are anticipated by the developers; see Fig. 17.2. Today's smart environments in the various research labs are usually built from devices and components whose functionality is known to the developer. So, all possible interactions between devices can be considered in advance and suitable adaptation strategies for coping with changing ensembles can be defined. When looking at the underlying software infrastructure, we see that the software engineers that have built this scenario have carefully handcrafted the interaction between the different devices, i.e., the "intelligence." This means that significant changes of the ensemble require a manual modification of the smart environment's control application. One may wonder whether the ensemble could simply learn how to use the new component. This however is hard as is argued in Sect. 17.4.

This is obviously out of the question for real-world applications where people continuously buy new devices for embellishing their home. And it is a severe cost factor for institutional operators of professional media infrastructures such as conference rooms and smart offices. Things can be even more challenging: imagine a typical ad-hoc meeting, where some people meet at a perfectly average room. All attendants bring notebook computers, at least one brings a projector, and the room has some light controls. Of course, all devices will be accessible by wireless networks. So it would be possible for this chance ensemble to provide the same assistance as the deliberate smart conference room above. Enabling this kind of self-organization, the ability of devices to configure themselves into a coherently acting ensemble, requires more than setting up a control application in advance. Here, we need software infrastructures that allow a true self-organization of ad-hoc appliance ensembles, with the ability to afford non-trivial changes to the ensemble; see also Servat and Drogoul (2002) for a similar viewpoint on this topic.

17.3.3 Significant Changes

The attentive reader might have begun to wonder what "significant" is supposed to mean here. After all, services such as *Universal Plug and Play* (UPnP) allow us to dynamically discover devices that provide a specific functionality, and even to rediscover alternative service providers once a certain device becomes unavailable. So, is the problem raised above not already solved? Let us consider an abstract example. Assume we need a service s named `moveFromAtoB`. And suddenly, this service has become unavailable. But there are other services, for instance p, named `moveFromAtoQ`, and q, named `moveFromQtoB`. So, to a human reader it might seem as if we could compensate for the missing service `moveFromAtoB` by combining the existing services `moveFromAtoQ` and `moveFromQtoB`, such that $s = q \circ p$. But this ad-hoc combination of services is a feature service discovery mechanisms are not designed for. The "intelligence" for service decomposition then needs to be provided by the requesting agent, rendering it private property and not a general capability of ensembles. Advanced multi-agent infrastructures such as the Open Agent Architecture of Martin et al. (1999) indeed look at providing mechanisms for service decomposition at the infrastructure level rather than at the individual agent level.

Now consider another situation: imagine, there are services named `moveFromAtoQ` and `moveFromQtoB`. Nobody so far has thought about a service named `moveFromAtoB`, whose *meaning* can be defined by the composition of p and q, which are denoted by the name `moveFromAtoQ` and moveFromQtoB,

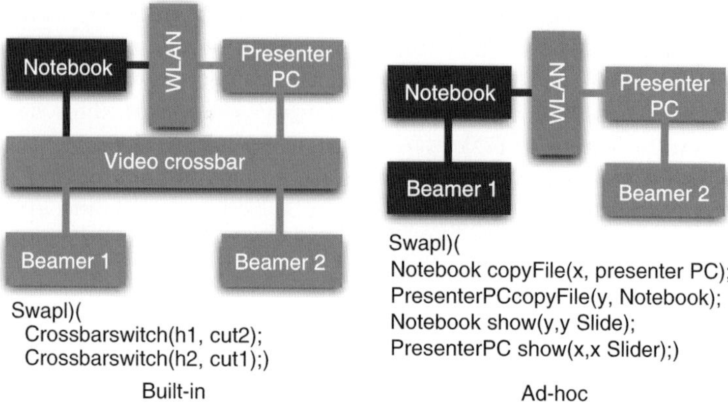

Fig. 17.3. Achieving the same effect with different ensembles

respectively. That is, the service s is available in principle, but we do not have a name to ask for it! In any ensemble, where p and q are available, s emerges naturally. But without a suitable name, we are not able to access it using conventional service discovery mechanisms that operate syntactically, on the *names* (more general: type signatures) of services. In order to cope with the discovery of emergent services as well as with service decomposition in general, we need a stronger mechanism for discovering and combining services, a mechanism that is based on the *semantics* of services rather than on their names.

An example from the conference room domain is outlined in Fig. 17.3, where the built-in infrastructure of two hypothetical conference rooms is displayed (greenish boxes). The room at left provides two beamers and a video crossbar, enabling a rather straightforward way for swapping two presentations. At right, the conference room just contains a single beamer; an attendee has presumably provided the second one. In both sketches, the reddish boxes denote components that have been added dynamically.

Clearly, both conference rooms require two significantly different strategies for realizing the user's goal of swapping two presentations. And, while in the built-in case one maybe could expect that the room designer has provided a suitable macro, this is not realistic for the ad-hoc situation: no designer of a smart room can be expected to anticipate every possible ad-hoc extension of the built-in infrastructure and to provide control strategies for every possible activity that could be performed with the thus extended ensemble.

When looking at the challenges of self-organization indicated above, we can distinguish the following two different aspects:

– *Architectonic Integration* – refers to the integration of the device into the communication patterns of the ensemble. For instance, the attachment of an input device to the ensemble's interaction event bus.

- *Operational Integration* – describes the aspect of making new functionality provided by the device (or emerging from the extended ensemble) available to the user. For instance, after connecting the second beamer + notebook to the right ensemble in Fig. 17.3, the capability of swapping presentations will now emerge from the ensemble. This can be termed the task of ensemble strategy generation.

Clearly, both aspects eventually have to be accounted for by a smart environment software architecture. First, we will look at the problem of strategy generation.

17.4 The Source of Strategy

How do the projects on smart environments outlined above handle the task of ensemble strategy generation? The following observations can be made with respect to this question:

- Reactive Video Conference Room. Relies on the system designer to define suitable strategies in the form of *divide demons*.
- Intelligent Classroom. Relies on system designer. Strategies are defined by providing models of user plans, as hierarchical finite state machines. The system keeps track of which plans are compatible with the current observation sequence of user behavior (*plan recognition)* and performs assistant actions appropriate for the current plan(s). A potentially more powerful system for plan recognition based on hierarchical Hidden Markov mdels has been investigated by Bui (2003).
- EasyLiving. Relies on system designer. Strategies are defined as condition–action rules in the system's rule base, which are matched against the current context data.
- Neural Network House/ACHE. Uses reinforcement learning (kaelbling et al. 1996), more specifically model-free Q learning, for learning a control policy that optimizes inhabitant comfort and energy efficiency. A neural net is used for learning (and predicting) the occupant's movement pattern.
- MavHome. Tries to learn an LZ predictor, more specifically an older k Hidden Markov model, for predicting the next user action from an observed prefix sequence. MavHome therefore essentially learns control strategies from the user. In addition, reinforcement learning is used for teaching a decision component if taking the predicted action makes sense with respect to user satisfaction and energy efficiency.

Now, from the discussion in Sect. 17.3.3, it is clear that we cannot expect the *system designer* to provide an ensemble with strategies that allow it to perform complex requests of behalf of the user. Therefore, approaches such as EasyLiving or the Intelligent Classroom are not viable for the case, where a dynamic ensemble provides the environment's capabilities.

On the other hand, the approach taken by MavHome, to learn strategies from the user, is not an option either: if a substantial set of devices is *invisible*to the user, they can obviously not become part of a control strategy the user might develop. Therefore, a system cannot learn from the user how and when to use these devices.

Since in dynamic ensembles neither system designer, nor system user have an overview over the complete ensemble and its potential, there is no human being that could provide strategies to this ensemble.

Either, the user has to be made aware of the available devices and their potential (pushing the responsibility back to the user), or the ensemble *itself* must become able to develop strategies on its own, based on the user's objectives. With respect to this, it should be noted that the systems developed in the above projects have *no* explicit notion of the user's objectives: they learn *procedures* from the user (or receive them from the system designer), but they have no concept of the *effect* of these procedures with respect to the user's objectives. This topic will be addressed in the next section.

17.5 Goal-Based Interaction

When interacting with appliances in everyday settings, we are used to think of interaction in terms of the individual functions these devices provide: functions such as "on", "off", "play", "record", etc. When interacting with devices, we select, parameterize, and then execute functions these devices provide. Upon execution, they cause an effect: a broadcast is recorded on videotape; the light is turned brighter, and so on.

But then, a user is not really interested in the function he needs to execute on a device – it is rather the function's effect, which is important.

This observation immediately leads to the basic idea of goal-based interaction. Rather than requiring the user – or a system designer – to invent a sequence of actions that will produce a desired effect (goal-)based on the given devices and their capabilities, we should allow the user to specify just the goal ("I want to see 'Chinatown' now!") and have the *ensemble* fill in the strategy leading to this goal. (Find the media source containing media event "Chinatown". Turn on the TV set. Turn on the media player, e.g., a VCR. Position the media source to the start of the media event. Make sure the air condition is set to a comfortable temperature. Find out the ambient noise level and set the volume to a suitable level. Set ambient light to a suitable brightness. Set the TV input channel to VCR. Start the rendering of the media event.) Goals allow services to be named by their *semantic*, i.e., by the effect they have on the user's environment – thereby evading the problems of syntactical service addressing outlined in Sect. 17.3.2.

Goal-based interaction requires two functionalities: *Intention Analysis*, translating user interactions and context information into concrete *goals*, and

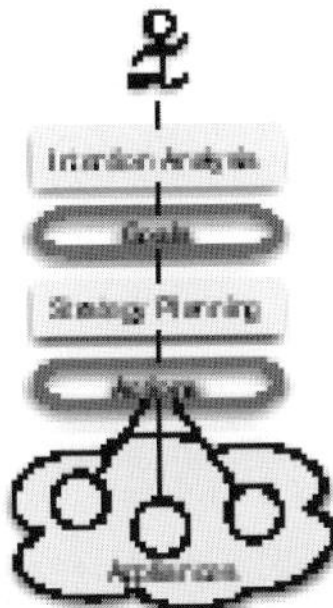

Fig. 17.4. Principle of goal-based interaction

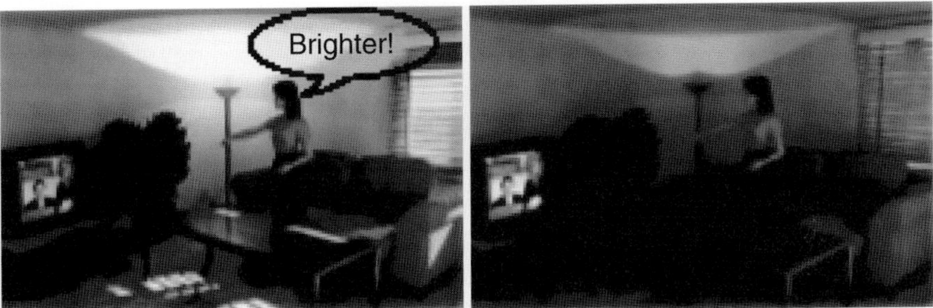

Fig. 17.5. Goal-based ensemble control

Strategy Planning, which maps goals to (sequences of) device operations; see Fig. 17.4.

Note that goal-based interaction is able to account for the operational integration called for above, while allowing the pitfalls outlined in Sect. 17.4: operational integration can now be realized based on an explicit modeling of the *semantics* of device operations as *precondition/effect rules*, which are defined over a suitable environment ontology. This environment ontology serves as a mechanism for explicitly representing user objectives. These rules then can be used by a planning system for deriving strategies for reaching user goals, which consider the capabilities of all currently available devices.

As example, consider the situation outlined in Fig. 17.5, left, where a user would like to increase the brightness of his TV set. Assuming the TV is already set to maximum brightness, the sensible reaction of the ensemble would be the one given at right: reduce ambient light. In order for an ad-hoc ensemble to arrive at this conclusion, TV set, lamp, and shutter must provide a description of their capabilities, similar to the one given below. For the sake of brevity, this capability definition has been very much simplified.

```
; lamp's action
(:action turn-down :parameters (?l - lamp)
  :precondition (luminosity ?l high)
  :effect (and (not (luminosity ?l high)) (luminosity ?l
low)))

;shutter's action
(:action close :parameters (?s - shutter)
  :precondition (and (time day) (open ?s))
  :effect (and (not (open ?s)) (closed ?s)))
(:axiom :vars (?s - shutter)
  :if (or (time night) (closed ?s))
  :then (luminosity ?s low))

;tv set action
(:action turn-brighter :parameters (?b - tv-set)
  :precondition (or (< (brightness ?t) max-brightness)
      (forall ?x (luminosity ?x low)))
  :effect (and (= (brightness ?t) max-brightness)
(brighter achieved)))
```

Then, based on a specific situation given by

```
(:objects l - lamp s-shutter - t tv-set)
(:init (= brightness t max-brightness) (time day) (open s)
(luminosity l high),
```

a suitable a plan for the goal `(brighter achieved)` could then be computed as `((close s) (turn-down l))`. Additional details on such a planning-based approach can be found in Heider and Kirste (2002b).

It should be noted that the reinforcement learning strategies used by MavHome and ACHE in some sense already employ an explicit model for effects: their environment model consists of two parameters, user satisfaction and energy consumption. Reinforcement learning is then used by the system to learn how the available operations affect these parameters. However, the approach proposed in this section enables strategy generation by the ensemble without a learning phase, which is crucial for dynamic ensembles. This specifically holds, when effect state spaces have to be considered that are more complex than just two environment parameters.

The crucial question for an approach based on an explicit effect model is of course the development of a suitably complete environment model that identifies the variables required for describing the preconditions and effects of operators. Here substantial additional research is required.

After looking at the topic of *operational* integration, we now briefly discuss the question of *architectonic* integration: the communication patterns in a dynamic ensemble that need to be supported.

17.6 Appliances and Event Processing Pipelines

17.6.1 Looking at Appliances

When developing a middleware concept, it is important to consider the objects that are to be supported by this middleware. For appliance ensembles, this means we have to look at *physical devices*, which have at least *one* connection to the physical environment they are placed in: they observe user input, or they are able to change the environment, e.g., by increasing the light level, by rendering a medium, or both. When looking at the event processing in such devices, we may observe a specific *event processing pipeline*, as outlined in Fig. 17.6, devices have a user interface component that translates physical user interactions to events, the Control Application is responsible for determining the appropriate action to be performed in response to this event, and finally the Actuators are physically executing these actions. It seems reasonable to assume that *all* devices employ a similar event-processing pipeline, even if certain stages are implemented trivially, being just a wire connecting the switch to the light bulb.

It would then be interesting to extend the *interfaces* between the individual processing stages across multiple devices, as outlined in the right side of Fig. 17.6. This would allow a dialog component of one device to see the input events of other devices, or it would enable a particularly clever control application to drive the actuators provided by other devices. By turning the private interfaces between the processing stages in a device into public *channels*, we observe that the event-processing pipeline is implemented cooperatively by the device ensemble on a per-stage level. In order to make ensemble dynamics transparent to message senders, these spread interfaces clearly need to support a kind of *content-based routing* mechanism that also is able to handle the competition between listeners for messages. Each pipeline stage is then realized

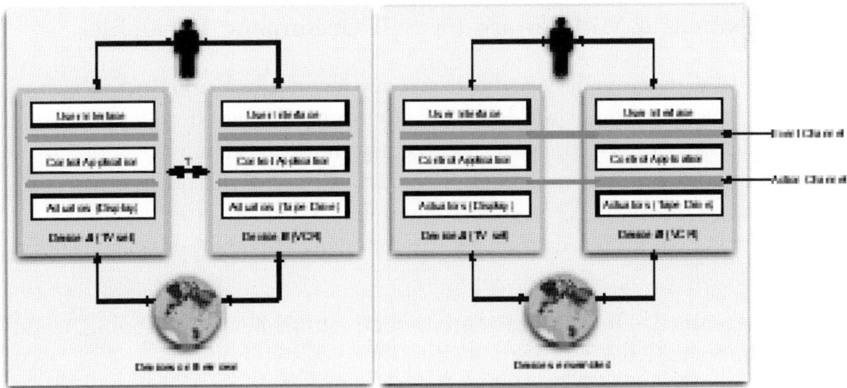

Fig. 17.6. Devices and data flows

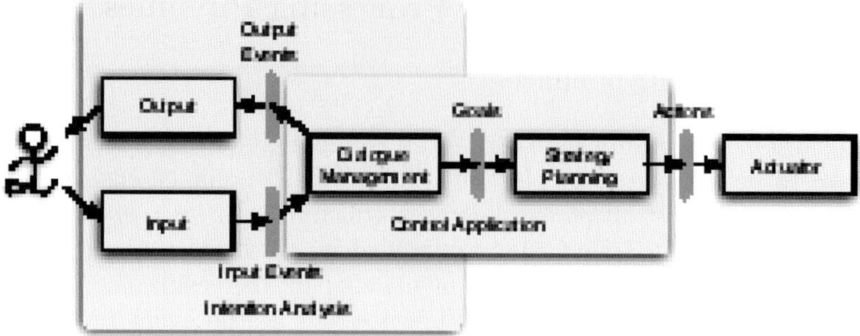

Fig. 17.7. Adding goals to the appliance pipeline

through the cooperation of the respective local functionalities contributed by the members of the current ensemble.

With respect to the information processing inside appliances as outlined here, the two functionalities required for goal-based interaction, i.e., intention analysis and strategy planning, can be interpreted as components of the control application, resulting in the extended processing pipeline shown in Fig. 17.7.

So, our proposal for solving the challenge of architectonic integration is to provide a middleware concept that provides the essential communication patterns of such data-flow based multi-component architectures. Note that the channels outlined in Fig. 17.6 and Fig. 17.7, respectively, are not the complete story. Much more elaborate data processing pipelines can be easily developed, as outlined by Heider and Kirste (2002b). Therefore, the point of such a middleware concept is not to fix a *specific* data flow topology, but rather to allow *arbitrarily* such topologies to be created ad hoc from the components provided by the devices in an ensemble.

17.6.2 Towards a Middleware for Self-Organizing Ensembles

In this section, we outline the basic properties of the SodaPop infrastructure, a prototype middleware for AmI environments based on appliance ensembles. SodaPop development is currently pursued within the scope of the DynAMITE project. For more details on the DynAMITE project the reader is referred to DynAMITE, online; Hellenschmidt and Kirste (2004).

The SodaPop model introduces the following two fundamental organization levels:

- Coarse-grained self-organization based on a data-flow partitioning, as outlined in Sect. 17.6.1
- Fine-grained self-organization of functionally similar components based on a kind of pattern-matching approach.

Consequently, a SodaPop system consists of the following two types of elements:

- *Channels*, which read a single message at time *point*and map them to multiple messages, which are delivered to components (conceptually, *without delay*). Channels have no externally accessible memory, may be distributed, and they have to accept *every*message. Channels provide for *spatial distribution* of a single message to multiple transducers. The specific properties of channels enable an efficient distributed implementation.
- *Transducers,* which read one or more messages during a time *interval* and map them to one or more output messages. Transducers are *not*distributed, they may have a memory and they do not have to accept every message.

SodaPop provides the capability to create channels – message busses – on demand. On a given SodaPop channel, messages are delivered between communication partners based on a refined publish/subscribe concept. Every channel may be equipped with an individual strategy for resolving conflicts that may arise between subscribers competing for the same message, i.e., the same request.

Once a transducer requests a channel for communication, a check is performed to see whether this channel already exists in the ensemble. If this is the case, the transducer is attached to this channel. Otherwise, a new channel is created. Through this mechanism of dynamically creating and binding to channels, event-processing pipelines emerge automatically, as soon as suitable transducers meet.

When subscribing to a channel, a transducer declares the set of messages it is able to process, how well it is suited for processing a certain message, whether it allows other transducers to handle the same message concurrently, and if it is able to cooperate with transducers in processing the message. These aspects are described by the subscribing transducers' utility, which encodes the subscribers' handling capabilities for the specific message.

When a channel processes a message, it evaluates the subscribing transducers' handling capabilities and then decides, which transducer(s) will effectively receive the message. Also, the channel may decide to decompose the message into multiple, presumably simpler, messages, which can be handled better by the subscribing transducers. Obviously, the transducers then solve the original message in cooperation.

How a channel determines the effective message decomposition and how it chooses the set of receiving transducers is defined by the channel's decomposition strategy. Both the transducers' utility and the channel's strategy are eventually based on the channel's ontology, the semantics of the messages that are communicated across the channel. A discussion of specific channel strategies for SodaPop is out of the scope of this chapter. It may suffice that promising candidate strategies for the most critical channels – the competition of Dialogue Components for Input Events, the competition of Strategists for goals on the Goal Channel, and the cooperative processing of complex output requests – have been developed and are under investigation; see Heider and Kirste (2002a) and Hellenschmidt and Kirste (2004) for further details on SodaPop.

17.6.3 Ensemble Organization by SodaPop

To summarize, self-organization in SodaPop is achieved by the following two means:

1. Identifying the set of channels that completely cover the essential message processing behavior for any appliance in the prospective application domain.
2. Developing suitable channel strategies that effectively provide a distributed coordination mechanism tailored to the functionality, which is anticipated for the listening components.

Then, based on the standard channel set outlined in Fig. 17.4, any device is able to integrate itself autonomously into an ensemble, and any set of devices can spontaneously form an ensemble.

Currently, SodaPop is available as experimental software from the DynAMITE web site (DynAMITE, online). Formation of an ensemble based on experimental SodaPop is outlined in Fig. 17.8. On the left, the individual appliances are shown, where the green boxes symbolize their hardware packages. This example contains a stereo and a TV set, both with standard WIMP-type user interfaces, a solitary speech input, a solitary display, and a solitary avatar, possibly on a mobile display. For all devices, their internal transducers and channel segments are shown. On the right, the resulting "ensembled" appliance set is shown, after the matching channel segments have linked up by virtue of SodaPop. Note how the vertical overall structure at left has been replaced by a horizontal overall structure. Note

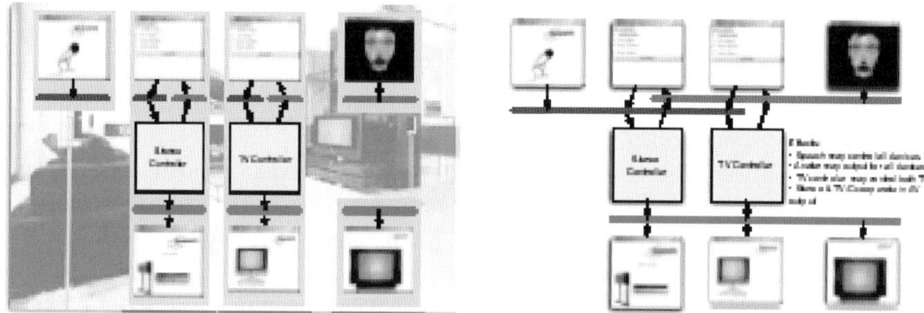

Fig. 17.8. A simple Smart Living Room Ensemble before assembly (left) and after topology (right)

also, that now stereo and TV both afford speech control, output may be done anthropomorphic through the avatar by all components, and the audio for a movie will be automatically rendered by the stereo system, winning competition with the TV-set's audio system by offering a higher quality.

17.7 Conclusion

Smart environments promise to enable ubiquitous computing technology to provide a new level of assistance and support to the user in his daily activities. An ever-growing proportion of the physical infrastructure of our everyday life will consist of smart appliances. In our opinion, an effective realization of smart environments therefore inherently requires to address the challenge of self-organization for ad-hoc ensembles of smart appliances.

We argue that a possible solution should be based on the fundamental concepts of goal-based interaction and self-assembling distributed interaction pipelines. Concepts such as SodaPop have given evidence that such an approach indeed is viable. We do not expect the solution proposal we have outlined above to be the only possibility. However, we hope that we have convinced the reader that there is *at least one* possible and sufficiently concrete approach towards solving the substantial challenges of dynamic ensembles, which are raised by the proliferation of ubiquitous computing technology.

A recent investigation of software infrastructures for ubiquitous computing (Endres et al. 2005) shows that there is a growing number of researchers looking at the software issues of ubiquitous information systems, from a wide range of different perspectives. It makes clear that the topics raised here are just a small fraction of the problems and challenges that have to be addressed in order to make ubiquitous computing as invisible and intuitive as it is called for in the well-known visions. Enabling truly spontaneous and smart cooperation between ubiquitous multimedia appliances will remain a major challenge for future research. In a similar vein, it remains an interesting question, how the user's mental model of the infrastructure influences the system acceptance and the perceived system quality. Chapter 14 looks at the option of using dedicated embodiments in user-interface robots of the system for focusing on user interaction.

18

Sensory Augmented Computing

B. Schiele

"Computing belongs in furniture and foot-ware much more than it does on the desktop."

MIT Media Lab 1995

18.1 Introduction

The vision of Ambient Intelligence is to embed computing and communication capabilities into nearly everything, namely the environment, objects, or even clothing. Great advances in mobile computing, communication, and device technology for example allow to access a large variety of computing services without the constraint of sitting in front of a desktop computer or being in a particular smart or intelligent environment. However, many research challenges remain. A particularly challenging research topic within Ambient Intelligence is the question of how to interact unobtrusively and in a seamless way with users. Quite obviously current desktop interaction techniques do not generalize well to the more versatile settings of Ambient Intelligence.

In order to realize the vision of Ambient Intelligence context awareness is often seen as a means to make the computing tasks sensitive to the situation and the user's needs. Ultimately, context awareness may support and enable seamless interaction and communication between human users and ambient intelligent computing environments. The notion of implicit interaction, for example, suggests to sense "an action, performed by the user that is not primarily aimed to interact with a computerized system but which such a system understands as input" (Schmidt 1999). That means, the user interacts with physical objects in a natural way, but a computer system also can extract inputs from these actions. Others such as Hinckley et al. (2000), Schmidt et al. (1999), Rekimoto (1996) and Harrison et al. (1998) propose physical interaction, e.g., tilting a device for configuring a device's functionality, as new and convenient forms of interaction for mobile user scenarios. System input generated from interaction with physical objects has been used for coupling physical objects with computer applications such as tangible user interfaces (Ishii and Ullmer 1997), computer-assisted furniture assembly (Antifakos et al. 2002), tracking a patient's medicine cabinet (Siegemund and Flotifakos 2003) or workflow monitoring in a chemical lab (Arnstein et al. 2002).

Empowering a computer system to process physical user inputs requires augmentation of today's computer nerve-endings, such as mouse and keyboard, by sensors: perception, reasoning, and interpretation of real world phenomena enable computer systems to observe the user's physical environment and serve the user in more appropriate ways than it is possible today. Current technology offers an impressive range of sensors and sensor modalities. Furthermore, it is also widely believed that many sensors will become so cheap and small that they can be deployed unobtrusively and in large numbers. Computing which has access and makes use of this vision of ubiquitous sensors is what I call sensory augmented computing.

This chapter is devoted to Sensory Augmented Computing since the accessibility to a large variety and diversity of sensor information has great potentials to change the way we interact with computers. The general vision is that the use of sensors and elaborate perception techniques will play an important role in order to derive interesting and high-level context. Using a multitude of sensors, distributed throughout the environment may enable applications to be aware of the situation of the environment and the users. Obviously, a computer interface which uses user models, contextual, and situational information to its fullest is a long-term research goal. The chapter starts with an overview and classification of interesting research in the area of sensing for Ambient Intelligence. Then we describe in more detail one of our own sensory augmented computing research projects, namely multi-sensor context awareness for proactive furniture assembly.

18.2 Sensing Opportunities for Ambient Intelligence

As mentioned above, we strongly believe that computers should have access to a large variety of sensors in order to see, hear, interpret, and eventually understand more about humans. Already today, there exist many sensors and sensing devices. Besides the prominent examples of vision and audio sensors there exist a large variety of other sensor modalities, which could be embedded in many objects and devices. This section gives an overview of some work related to sensory augmented computing for Ambient Intelligence. For the discussion we will characterize sensing opportunities with two criteria: the logical view and the physical view.

18.2.1 The Logical View: Dimensions of Sensing

The first criteria we use to characterize various sensing technologies is the type of information that can be extracted from them. Since we are particularly interested in information those sensors contain about humans we concentrate on various aspects of "human-sensing." We have identified the following six different aspects (Michahelles and Schiele 2003): human ID, object usage, location, bio signs and emotions, human activity, and interaction among different humans.

Human or user *ID* has been widely used, e.g., for customizing and personalizing services without requiring explicit user inputs (Richardson et al. 1994; Bohnenberger et al. 2002). In fact, we use a more general definition of *ID* ranging from differentiating people to actual person identification. The second dimension is *location*, which is the most prominent and widely used form of context information (Abowd et al. 1999) used in ubiquitous computing applications such as Want et al. (1992) and Davies et al. (1998). It does include 3D coordinates but also semantic location descriptions. The third dimension, *activity*, describes the task the user is performing which ranges from simple moving patterns (Van Laerhoven et al. 2001) to precise job descriptions. The fourth dimension, *object use*, comprises collocation of a user to an object (Richardson et al. (1994), carrying an object (Langheinrich et al. 2000) and the actual use (Antifakos et al. 2002). The fifth dimension, *bio signs/emotions*, describes the internal state of the user. Research in this area is still in its infancy. First results could be obtained with heart rate and skin-resistance, for reasoning about a user's affects (Picard and Klein 2002). The sixth dimension, *human interaction*, characterizes the relationship between humans including simple collocation, listening to a speaker, gazing, and actual interaction as discussion.

18.2.2 Physical View: Placement of Sensors

We differentiate four different sensor placements. *In environment* refers to stationary installed sensors, e.g., in the floor, walls, where placement can only be changed with effort. Whereas *in environment* installations work with all users at the stationary location *on human* has the opposite characteristics: only users wearing the sensors can participate, therefore they are not bound to a location. *On object* is in between the two previous categories, as objects can be personal and can be carried by a human, such as a key, but also stay at a certain location, e.g., chair. This distinction depends on the object. Additionally, *mutual collaboration* defines sensing system that always require more than one unit in order to operate properly, e.g., triangulation of signal strength for localization.

Using the six logical dimensions of sensing and using the four sensing placements (*in environment*, *on object*, *on human*, and *mutual collaboration*) the following discusses some work of sensory augmented computing for Ambient Intelligence.

18.2.3 Sensors in AmI Research

For recognizing a person's *ID* the best results can be achieved with biometric sensors (Wayman et al. 2003), such as fingerprint or iris scan. Methods based on vision (Donato et al. 1999), audio, or load-cells embedded into the floor (Cattin 2001) deliver less quality. Inertial sensors placed *on object* and *on human* can be used to sense typical movements, e.g., perceiving the signature at a pen, for identification. Scheirer et al. 1999) report on using vision. Kern et al. (2002) report on using audio worn *on human* for people identification. Location systems as described by Hightower and Borriello (2001) can also be used for identifying people at different locations. These systems require both sensors worn by humans and installed base stations.

For detecting *object use* load-cells (Schmidt et al. 2002) have been proven useful installed both *in environment* and *on object.* Object classification with vision is well established in static settings, occlusion during dynamic use can be challenging. Audio is another option, if the *object use* generates characteristic sounds. Inertial force sensors placed *on object* have been successfully used for *object use* as reported by several authors including Hinckley et al. (2000), Schmidt et al. (1999), Rekimoto (1996), and Harrison et al. (1998). Obviously, motion during *object use* can be also sensed *on human* but with less quality. Audio *on human* is also possible (Lukowicz et al. 2002) but is an indirect measurement compared to *on object* placement. Location systems can give hints as well for *object use*, e.g., teleporting X Windows to user's current location (Richardson et al. 1994).

Location is the most explored sensing dimension in ubiquitous computing. Load-cells (Schmidt et al. 2002), vision (Brumitt et al. 2000), and audio (Darrell et al. 2001) have been explored in different projects. Coarse location can be also gained through passive-infra-red sensors, mechanical switches, or IR-barriers. *On object* and *on human* the primary outdoors is GPS, more low-level information delivers humidity, inertial, or pressure sensors (Vildjiounaite et al. 2002). The variety of location systems based on *mutual collaboration* is huge: differential GPS, ultra-sound, radio, etc. There exist various systems that integrate several of the standard techniques such as GPS, GSM, or WLAN; see for instance LaMarca et al. (2004) and Fox (2003).

Sensing *bio signs/emotions* with *in environment* sensor-settings is difficult: Donato et al. (1999) and Fernandez and Picard (2003) report on vision and audio for reasoning on user's *bio signs/emotions.* Augmented objects measuring force and touch (Ark et al. 1999) can give some hints about *bio signs/emotions.* However, most promising are *on human* measurements such as reported by Healey and Picard (1998) and Michahelles et al. (2003).

Activity can be well sensed with special purpose system, such as commercially available smart white boards. Load-cells, passive infrared, pressure and capacity sensors can be used for low-level detection only. *On human* sensing has been well explored for motion activity (Farringdon et al. 1999). Location system can give hints reasoned from semantical location descriptions.

Interaction among humans has not been explored very well. *In environment* sensing systems based on vision, load-cells, and audio could help to perceive characteristics of interaction, such as collocation, gestures, or speech. The *on object* field is blank, as objects are not involved here. *On human* the same sensors can be used as for *activity* if measurements are correlated among interactors. Location systems do not really help here, as collocation is not significant for interaction.

18.2.4 Discussion

Quite interestingly each sensor placement *in environment*, *on object*, *on human* and *mutual collaboration* is meaningful for at least one of the six sensing dimensions. *In environment* placement is the primary choice for *ID* sensing. Regarding the other five sensing dimensions the power of *in environment* placement is mainly based on video and audio methods. However, the perception quality relies on computational expensive methods. Nevertheless, once an environment has been augmented with sensors, e.g., Smart Rooms, applications work without additional instrumentation of users or objects. It also can give hints for human–human interaction. As our focus is on human sensing it is obvious that *on object* points out useful for *object use*. As physical interaction with everyday objects mostly involves movements, such as grasping, moving, or turning the dominant sensor choice for *object use* are inertial sensors and force strips to a certain extent. *On human* is suited for direct measurements of human-centric sensing aspects, such as *bio signs/emotions* and *activity*. Applicable sensors include inertial sensors, audio, biosensors, and also video to a certain extent. *Mutual collaboration* sensors, such as the location systems, have similar characteristics as video and audio with even lower quality: location can provide coarse information about *object use*, *activity*, and *in environment* due to the strong implications of physical location. However, in direct comparison with *on object* and *on human* sensing location system are in an inferior position.

18.3 Proactive Furniture

As an example of sensory augmented computing this section describes a context-aware system enabling proactive instructions for assembly. We demonstrate the system with flat pack furniture but the approach generalizes to a variety of cases where there exist instructions that need, should, or could be followed by a person.

Proactive instructions are taken as example for various reasons. First of all, many of today's instructions, handbooks, and even reference manuals are rarely used even though many of us would and could profit from getting the appropriate instructions at the appropriate moments in time. Secondly, the level of instructions required varies depending on particular person. So the system should proactively adapt its instructions to the current user. Thirdly, most of today's instructions are mostly linear in the sense that they do not model and allow variations in the way or the order people perform actions. This is a common problem for many paper-based but also computer-based instructions. And fourthly, since instructions are typically detached from the physical object the users have to make the connections between the "virtual instructions" and the physical objects and actions themselves. For all those reasons modeling the various states and actions of an assembly, recognizing them using multiple sensors embedded in the involved physical objects, and giving appropriate instructions and feedback depending on the actions performed by the user is a highly promising approach to overcome many problems with today's instructions.

As the running example for proactive instructions or more generally for proactive guidance we chose the example of presenting instructions during the assembly of *Do-It-Yourself* furniture. We use the parts of the assembly as the interface to the instructions, by sensing what the person is doing and how far he or she has got with the assembly. Quite obviously the idea of perceiving the user's actions and presenting instruction based on the actions applies to many other applications. The Labscape Project, described by Arnstein et al. (2002), is such an example, where the actions in a biological laboratory are monitored. Here, both a logbook of daily activities is created, and instructions are presented in situ. Other examples in the field of aircraft maintenance have also been discussed; see for instance Lampe et al. (2004). More examples will be presented later.

The two main questions and challenges we are addressing in the following section can be formulated as follows: (1) Can sensing be implemented reliably enough, so that the user can interact with such systems? (2) How can feedback be given to help the user understand what the system is doing? In our example, the second challenge translates into, how can the assembly instruction be presented. Section 18.4 gives an overview over the sensing task and the technology used. Section 18.5 presents our vision of situation-aware affordances and summarizes a user study in which we compare our prototype with traditional paper-based instructions.

18.4 Sensing a Furniture Assembly

This section starts off with a brief overview of the assembly instructions provided by IKEA. We propose our own assembly plan, offering more possible solutions of assembling the wardrobe. How this plan is used and which user actions need to be perceived is explained and demonstrated with experiments.

18.4.1 The IKEA PAX Wardrobe

The PAX wardrobe is a simple wardrobe that can be used for many different purposes. Many types of shelves can be inserted at different heights. Our discussion will be concerned with the assembly of the main wardrobe without its shelves.

The wardrobe consists of six wooden boards, two metal corners, cams, cam-bolts, dowels, screws, and nails. For a standard assembly of the wardrobe, a screwdriver and a hammer are the only tools required. If the wardrobe is mounted to the wall a drill is also needed. In Fig. 18.1 Steps 1 up to 6 of the assembly instruction are depicted. Steps 1 and 2 show the preparation of the

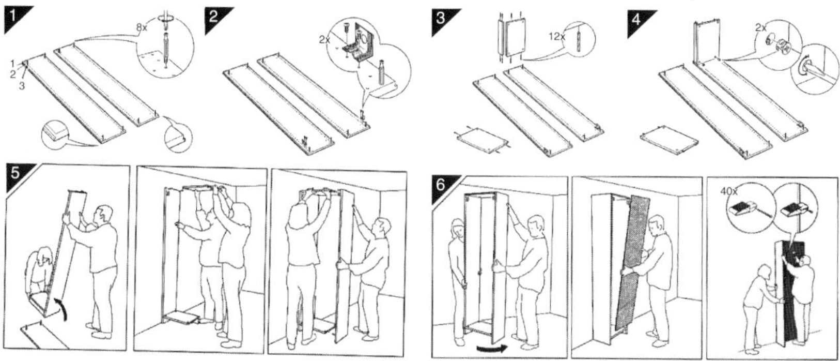

Fig. 18.1. Steps 1 to 6 of the IKEA assembly instructions. Steps 1 and 2 (upper row, left) show the preparation of the two sideboards. Steps 3 and 4 (upper row, right) show how one horizontal board and the base strip are attached. Steps 5 and 6 (lower row) of the assembly instructions show how the compound is lifted into an upright positions, the remaining sideboard and horizontal board are attached, and the back panel is nailed on (Reproduced with the permission of the IKEA corporation.

two sideboards. The first step is to insert the four cam-bolts in each board at the right positions. Then the two metal corners have to be attached. Steps 3 and 4 show how one of the horizontal boards and the base strip are attached to a sideboard. Before attaching these boards they need to be prepared with dowels. The last step is to tighten the cams to fix the board. Step 5 shows how the compound part from Step 4 is lifted into an upright position. It is important to lift the wardrobe into an upright position before continuing with the assembly, because in rooms with low ceilings it is not possible to lift the fully assembled wardrobe. After lifting the wardrobe upright the top (horizontal) board subsequently the remaining sideboards are attached. Step 6 shows how the back wall has to be nailed on by using the nail-holder to position the 40 nails correctly.

Looking at the instructions offered by IKEA it is clear that they are well optimized. However, they only represent linear sequence of actions similar to many other types of instructions. In its current paper format, the user has to make a connection between the physical world of the wardrobe and the virtual domain of the instructions explicitly. Making this connection is typical burden, which is not necessary. To overcome these problems we suggest an assembly plan, which models all possible paths the user can take, in the following section.

18.4.2 Assembly Plan

To successfully assist the user the assembly plan has to modeled. Here we present an assembly that models the different states of the assembly as well as the different actions performed by the user. Those actions are modeled as state-transitions.

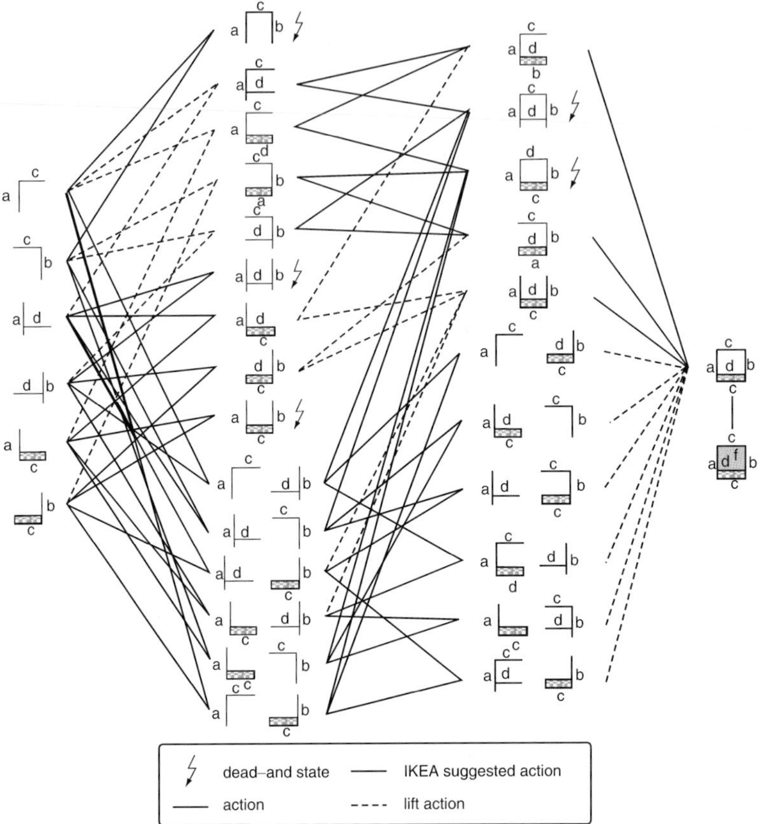

Fig. 18.2. Assembly plan for the IKEA PAX wardrobe

To create such a plan, we gave the wardrobe parts the following identifiers: sideboards a and b, horizontal boards c and d, base strip e and back panel f. Figure 18.2 shows the main part of the full assembly plan. The graph consists of icons representing partial states of the wardrobe and interconnecting lines. These lines describe actions that need to be completed to move from one state into the next.

Actions always consist of joining a not previously used board to the compound depicted or joining the two compounds together. The actions of preparing the boards (adding dowels, cams, or screws) are not shown in the graph. The only restriction for those actions is that they have to be completed before the board is used. We had to distinguish between the situations of connecting the sideboard a and the one of connecting sideboard b to the horizontal board. This is important since the board orientation matters.

The graph is read from left to right. The wardrobe can be in any one state in one column at each step in time. An action transfers the wardrobe to the next column on the right or to the end state. The dashed lines in the graph symbolize actions in which the user has to lift up one of the compound parts before adding the extra board or joining the two compounds. This is due to the restriction that the compound has to be lifted before the two horizontal parts are added to one sideboard. States from which one cannot continue are marked as dead-end states with a lightening-bolt. If the user reaches such a state he has to go one step back before he can continue.

The plan offers a total of 44 possible assembly sequences, and shows 14 sequences leading to dead-end states. The four sequences marked with thick lines are the ones proposed in the assembly instructions by IKEA. It is worth noticing that the lift action occurs in the second step in all the IKEA sequences. Besides the IKEA-sequences there are four other sequences that also have the lift action as the second step. From our experience we can say that these sequences are just as simple for the user to set up as the ones proposed by IKEA. Knowing with which state the user is occupied, it is possible to implement a variety of different types of assembly instructions. The user can be given information about the best action to take next. Alternatively, the user could only be informed when he or she has arrived in a dead-end situation. For quality monitoring reasons, simply noting all states the user passed through may be of interest. The following is concerned with how the assembly state is inferred using sensors.

18.4.3 Observing the User's Actions

The following presents a sensor-based approach for perceiving the actions in the assembly plan. First, we show how the actions can be subdivided into partial actions. We then show which sensors can be used to recognize the partial actions.

In Table 18.1 the actions that have to be recognized to trace the full execution plan are listed. Preparing the different parts, lifting a compound part into an upright position, joining a sideboard or the base strip to a horizontal board have to be detected. Furthermore, we distinguish the case when the second sideboard is added to the compound and the action of nailing the back panel to the rear of the compound resulting in the finished wardrobe.

Most of the actions in Table 18.1 can be subdivided into partial actions. These actions are relatively simple and self-contained such as tightening a screw, hammering in a dowel or a nail, turning a board, or joining two parts together. In the following we will show how these simple partial actions can be detected using the appropriate sensors. In the third column the table gives the sensor configuration we used in our experiments. The fourth column provides some sensor alternatives, which may influence the precision of the perception and the total sensor cost.

Taking a look at the simple example of preparing a horizontal board (inserting four dowels), we see that this can be recognized using only one accelerometer attached to the board itself. Alternatives would be to enhance the hammer with an accelerometer or to use an electric contact that reacts

Table 18.1. Assembly actions and possible sensor configurations

Action	partial actions	our sensor configuration	alternative sensor configuration
prepare sideboard	screw 4 cam bolts screw 2 screws	screwdriver with gyroscope	contact sensors
Prepare horizontal board/base strip	hammer in 4 dowels possibly turn board	accelerometer on board	accelerometer on hammer dowel contact sensors
lift compound part		accelerometer on board	
join sideboard and horizontal board	join parts tighten 2 cams	force sensors screwdriver 2 accelerometers	distance sensor contact sensors
join base strip to sideboard	Join may be hammer	force sensor 2 accelerometers	
nail wall to back	hammer 40 times	accelerometer	

as soon as the dowel has been fully inserted. What makes this example interesting is that one can insert the four dowels in many different ways. As the dowel insert-points lie on opposite sides of the board, one usually has to turn the board during its preparation. This action of turning the board can also be easily recognized using the accelerometer attached to the board. How many times and when exactly the board is turned, however, can be varied by the user. Next we discuss an approach for how such different sequences of partial actions can be incorporated into the perception process.

18.4.4 Sensor Experiments

In this section we show how sensors can be used to detect the partial actions presented above. After that, we describe how Markov chains can be used to detect sequences of these partial actions. For our experiments we used off-the-shelf sensors. The available sensors were attached to parts of the wardrobe and to the tools used during assembly. Data was collected on a standard PC. In a second step we then developed an interactive prototype using wireless communication technology developed in the Smart-its project (Smart-Its, online; Beigl et al. 2003).

Figure 18.3a shows two 2d-accelerometers connected to the sideboard and horizontal board of the wardrobe. The accelerometer used is the MEMS accelerometer ADXL202 from Analog Devices on the evaluation board.

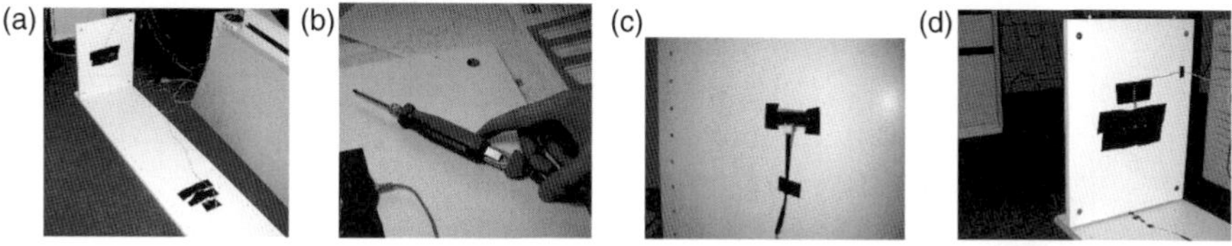

Fig. 18.3. (**a**) Horizontal board and sideboard both equipped with 2d-accelerometers. (**b**) Screwdriver enhanced with gyroscope. (**c**) Sideboard with attached distance sensor. (**d**) Horizontal board with accelerometer and sideboard with attached force sensor

Figure 18.3b shows a screwdriver enhanced with the gyroscope ENC-03JA from Murata. Furthermore, Fig. 18.3c shows the Sharp GP2D12 infrared distance sensor attached to the sideboard. In Fig. 18.3d a horizontal board being joined to a sideboard is shown. The horizontal board is equipped with an accelerometer and the sideboard is enhanced with a standard force-sensing resistor (FSR) to measure when the boards are joined.

18.4.5 Detecting Partial Actions

It is worthwhile going through a few of the experiments to see how the partial actions can be detected. For example, the preparation of the sideboard consists of screwing in four cam-bolts and two screws. In Fig. 18.4a the output of the gyroscope enhanced screwdriver is plotted over a time period of 5 min. By calculating the standard deviation over a time-window one can easily recognize that the user was using the screwdriver six times. We also conducted experiments to detect whether a user is opening or tightening screws. It shows that these actions are also easily distinguishable as the standard deviation of the gyroscope signal is clearly negative when tightening a screw and clearly positive when opening a screw.

The action of joining a horizontal board to a sideboard is shown in Fig. 18.4b. This plot incorporates the output signal of the gyroscope-enhanced screwdriver, the force sensor, and the two dimensions from the accelerometer attached to the horizontal board. One can reconstruct that the horizontal board was moved into place, and then the screwdriver was used to tighten the cams, which in turn increased the pressure on the force sensor. In another experiment we included the infrared distance sensor mentioned above. We used this sensor to detect the orientation of the horizontal boards with respect to the sideboards. To do this, IR-receivers were placed on both sides of the horizontal boards.

Beyond the sensors described above, Table 18.1 presents a variety of alternatives. Depending on the required system reliability and the product cost, different design choices can be made. For example, metal contacts could be used as a very cost-efficient sensor, to detect when two objects are connected.

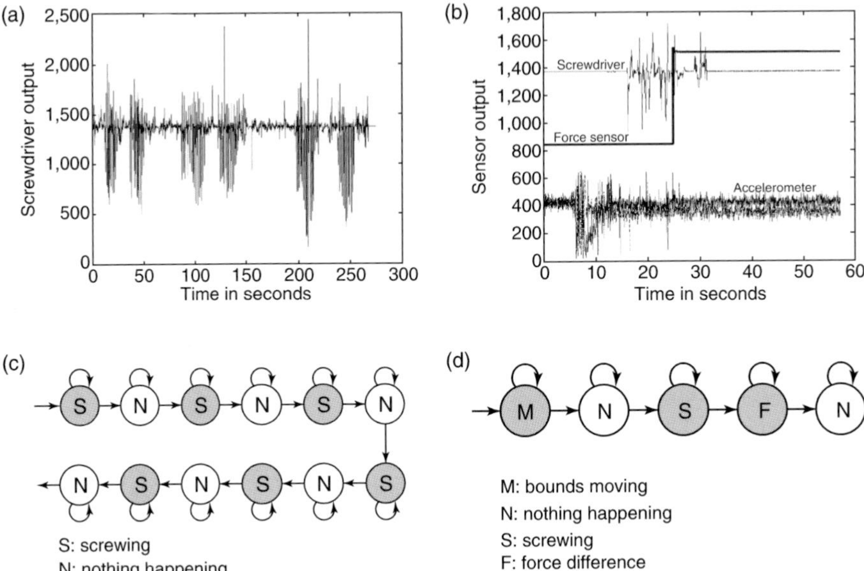

Fig. 18.4. (a) The output signal of the gyroscope enhanced screwdriver during the preparation phase of a sThis action consists of tightening four cam-bolts and two normal screws. **(b)** The outputs of the screwdriver, force sensor and accelerometer are plotted during the action of joining the horizontal board to the sideboard. **(c)** and **(d)** The Markov chains for the sideboard preparation and the joining actions, respectively

From various experiments we can conclude that all partial actions necessary can be perceived quite easily and reliably; see also Table 18.1. There are surely still problems and ambiguities, for example one cannot differentiate if someone is hammering in a nail or a dowel by using only an accelerometer. Similarly one cannot precisely count how main nails/dowels have been hammered in, as the user might start hammering, take a break, and then continue hammering in the same part. Nevertheless, one must say that the partial actions being recognized can be distinguished to an adequate degree. Following, we present a method that allows us to detect the complete actions by modeling valid sequences of possible partial actions.

18.4.6 Detecting Complete Actions

One approach for detecting actions with a higher confidence is using Markov chains to model sequences of partial actions. With this technique the chronological order of the partial actions can be taken into account. Simple Markov chains were designed for each action. Figure 18.4c shows the Markov chain for the action of preparing the sideboard. It incorporates the partial action of tightening a screw, six times.

Similarly the Markov chain for the joining action is shown in Fig. 18.4d. Here we model how the pressure on the force sensor rises when the screwdriver is used.

To test our Markov chains we performed the actions described in Table 18.1 several times and recorded them using the described sensor configuration. Applying simple classifiers to the data, such as Bayesian or threshold classifiers, we generated sequences of partial actions. These sequences were then fed to the Markov chains for recognition. The actions were all easily recognized. This is due to the fact that the detection of the partial actions is quite reliable and that the order of partial actions is distinctive for each action.

18.5 Situation-Aware Affordances

The previous section showed how the actions modeled in the assembly plan can be divided into partial actions and how the necessary action sequences can be recognized using Markov chains. As a result the system can recognize the various states as well as the actions of the assembly. In this section we present the concept of situation-aware affordances as a technique to make physical objects more interactive. We start with a general discussion of the concept and then show, how the concept can be implemented for the specific example of proactive furniture.

18.5.1 General Concept

Interactive environments such as the Aware-Home (Abowd et al. 2000) and smart offices (Johanson et al. 2002) introduce new and diverse tools into everyday life. It is crucial to design these environments in a way that people can explore, understand, and predict functionality and effects. Beyond training and instruction manuals, appropriate design has proven essential to make such systems more usable and intuitive (Van Welie 1999). One way to approach this is to provide objects with clear affordances as have been defined by Gibson (1986) and made popular later by Norman (1988). Affordances give people visual cues about how to use objects and thus offer a simple form of instructions. For example, buttons are here to be pressed, and a coffee-cup handle is here to lift up the coffee-cup.

Although carefully designing objects with discernible affordances can lead to better results in many cases, affordances are mostly static and bound to a single object. In contrast to that, in interactive environments objects can adopt different roles at different times and may be involved in multi-step tasks. Classic object affordances can display information regarding the use of single objects. However, what would be needed for interactive environments is a situation-aware notion of affordances that can also reflect relations among several objects and changes in the environment.

Object affordances are closely related to usage instructions. Several guidelines for designing instructions have already been proposed. Actually, an ideal design of objects should not require any instructions at all: the user should be able to guess and understand the functionality at a glance. However, it is hard to eliminate instructions in general. It would already be an achievement to integrate them into the related objects. Instructions could be split into smaller portions – hints – that subtly but infallibly guide users towards correct conclusions. These hints should be tailored to the users momentary task. Each hint helps in one dedicated situation in contrast to manuals covering all error cases.

As an overall requirement, successful interactive instructions have to follow three principles described by Constantine and Lockwood (2003): explorability, predictability, and intrinsic guidance. Explorability enables users to explore, experiment, and discover functionality without penalization of unintentional or mistaken actions. Predictability builds upon intuition: a user can draw conclusions based on first impressions without having to understand all details. Intrinsic guidance is integral and inseparable of the user interface. Instructions are provided as needed without requiring any special action or initiative on the part of the user.

The following presents a specific solution on how an implementation of situation-aware affordances may look like for the example of proactive furniture.

18.5.2 Specific Solution

Our approach is to show and evaluate how the notion of affordances can be extended to situation-aware affordances including dynamic cues, so that workflows, and relations among objects can be presented. Due to the physical nature of the assembly, the symmetry of boards, and the interchangeable activities, several sequences of assembly steps are possible. However, steps depend on each other, such that previous steps constrain the assembly of the parts in consecutive steps. Thus, the role of parts changes during the assembly process and have to be visualized to the user.

In Sect. 18.4 we showed how the states of assembly could be sensed using integrated sensors. Knowing the state of the assembly, instructions can be given to the user at any time. Displaying instruction on a computer screen does not overcome the disadvantage of conventional paper instructions. By looking at the instructions the user is distracted from his original task of assembling the furniture. The flow of action is disturbed. Augmented reality presents one way of bridging the gap between the instructions and the real world. It has been used to integrate information into a user's physical environment often (Tang et al. 2003; Zauner et al. 2003). However, AR is cumbersome and, typically, computationally expensive. Audible instructions offer a cheaper way of immersion but have to tackle the problem of addressing the appropriate parts by a vocabulary the user is familiar with or has to learn in advance.

Our vision and aim is to integrate instructions directly into objects. We study how affordances of physical objects can be exploited and enhanced by dynamic cues resulting in situation-aware affordances. In particular, we evaluate the effectiveness of LEDs attached to objects as a way of extending static affordances. We use LEDs attached to the furniture parts in the furniture assembly task to guide users through the assembly process.

For the example application of proactive furniture, a set of video based mock-ups was designed. Figure 18.5a shows a screenshot from one of the videos produced. With these we investigated the feasibility of visual markers on objects for enhancing affordances. After showing the different videos to several people at multiple occasions, we defined a set of visual guidance principals we found appropriate. To guide the user through the furniture assembly we identified the following five types of feedback:

1. Direction of attention
2. Positive feedback for right action
3. Negative feedback for wrong action
4. Fine grain direction
5. Notification of finished task

Users unwrap the furniture package and their attention gets directed immediately (feedback type 1) to the parts they are supposed to start with. User's actions, such as turning and moving boards are sensed and blinking green light patterns indicate which edges have to be connected in which manner.

(a)

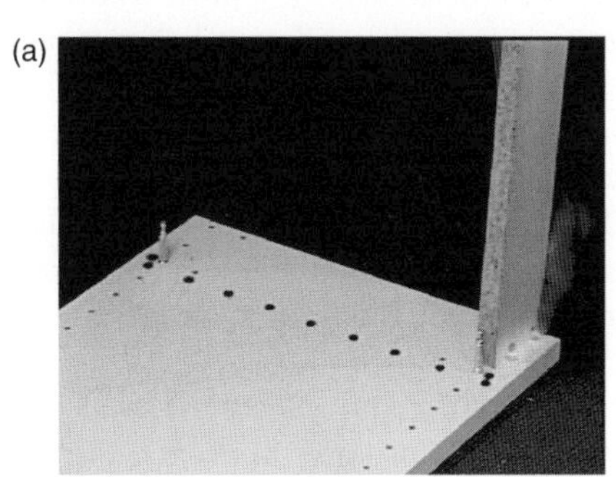

(b)

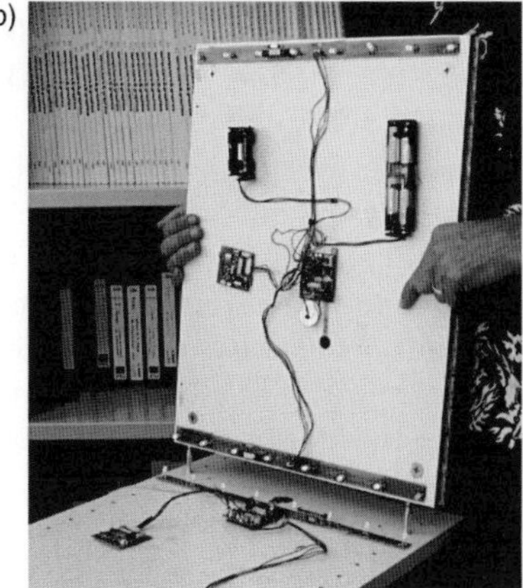

Fig. 18.5. (a) A short video showing the instructions in use was produced. Different colors and flashing sequences were tried out. The video ws showed to a side public to gain experience. The figure shows red flashing lights, while someone is trying to assemble the parts in the wrong way. **(b)** A person assembling two boards with the LED instructions. The short board shows that its orientation is wrong by flashing the lower strip of LEDs red

If boards are aligned in the proper way, a synchronized green light pattern on both edges indicates a well-performed action (feedback type 2). If the user takes a wrong action, a red light pattern appears representing a mistake (feedback type 3, see Fig. 18.5b). Additionally, a green flash pattern shows the alternative. After boards have been aligned together in the right way, individual green lights direct user's attention to the holes where the screws have to be inserted and tightened (feedback type 4).

Coding information with colors has to be done with care as different cultures map meanings to colors in different ways. Even so, Helander (1987) points out that red, yellow, and green should be reserved for "Danger", "Caution", and "Safe", respectively. In accordance with these guidelines we use red to signal an error or wrong position in assembly and green for a correct position. Compared to Tarasewich et al. (2003) we do not evaluate the information rate of LEDs as such, but use them to display instructions. In contrast to tangible bits by Ishii and Ullmer (1997) or the work presented by McGee (2002), we introduce and evaluate the concept of situation-aware affordances to visualize usage of objects rather than using objects to facilitate new forms of human–computer interaction.

18.5.3 Summary of a User Study

To evaluate the use of LED-based instructions we used our interactive prototype. To present instructions to the user we have developed a custom layout board carrying eight dual green/red LEDs; see Fig. 18.5b). These LED strips are connected to the described sensing hardware and are attached to the ends of all three boards to give instructions to the user. For more details on the used hardware, see also Beigl et al. (2003), Holmquist et al. (2003), and Michahelles et al. (2003). Besides only presenting information using the LEDs we have the possibility to provide visual and auditory instructions on a laptop computer.

The study was carried out with 20 participants with different backgrounds. Fourteen of the participants are male, six are female. The average age of the participants is 26.16 years with a standard deviation of 1.64 years. Five of the participants had computer science backgrounds, four were from engineering disciplines and nine from other fields. The participants had different levels of experience with flat-pack furniture assembly. All participants reported normal or corrected-to normal vision. For more details of this study please refer to Michahelles et al. (2003).

The overall goal was to compare the usability and effectiveness of classic paper instructions with our situation-aware affordance approach. To this extent, the user study was conducted in two phases. In the first phase the assembly time between an assembly conducted with classic instructions and one with LED-based instructions was compared. In the second phase of the experiment, participants were encouraged to perform the setup again three times using instructions presented in different modalities. The first modality employed only the LEDs to display the situation-aware affordances. The second modality displayed interactive instructions on a computer screen. The information for the instructions was based on the same sensor setup as with the LEDs. The third modality extended the LEDs with auditory spoken instructions. Overall, the user study presented by Michahelles et al. (2003) revealed that there is a measurable time gain when using LED-based instructions. Beyond that, we saw how errors during assembly can be reduced using instructions in the right place. Designing the sensors and instructions for this purpose, it may even be possible to totally prohibit errors during assembly. For applications beyond furniture assembly, such as airplane or power plant maintenance, this is a critical issue.

Besides these performance-related gains we found other problems solved through the LED-based instructions. The questionnaire showed that determining *which part fits where* is one of the main problems using today's instructions. Interestingly, 75% of the participants found that the LED-based instructions help with exactly this problem. Because the LEDs light up on both boards that need to be joined, finding out what goes where becomes a straightforward task. The instructions do not need to be mapped to the objects anymore, as they are simply integrated into them.

The comparison of instructions presented in different modalities led to further insights. Participants stated that instructions presented on the computer screen helped for orientating the boards correctly, but the screwing direction remained unclear. The various participants received presenting instructions with audio in the form of spoken words differently. About half of the participants found spoken instructions useful. About a quarter found them disturbing. The general opinion is that spoken instructions have to appear at exactly the right time, in order not to disturb too much. In the questionnaire the participants mentioned that they could be useful for instructions that cannot be presented visually. Getting the screws was mentioned as an example.

18.6 Conclusions and Discussion

In this chapter, we have argued that perception and context awareness have great potential to change the way we interact with computers in general and in the context of Ambient Intelligence in particular. The first part of the chapter gives an overview of various sensing and perception opportunities classified by the physical placement of the sensors and by the logical view of the information sensed. The second part of the chapter then describes in more detail an example from our own research for proactive instructions for the running example of a flat pack furniture assembly. For this example we described the modeling, recognition, as well as a proposed feedback mechanism.

We believe that the example of proactive furniture together with the concept of situation-aware affordances can be generalized to several other applications. A wide range of assembly and maintenance tasks could benefit from embedded sensors and proactive instructions. Sensors could be used to monitor the assembly of aircraft or the installation of roof racks on cars. Integrated into machine parts sensors could then be used to continuously monitor system performance. Proactive instructions can offer the user freedom in his choice of actions, while still guaranteeing a correct assembly. Generally, security-critical applications could benefit from information presented at the right time and at the right place. Beyond today's applications we believe that situation-aware affordances have the potential to let the user explore the functionality in future interactive environments in a more intuitive way. Smart homes and office environments will need simple and effective ways of letting the user know how they can be controlled.

As stated above, modeling and recognizing context information is often seen as one of the most important ingredients of Ambient Intelligence. However, the number of examples where context information really is used is rather limited and restricted, for example, to location information, which can be sensed with a predictable or measurable accuracy. The main reasons for this may be characterized by the fact that context information is often too uncertain, ambiguous, and cannot be extracted reliably. One might argue that today's technology is not good enough yet to overcome these problems but as I will argue in the following there are fundamental and inherent problems which are either very difficult if at all solvable. In my opinion, there are at least the following five fundamental issues for context-aware systems:

- *Unobservable information.* One of the most fundamental problems is that not all relevant information is measurable or observable. For example, a human's mental state is not observable or the personal interests and objectives are difficult to reliable estimate for a computer or even for another human observer.
- *Missing information.* Even for observable information in most circumstances a context-aware system will not have access to all relevant information such as complete history information. Even when a particular piece of information is in principle observable this information might not be available since it may not be stored or measured.
- *Unpredictable behavior.* Humans are notoriously unpredictable and change objectives, goals, and motivations often. This poses a great challenge for context-aware systems.
- *Ambiguous situations.* Clearly, many situations do not have a single interpretation but do have multiple interpretations. Those do not necessarily have to contradict each other but they leave enough room for drawing different conclusions.
- *Context is changing constantly.* An interesting but often overlooked issue is that context is changing constantly with every interaction or communication we might have with a particular environment or device. This is probably most obvious in the case of human-to-human communication where every single discussion or communication we have with a person changes our knowledge and understanding of the particular topic as well as of that person so that any subsequent discussion is influenced by that

change in context. However, I do not know of any context- aware system today that does take this into account effectively.

The above list of fundamental issues suggests that context-aware systems should be designed and evaluated much more carefully than it is done today. While these problems really are fundamental and important they are seldom raised and discussed with respect to context-aware systems.

In the following I would like to briefly discuss which and how fundamental challenges are addressed in the described example of proactive instructions for furniture assembly. The issue of unobservable and missing information was alleviated by the fact that we explicitly reduced our modeling and recognition task to the furniture and the actions performed with the various objects involved. This consideration of modeling only what can be modeled reliably clearly helps in many circumstances. Nevertheless, this separation of what can and cannot be modeled is often not done well. In order to reduce the ambiguity of situations to a minimum we basically used a set of sensors that enabled the recognition of the different actions and states with a close to perfect reliability. The unpredictability of human behavior was not an important issue here since we did not aim and need to predict human behavior. Rather than to consider the fact that context is changing constantly over time we allowed the system to adapt its behavior depending on the actions performed by the user. While this does not really address the issue of constantly changing context directly this appears to be sufficient for simple applications such as the one described here.

In summary, we can state that context-aware systems using perception techniques certainly have the potential to change the way we interact with computing in general and with ambient intelligent computing in particular. A first challenge is to make sensing more robust and reliable, for example, by fusion of multiple sensor modalities. A second challenge we have pointed to is the fact that there exist various fundamental challenges inherent in the use of sensing information and context information. While the first issue is well known the second issue seems to be largely underrepresented in our community and we hope that future research will enable us to deal better with these fundamental challenges.

19

Experience Design

B. Eggen and S. Kyffin

"Design explorations take the next step of considering how computation is to be manifest when it moves into the physical environment, and recognizing that this move makes the physicality of computation central."

Paul Dourish 2002

19.1 Introduction

One way of looking to the future is to take the past as a point of departure and in passing by the present-day extrapolate to things to come. Actual practice has demonstrated that this type of forecast can be highly accurate when it comes to the advancement of core AmI technologies. For decades Moore's law has correctly predicted the exponential growth of computer processing power. And in its generalized form it is also properly indicating performance improvements in the areas of wireless connectivity, storage capacity, and power battery technology. But despite the evident empirical observations on which this law is based, uncertainty will inevitably creep into the equation the further we look ahead in time. Taking a point of reference that is predominantly monodisciplinary in nature might cause other restrictions of "extrapolated" visions of the future. For example, by focusing primarily on technological developments danger exists that complementary disciplines in the arts and sciences will remain underexposed. This situation seems undesirable, especially as true innovations seem to happen on the borderlines between different scientific disciplines.

Alternatively, current trends can be ignored in order to freely, i.e., without any restrictions emanating from the past, think about the future. Realists call this way of looking to the future "dreaming" and attribute little predictive power to the resulting visions of the future. Visionaries, on the other hand, argue that the intrinsic needs and desires of people, but also the physical, mental, and social capabilities of people, only very slowly change over time. A change that is much slower, in fact, than the time scale typically used to forecast trends. These so-called normative visions of the future put the focus of attention on people's everyday life and top down inspire different disciplines to bring about new insights through collaboration.

In practice, both ways of looking to the future, extrapolated and normative, are hardly ever encountered in their purest form. It is the continuous exchange of these two directions of view, "*past* → present" and "present ← *future*," respectively, that should characterize and determine successful strategies for exploring the future of electronically enriched living environments. In fact, the intertwining of both approaches naturally puts the spotlight on the present. And it is precisely the *present* the authors of this chapter want to focus on when it comes to the "*true* visions" theme of this book.

To fully appreciate why the present is put in the center of interest, it is important to know that, from the early days on, the authors have been approaching the area of Ambient Intelligence from an *Industrial Design* perspective. In short, this means that the main objective of our design research efforts has been to explore the possibilities of Ambient Intelligence through concrete experience prototypes. The concept generation processes that underlay these prototypes were always grounded in sound technological, societal and cultural trend analyses, and the actual realizations often made use of just-beyond-the-state-of-the-art technologies only available in a laboratory setting. In addition to this, potential end-users were systematically involved to evaluate their experiences while interacting with the prototypical implementations of future intelligent products and systems. In our view, the real-time and physical characteristics of these interactions are indispensable for the creation of true points of contact between extrapolated future technological possibilities and normative long-term psychological and sociological human capabilities.

Although true encounters can only happen in the present, they can be captured in time, and further analyzed and generalized into knowledge relevant for the design of future intelligent products and systems. In this chapter, we will review some insights we consider relevant for designers of AmI environments. These insights originate from the numerous encounters between people and the experience prototypes that were created by our co-workers and us over the last couple of years. In the remainder of this chapter, we will first briefly review relevant technological developments and expectations. Next, we will summarize the design relevant knowledge we acquired from the experience design research activities we conducted within the area of Ambient Intelligence. Finally, we discuss new directions in design research that we consider crucial to truly realize the exciting possibilities offered by the AmI vision (Aarts et al. 2001; Aarts and Marzano 2003).

19.2 Looking to the Future

19.2.1 Disappearing Technology

In 1991, Mark Weiser and his colleague's at Xerox Palo Alto Research Center (PARC) concluded that technologies that have the greatest impact on people's everyday life are those that in the end perceptually disappear in the

Fig. 19.1. GB Microdrive technology "hidden" in everyday objects (© 2004 Hitachi Global Storage Technologies)

environment (Weiser 1991). They stated that the same would hold for information and communication technology, which eventually will only come out well when it fully merges into people's everyday life. One way to become invisible is to literally, i.e., physically, disappear out of sight of the end-user. Recent developments in the area of miniaturization have reached a point where we can actually start to "hide" electronics in the environment; see for instance Fig. 19.1.

Another way for information technology to disappear is to become transparent. In this case, the user is no longer aware that he or she is operating technical equipment, but attention is fully allocated to the experience of the interaction itself; technology does no longer put up barriers.

Making information technology transparent for people poses new challenges to design research that clearly go beyond pure technical specifications. Transparency characterizes the interface between the information processing system and its users, or, when the system has physically disappeared into the environment, between the environment and the user; see for instance Fig. 19.2. In practice, this means that in addition to knowledge on advanced interaction technologies, knowledge about the user and his or her environment is vital to guarantee transparent interaction. We believe that system intelligence is a key enabler for transparency. After all, when the system is aware of its context, the user of the system no longer has to feed the system with explicit information about the environment. Reducing explicit input not only increases interaction speed, but it also reduces the cognitive workload of the user and the chance of making mistakes (Lieberman and Selker 2000). At the output side, context awareness enables the system to decide about the best moment in time and place to present the information to the user in a way that optimizes information transfer while minimizing information overload. In summary, a context aware system that shows meaningful and appropriate adaptive behavior enables users to primarily focus on the basic intention that in first instance triggered the user–system interaction.

Fig. 19.2. Transparency in Ambient Intelligence – communicating with distant family members objects (© Philips Research)

Future scenarios are frequently used to systematically explore the possibilities of these kinds of intelligent environments. In 2001, the Information Society Technologies Advisory Group (ISTAG) of the European Community published four scenarios that describe in detail how people's everyday life in 2010 would look like assuming Ambient Intelligence has pervaded the living environment by that time (ISTAG 2001). The next excerpt summarizes a part of the ISTAG "Dimitrios" and the "Digital Me" (D-Me) scenario, which puts the emphasis on system-mediated communicative acts supporting human relationships:

> *Dimitrios is wearing, embedded in his clothes (or in his own body), a voice activated "gateway" or digital avatar of himself, familiarly known as 'D-Me' or 'Digital Me'. At 4:10 p.m., following many other calls of secondary importance – answered formally but smoothly in corresponding languages by Dimitrios' D-Me with a nice reproduction of Dimitrios' voice and typical accent, a call from his wife is further analyzed by his D-Me. In a first attempt, Dimitrios' 'avatar-like' voice runs a brief conversation with his wife, with the intention of negotiating a delay while explaining his current environment. [... time goes by in which Dimitrios gets entangled in other activities ...]. Meanwhile, his wife's call is now interpreted by his D-Me as sufficiently pressing to mobilize Dimitrios.*

The ISTAG scenarios do not only aim at sketching a provoking view on a possible future, but they are also especially intended to systematically map developments in the area of information and communication technology, and to get a discussion going on related economic, social, and public factors. The scenarios are called "human centered" by their creators, because people, the individual is put in the center of our future information society. Also in the

Dimitrios scenario, human relationships play a central role. However, after reading the scenario, the question easily arises whether you and I would positively judge the way in which these relationships are maintained and supported by technology. Despite the fact that the advanced speech and language technologies and the artificial intelligence needed to realize these scenarios are impressive, at the same time, these technologies to a large extent give color to the personality of D-Me. It is literally impossible for Dimitrios' wife to get round D-Me. What does she think of the way D-Me behaves? And how do these situations influence her relationship with the real Dimitrios? In case these questions will be answered negatively, will it be possible in the long run for Dimitrios to identify himself with his Digital Me, or is D-Me urged to disappear from Dimitrios' sight once and for all?

In a more recent publication the ISTAG working group "Experience and Application Research" has concluded that a new approach is needed to ensure the successful application of research and development efforts in the area of Ambient Intelligence (ISTAG 2004). This new approach should be based on an extensive involvement of those people who will be confronted with the advantages and disadvantages of Ambient Intelligence. In the next section, we will elaborate on the role of end-users in the design of intelligent products and services.

19.2.2 The Future Seen from a People's Point of View

As indicated before, it is considered of utmost importance that a designer of future intelligent systems integrates technical knowledge with knowledge of the user and the environment into innovative solutions for relevant problems. User involvement in the design process is a crucial element for a number of reasons. Firstly, interactions with end-users can be employed to define and validate the design problem from the viewpoint of the individual, his or her social context, or the public interest. At the same time, user involvement can be instrumental in mapping out the characteristics of the target group that will use the system, the context in which the system will be used and the suitability of available technology. This leads to a set of design requirements the intelligent system should comply with. After solutions have been generated and implemented, people can once again be brought into action to evaluate to what extent the design meets the requirements.

Over the last decade, we have been actively involved in the set up and execution of numerous *user-centered* design research projects within the AmI programs of Philips Research and Philips Design; for an overview see, for example Aarts and Aarts and Marzano (2003). More recently, this collaboration was continued in the context of the AmI research program carried out at the department of Industrial Design of the Eindhoven University of Technology (TU/e) (Lundqvist 2004). The vision of Ambient Intelligence always formed the basis of the considerable efforts in design research carried out in these projects. In the majority of projects, experience prototypes were built which created *true* points of contact between people and future technological possibilities.

19.3 The Home Experience

To better understand the physical, social, and cultural context in which AmI technology will be used and its implications on daily life, early on in the research program we conducted a user study on people's home experience. To acquire this knowledge, we left our laboratory and went out to talk to consumers. We visited ten carefully selected Dutch families at their typically Dutch homes as shown in Fig. 19.3 (Eggen et al. 2003).

During interactive family sessions, various techniques, such as telling, drawing, writing, taking pictures, free association, and family members interviewing one another were employed to stimulate the families to express their experiences in the present home. We also learned about their attitudes towards so-called "Smart Homes" and towards their Dream Home of the future. Although a Smart Home is a future concept, it should be noted that people's ideal Dream Home is not necessarily smart. The main findings of our study can be summarized as follows.

- Home is a feeling. It is a cozy, trusted, and safe place, a place to return to; where you can be yourself and do what you want; where your own things are; where you meet the people you love and like.
- People have mixed feelings about a Smart Home. The families see a Smart Home as a home where technology is applied to make life easy and to save time that can be used to do things one really likes to do. It provides comfort and luxury, and overcomes the drawbacks of the current home by taking over the unwanted tasks. When talking about a Smart Home, people start to talk about fears and worries they have regarding this Smart Home. They fear a cold and emotionless place where boredom and laziness will take over. So, besides positive effects, people explicitly discuss negative aspects. Some people do not even believe in a Smart Home at all. They

Fig. 19.3. Family studies – the home experience (© Eindhoven University of Technology)

think it will never be possible: "people are smart; a home will never be smart.". Given the anticipated disadvantages and dangers of the Smart Home, people start to formulate requirements: "If you build a smart home, make sure I can trust it. It should be easy to use and integrated. I want to be in control."
- The Dream Home equals home as it is now, but better. In the future the feeling of home should remain the same; the core values of the home must remain untouched. What people would like to have are the benefits of the Smart Home, and the positive experiences and values of the current home. People's expectations of the benefits of the Smart Home are limited to what they think will be technically feasible. Because they sometimes underestimate what is possible, potentially interesting options are not mentioned. This is why we asked the families to imagine that anything is possible and to tell us what they would like. In general, the role of a future home would be one of an assistant. It would give advice, create the right conditions, and support the family in their activities and with the things that have to be done.

In the remainder of this section, we generalize some of the lessons we learned from these interactions with people into a number of insights. These insights are not only based on the family studies mentioned above, but also stem from many additional user-centered studies carried out in the AmI research programs at Philips and the Eindhoven University of Technology.

19.4 Design Insights of AmI Systems

For the purpose of this paper we only selected insights that we consider most relevant for the design of future intelligent products and systems. Although many projects were carried out in the context of the home environment, additional projects were done in the office, public, automotive, and medical domains. We therefore like to stress that the relevance of the insights described below definitely go beyond the home domain and hold for other domains too.

Insight 1. Intelligent Products Should Not Offer Predefined Experiences to People, But Only Create the Right Conditions to Enable and Support a Personal or Social Experience

From a business perspective, quality is no longer the only differentiator between competitive products. In their book "The Experience Economy," Pine and Gilmore have convincingly argued that the "user experience" offers new ways to create competitive advantage (Pine and Gilmore 1999). The user experience can be realized by making the interaction between the user and the product unique and personal. Consumers are willing to pay for this experience. It is the task of the designer of intelligent products to uncover such unique and personal aspects, and take them as the starting point for the design of new products or systems. In many of the projects carried out by the authors and their co-workers, end-users have repeatedly indicated which product-interaction aspects they really valued. An illustrative example is provided by the "Phenom" project shown in Fig. 19.4 in which we developed a system to support the recollection of memories in the context of an AmI living room (van den Hoven and Eggen 2003). Initially, the project focused on fast and efficient retrieval of photos stored in a digital home archive. The photos were considered digital representations of memories. Interactions of users with early prototypes showed that the photos only act as a trigger to start the recollection process. The real memories emerge and become tangible in the form of stories told by the users. This insight caused the project to shift its focus from "information retrieval" to "storytelling" and the role various artifacts like souvenirs, photo albums, and physical aspects of the environment play in the experience of recollecting. This change of direction led to many new technological challenges concerning the localization and tracking of objects and persons in the environment, the synchronization of decentralized digital databases, the development of intelligent algorithms, and the application of advanced network, display, and user–system interaction technologies.

Insight 2. The Behavior of Intelligent Products or Systems Should Fit the Rhythms and Patterns of Everyday Life

During the many interactions we had with people, time after time, it turned out that the use of electronic systems should be seamlessly integrated in everyday activities that at closer inspection seem to be driven by hidden personal and social user needs. The primary task or activity the product was originally

Fig. 19.4. Phenom – an intelligent living environment supporting the recollection of memories (© Philips Research)

designed for is carried out subconsciously focusing the "true" user experience on the underlying user needs. Family members, for example, indicated that daily household tasks like vacuum cleaning or ironing are often considered boring, but that, on the other hand, the nature of these activities enables them to create precious moments in time where they can daydream and reflect on or escape from their daily worries (Eggen et al. 2003). People realize such moments are of great importance for their personal well-being and explicitly question what the personal and social consequences will be when the future Smart Home will "relieve" them from these daily household chores.

The designer of future intelligent systems should be aware of possible interferences between the use of the system and the rhythms and patterns of everyday life. Recent investigations, for example by O'Brien (1999), have shown that in case of interference, the user will not accept the new system and the system is doomed to fail.

Insight 3. Intelligent Products Should Only Explicitly Attract People's Attention When This is Meaningful and Appropriate

Based on the present situation, many people tend to worry about a future in which the number of "smart" products is said to grow exponentially. People indicated that, already now, they often feel overloaded with information that is inadvertently pushed onto them. People demand from intelligent systems that this problem is solved, or at least that it is severely reduced in order to restore a situation in which they could experience a certain degree of "freedom from choice" again. In such a situation people will only read, listen, look, choose, and act when meaningful information is presented in an appropriate manner and at a suitable moment in time.

According to Weiser and Brown (1996), *Calm Technology* that makes better use of the periphery of the human perceptual and cognitive system needs to be developed. People are only subconsciously aware of what happens in the periphery of the perceived situation, but when desired or necessary this information can immediately be put in the center of attention. Calm technology prevents overloading people with information by presenting this information in the periphery. At the same time this peripheral information gives people the feeling that they know what is happening around them and that they are fully in control when it comes to decide whether or not and when they should act. Within the "Home Radio" project this approach was investigated in more detail. A system was built that supports family members to stay in touch with each other and their home (Eggen et al. 2003). The nature of this contact can be characterized by three modes of interaction which seamlessly can change into one another: in the "ambient" mode audiovisual means are used to unobtrusively display information in the environment to create an ambience informing the user that everything at home is fine, in the "attention" mode the user consciously pays attention to the audiovisual information display, and in the "interaction" mode the user performs explicit actions to

Fig. 19.5. Home Radio – ambient awareness (© 2003 Philips Research)

extract additional information from the system about the home and/or its occupants. Experiences gained within projects such as the Home Radio project, shown in Fig. 19.5, underscore the importance of the development of calm technologies within the context of Ambient Intelligence. Designing calm interaction styles implies the integration of knowledge of information modeling and multimodal input and output technologies with psychological knowledge about cognitive processes that regulate human attention management.

Insight 4. Intelligent Systems Should be Trustworthy

People acknowledge the fact that systems can only behave intelligently when they know something about their environment. However, only a small portion of this knowledge can be preprogrammed in the system when it is bought. The greater part of the necessary knowledge has to be learned from the user, from the environment in which the system is supposed to operate, and from the actual use. People indicate they are willing to invest in this training process only when the use of the resulting intelligent system is guaranteed for long-term use. Negative experiences with computer crashes and "upgrades" to new operating systems, in particular, seem to worry people most in this respect. The user should also be able to trust the system that privacy sensitive information is in good hands. This means, for example, that personal information needs to be protected against "hackers,", commercial and/or government institutions. In the Home Radio project it turned out that at the sender site there not only was a need to secure privacy sensitive information from unwanted disclosure to close relatives, but it was also found that the receiver of personal information should be able to indicate the kind of information that he or she considers appropriate to receive. Unwanted disclosure of privacy sensitive information might negatively influence the receiver's peace of mind or, in some cases, might even burden the receiver's sense of responsibility. In general, people indicate they want to have full control over the kind of information a system is allowed to know about them and which part of this information can be shared with other people and other systems. These requirements put new challenges to designers of intelligent systems to develop tools that enable users to effectively manage their stored personal profiles.

Insight 5. People Should Always Stay in Control of Intelligent Systems

In many projects people have indicated that if needed or wanted they should be able to regain system control. This requirement has always played an important role and been a determining factor in the design rationale for favoring semiautomatic design solutions over fully automatic ones (Van de Sluis et al. 2001).

Insight 6. Building and Testing Experience Prototypes in a Realistic Setting Represents a Crucial and Necessary Phase in the Design of Intelligent Systems

In most design research projects from which the "lessons learned" described above were derived, big efforts were spent on the implementation of design concepts into working prototypes. Although these demonstrators clearly did go beyond supporting tasks, they also had their limitations. We have learned that one should go to considerable length in allowing end-users to realistically interact with and experience an application in the proper context of use. It turned out this is of particular importance for the multimodal and adaptive interaction components of the intelligent user interface. Multimodal interfaces aim to improve the naturalness of user–system interaction. True naturalness can only be achieved in the actual physical, social, and cultural context of use. We also learned that valid evaluations of adaptive systems (from a user point of view) should be performed in a realistic "everyday life" setting where adaptation of the system to the user (and vice versa) is done under "real" space and time conditions. Finally, it was found that the investigation of smooth transitions from background to foreground activity, in particular, requires believable and realistic environmental conditions.

Not only to fulfill these experimental research conditions, but also to better support investigations into the feasibility and integration of advanced technologies and to study the practical, psychological, and social implications of Ambient Intelligence, the HomeLab was built (Eggen and Aarts 2002); see also Fig. 19.6. HomeLab is a research laboratory comprising a fully furnished modern one-family home, complete with living, sleeping, and kitchen facilities. HomeLab is equipped with a distributed embedded infrastructure in which Ambient Intelligence can be developed and investigated.

Fig. 19.6. HomeLab – a peek at the living room taken from the observation cameras mounted on the ceiling (© 2003 Philips Research)

The insight that experience prototyping should be a necessary phase in any design process of intelligent products and systems concludes this section in which lessons learned were presented that are considered of crucial importance for the design of future intelligent systems. These understandings help us to set out new directions for future design research. In the following section, four different research directions are discussed that will deliver us more detailed and specific knowledge necessary to take the next concrete steps towards real-world AmI applications.

19.5 New Directions

19.5.1 Multimodal Interaction

One of the goals of multimodal interaction research is to increase the bandwidth of the communication channel available for the interaction between people and systems. By integrating multiple sensory modalities, not only when processing user-generated input, but also when generating system output, multimodal interaction better fits the human communication capabilities. This increases the naturalness of human–system interaction; see, for example Bongers et al. (1998). We believe multimodal interaction will only become more important in the context of the design of intelligent systems. Seen from the viewpoint of the system, a rich interaction with the environment is a first and necessary requirement to be able to show intelligent behavior. Multimodal-interaction technologies enable the system to extract more relevant information from the environment and to subsequently, based on explicit knowledge of the context-of-use (who is where, is doing what, with whom, etc.), take appropriate action. We are currently working on multimodal interfaces that enable systems to better identify when somebody is directing explicit attention to the system. In contrast to the often-studied "single user–single system" scenario, we investigate situations where the presence of multiple people is highly likely and the interaction with the system is interleaved with interaction between people. In such more realistic scenarios it turns out that besides the use of verbal cues (e.g., explicitly addressing the system or fitting the incoming utterance to a language model) nonverbal cues like the eye gaze of both the speaker and the listener can improve the identification of the intended addressee in mixed human–human and human–computer interaction (Van Turnhout et al. 2005)

In future AmI environments where many intelligent objects will be present, it seems inevitable that the various objects have to communicate to people what functionality they offer or what interaction possibilities they support. As mentioned above (Insight 3), this "information push" should preferably only be instantiated by the objects if it will be judged meaningful and appropriate by people. From a system's perspective this dilemma poses a complex problem that seems extremely difficult to solve by means of computational intelligence. Multimodal approaches to calm technology in which various output modalities are combined offer alternative ways to address this problem. Unused sensory channels can be allocated in parallel to the modality supporting the primary activity. These complementary modalities can be used to communicate background information that people subconsciously perceive but that can be easily put in the foreground when necessary. We are currently exploring the possibilities of haptic devices for enhancing person-to-person interaction over the Internet. More specifically, we are investigating the usability and fun of foot interaction enabled by real-time haptic signals; see Fig. 19.7. First experiments in which the Foot IO devices were used in an Instant Messaging application showed that users abundantly used "hapticons" and experienced the addition of haptics to the IM application as "more fun" (Rovers and Van Essen 2004; Van Essen and Rovers 2005).

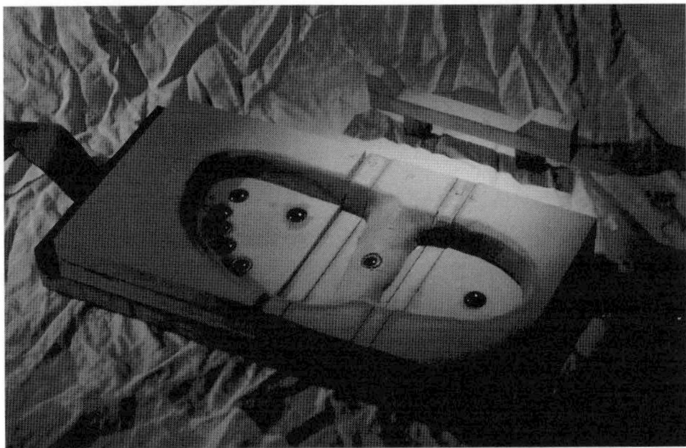

Fig. 19.7. Foot IO prototype (© Eindhoven University of Technology)

19.5.2 Interaction with Intelligent Tangible Objects

Today's electronic devices to a large extent integrate similar basic functional building blocks: audiovisual displays to present system feedback, storage capacity and processor power to store, process and generate information, and user interface components to enable users to directly interact with the system. New developments in the area of wired and wireless network technology have allowed these functional components to separate and migrate into the environment. Audiovisual displays, for example, can be integrated in tables, walls, windows, and even in clothing and storage capacity and processing power can disappear out of people's sight by being hidden in the environment. And, although, sensors, cameras and microphones, in principle, provide system designers the means to replace physical interaction elements with a perceptive environment, we strongly believe that also in the future tangible interaction objects will remain important. In the recent past, in many Ambient Intelligent projects it has been demonstrated that people like to grab and use physical interaction objects and that tangible interfaces can be very effective in supporting people to fully immerse in the creation of valued experiences (Insight 5). Examples of projects include the WWICE project in which physical tokens were designed and used to move multimedia activities in the social context of the home environment (Van de Sluis et al. 2001) and the Phenom project in which souvenirs were used to support the recollection of autobiographical memories (van den Hoven and Eggen 2003). Recent work by Fels has shown that the actions performed on an object and the resulting system behavior should subjectively match in order to guarantee that the emotional content of a message is effectively transmitted (Fels 2000). When this subjective match is sufficiently strong the object will be experienced as an extension to the human body instead of being treated as a "separated" interaction tool; the interaction has become "embodied" (Dourish 2002). When embodied interaction happens, it is the interaction itself that satisfies the user and no longer the achievement of the final goal only. Further research is needed to uncover the design requirements that tangible interaction objects need to meet in order to bring about embodied interaction.

A second line of research currently being explored concerns making tangible objects "smart" by empowering them with built in sensors and actuators, processing power, and communication possibilities. Within the department of Industrial Design of the Eindhoven University of Technology we recently started investigations into ways objects can express basic emotions through movement (Kyffin et al. 2005); see also the emotional walnut of Fig. 19.8. To explore options for adding behavioral expressions to existing movement possibilities of objects and to study the mediation of behavior a new approach called 4D sketching was developed. This method includes sketching in 3D space with the active behavior adding a fourth, temporal dimension. Easy-to-use materials for spatial sketching like foam and cardboard are used in addition to microprocessors, servomotors, and sensors to sketch the active

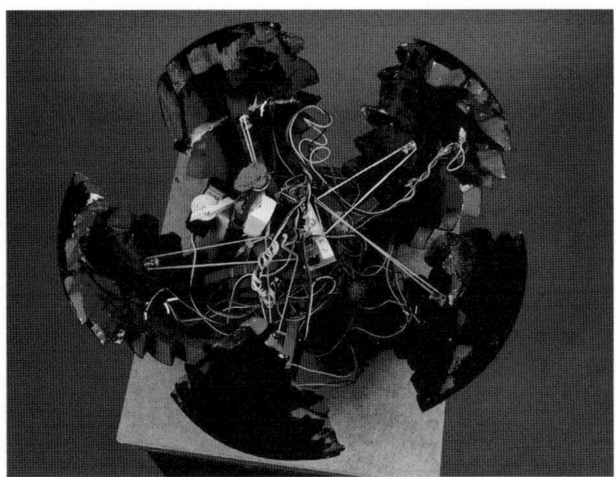

Fig. 19.8. Crack, the angry walnut machine (open) (© Eindhoven University of Technology)

behavior. Early results show that objects indeed can express at least basic emotions through movement, but also that it is particularly hard to separate form, color, motion and sound, and that an integrated multimodal design approach is needed.

19.5.3 Ambient Culture

Connectivity is one of the core functionalities of future electronic systems and offers new opportunities to connect many simple smart objects through a network. These objects can then collaborate resulting in emergent group behavior that fulfills a specific user need or supports a desired user experience. Watching television could, for example, become a total new experience when different electronic systems in the living room like lighting, sunblinds, central heating, telephone, electronic agenda, furniture, and audiovisual consumer products become equipped with "eyes and ears" and rules that define how signals in the environment should be interpreted, and the resulting behavior of the individual systems should be adjusted accordingly (Diederiks et al. 2002). It is important to realize that this collaboration does not necessarily result from explicit planning, but emerge from the relative simple interactions between the individual devices without one of them taking the central "lead." We have successfully applied such a "decentralized systems" approach in the software domain to automatically select audio tracks from a music database to generate "playlists" that fit particular moods like exciting music for a party or more relaxed music for a romantic dinner (Pauws and Eggen 2003). We are currently investigating the possibilities to embed these software agents in physical objects that can collaborate in an intelligent way with people to support the co-creation of desired user experiences (Insight 1). In general, we are interested in the ecology, the *Ambient Culture*, which will arise from the interactions between people and these smart objects and from the interactions between the objects themselves. See Chap. 3. Such communicative acts eventually will lead to a dynamic set of shared attitudes, values and goals defining the quality of the relationship of people, and their smart environment.

19.5.4 User-Centered Design

Over the years and throughout this paper we have advocated and demonstrated the involvement of end-users in the design process of intelligent systems. This so-called *User-Centered Design* approach was recently sustained by the working group "Experience and Application Research" of the ISTAG of the European Community (ISTAG 2004). We believe successful introduction and acceptation of intelligent products and systems in daily life will be strongly influenced by the way in which these products will enable people to shape and enhance their everyday social and cultural experiences. This calls for a continuous effort into the development of new design and evaluation methods that explicitly addresses these social and cultural aspects of intelligent product environments. Below we will elaborate four different lines of research.

We again want to stress the importance to better know and understand people's daily routines. People have indicated these routines greatly influence their personal well-being; see for example Eggen et al. (2003). It should be noted that, although the designers could not create the routines themselves, the electronic products they design and that take part in these routines could. Of course, any product should in first instance just offer well the functionality it is supposed to deliver. But, in addition, the designer should explore new dimensions of the user experience that enable the user to seamlessly integrate the product in existing or new routines (Insight 2). Ethnomethodology represents a direction within social psychology investigating commonsense routines people apply to manage and organize everyday life (Dourish 2002). We are currently developing new design methods based on and inspired by ethnomethodology that will be used to uncover commonsense routines that are of importance for the area of health and well-being.

The design of decentralized intelligent systems deserves special attention. As explained before, the overall behavior of a group of connected devices emerges from the interactions between the individual devices. The designer of this kind of system faces the difficult challenge to design the situational behavior of the individual product, while, at the same time, keeping an eye on the group behavior of the community of products. Currently, we are developing user-centered design methods that involve people to solve this complex design problem; see Fig. 19.9. In a "Wizard-of-Oz"-like experimental setting, persons temporarily put themselves in the position of the smart devices. In an iterative "playing–reflection" process the local roles for each "human-as-

Fig. 19.9. User-centered design – "gaming" focus group with children (© Eindhoven University of Technology)

device" are defined leading to emergent solutions that solve the decentralized design problem.

The issues of privacy deserve special attention (Insight 4). Personalization will become a key characteristic of the interaction of people with future intelligent systems. In practice, this means a system has to build up a user profile over time. The use of such a profile is especially critical when it comes to applications that support social relations between people as well as in situations in which the profile autonomously carries out actions on behalf of the user. In these situations the user should be able to trust the profile. Trust in the system needs to be building up over time on the basis of experiences obtained in daily interactions of the user with the system. Also, the way in which the user is involved in the actual building, the manipulation, the reconstruction, and the application of the distributed user profile determines the level of trust the user is willing to attribute to the system. Within the faculty of Industrial Design, we are currently conducting research on cost–benefit aspects of the disclosure of privacy sensitive information (Perik et al. 2004).

So far, the stakeholders directly involved in the development process of AmI applications mainly belonged to the design and research departments of our organizations. The next stage of design research calls for a more collaborative and interdisciplinary approach than ever before including business development. This is compounded, on the one hand, by the need to innovate more rapidly in a dynamic and shifting market, and on the other hand, by the need to differentiate in a saturated market by more fully incorporating end-user insights and providing true value. Within this context, we propose that the envisioning, designing, and development of innovative solutions require a multiple "creative process" which encompasses all the contributory issues, from people insights and context to software and platforms, simultaneously. Within Philips we have applied this approach in 2004 for the first time through the establishment of a mutual agreement among various business, design and research groups to work in partnership towards a joint vision demonstrator. This collaboration has led to its first result, the Intuitive Connected Home II demonstrator targeted at aspects of lifestyle, healthcare and well-being in the home and on the move. In the future, this approach may be adapted to enhance the way the company works with external partners to the benefits of all (Andrews et al. 2005).

19.6 Conclusion

In our work described in this paper we have always taken experience design as a vehicle to create true points of contact between people and future technological possibilities. The many AmI applications that were implemented throughout the work served as true encounters of the invisible future from which we were able to derive a number of key insights that are of prime importance for the design of intelligent products and systems. These validated insights helped us to focus our research efforts into new directions that eventually will deliver the missing detailed design relevant knowledge that we need to build the next generation of successful real-world AmI applications.

20

Experience Research

E.T. Hvannberg

"Of all affairs, communication is the most wonderful."

John Dewy (1859–1952) in Experience and Nature, 1925

20.1 Introduction

Research is meant to explain, create, and evaluate entities to advance knowledge. we ask questions about relationships between entities, cause, and effect. We invent purposeful entities that are valuable, and explore whether entities have desired characteristics. Morrison and George (1995) describe research approaches as formative, descriptive, evaluative, and developmental. Basili et al. (1986) list four types of purposes in the scope of software engineering experimentation: to characterize, evaluate, motivate, and predict. The underlying motivation may be to understand, assess, manage, engineer, learn, improve, validate, and assure. Wynekoop and Conger (1990) describe research purposes, based on the work of Basili et al., as understanding the meaning of entities studied, engineering the original development of a prototype, re-engineering an existing entity, evaluation of an entity, and describing or defining entities.

Ambient Intelligence implies that technology is *embedded* in everyday objects, such as walls, furniture, even nature, and surrounding humans in an *intelligent* way enabling them to participate in a networked society. The AmI space is open, allows evolution and extendibility. It is dynamic to allow for reconfiguration (ISTAG, 2003). Technologies in AmI space must have qualities such as safety, reliability, security, privacy, usability, and interoperability. People must have trust and confidence that technologies bear these qualities. In Ambient Intelligence the user is in control and not controlled.

New technology needs to be developed to reach the goals of AmI space: on one the hand, basic system components, such as smart materials, micro/nano-electronics systems, I/O technology, embedded systems, and ubiquitous communication, while on the other hand multimedia, natural interaction, context awareness, emotional computing, multi-modality, and computational intelligence will serve the needs of the user. To connect these two sides, i.e., the ambient and user technologies, researchers and developers need an integration research platform.

The AmI vision motivates us to consider whether researchers and developers need different types of methods to invent and evaluate technology that realizes Ambient Intelligence, than is already used. A plethora of methods for research have been used in the fields of computer science, information system, and software engineering, but with different emphasis (Glass et al. 2004). Computer scientists inventing technology, information systems scientists evaluating technology, and software engineers studying processes and methods, reuse, measurement, quality, and trade-off of options.

Without specifying in detail the characteristics of the type of research needed, it is clear that in order to realize Ambient Intelligence, both inventive and evaluative researches are required and a strong bridge, or an interleave, between the two types of research is imperative. Our problem will not only be providing interaction between the above three related fields of information technology, but to involve a broad range of scientists, e.g., from natural, human, social sciences, engineering, biological sciences as well as people knowledgeable about the application domain. Current research platforms tend to stimulate invention or evaluation, innovation or adoption, but we need an environment that facilitates both sides, aplatform that is close to the situation and the context, yet that accommodates models of information technology, which have not been implemented yet, into robust systems, components, or methods.

Increasing cooperation between developers and researchers allows for larger scale experiments, observations, and analytical studies of different interfacing technologies. Teamwork aids engineers with reusing designs, trading off different design options, and validating quality solutions.

As the AmI space will evolve and alter the ecology in an unforeseen manner, researchers need tools that can predict changes in needs and assumptions. These types of tools could enable developers to build technologies that can adapt to the change.

Experience research entails that researchers allow people to experience life through innovative technology, observing the interaction in the problem domain: people interact with technologies, people interact with people in the presence of technologies, and technologies interact with other technologies. Researchers observe technologies' affect on people: on people's performance, emotions, feelings, moods, health, or on any other measurable characteristics. During these observations, researchers discover the technologies' facilitators and hindrances. The researcher is not only a presenter of technology but also an apprentice to the application domain, exploring people's experiences, not only eliciting but also predicting needs and values in an evolving life, and improving technologies. In user experience research, people need not only be subjects of research but can be active participants, taking responsibility and influencing their future. Experience research is a holistic approach capturing the total experience, yet focusing on a certain aspect and its interaction with the outside.

20.2 Goals of Experience Research

The goals of the experience research platform that integrates the two previously mentioned sides of Ambient Intelligence, i.e., the components in the background and the people in the foreground, can be formulated as follows:

- allow research for Ambient Intelligence, i.e., embedded, transparent, mobile, intelligent, inclusive, open, evolving, and extensible,
- innovate at the boundaries by relating phenomena from different scientific areas, both theory and practice,
- provide different levels of experiences to users, researchers, and developers from abstract, primitive, limited in scope and span, through several iterations of refinement, to detailed, comprehensive and robust systems,
- enable research for large-scale use of diverse scope,
- bridge the gap between invention and evaluation, and
- offer research platforms for industry or academia alike for researching technologies aiming for different values from improved family life to increased competitiveness.

To reach the above goals we introduce an Experience and Application Research Centers (EARC) life cycle model in the next section, but first we review current research methods briefly and discuss expectations of future research methods.

Research methods can be categorized according to where they are used, e.g., in a natural environment or in a laboratory, on one subject or many, whether the methods are used in applied research for a specific goal or basic research where there are no restrictions on solutions. Action research, case studies, and field studies are examples of research methods that are conducted in a natural setting. In action research the practitioner is the researcher; phenomena are observed and improved in iterative cycles where practitioners reflect on their practices (Dick 2000). Case studies and action research are applied to a single or a few entities but field studies and field surveys are usually applied to many entities. Experiments can be conducted in a laboratory or in the field. In-depth interviews and observations are examples of methods of ethnography where studies are always carried out in real setting, take a holistic approach, and are often longitudinal. Conceptual analysis and mathematical proof are analytical methods.

Emerging research platforms need to take into account both analytical and empirical research. The former is not limited to the current practical realm; the latter creates data in actual surroundings among people, outside in open air or inside, in a private, social or organizational contexts. Both arms of research induce and deduce knowledge based on the data collected. Various types of data are collected, e.g., data that can be converted to numerical data that are processed quantitatively and descriptive data, people's expressed opinions and behavior, that is analyzed qualitatively (Taylor and Bogdan 1998).

20.3 EARC Life Cycle Model

Polanyi (1962), the philosopher of tacit knowledge, states that "The art of scientific research defies complete formalization; it must be learned partly by examples from a master whose behavior the student trusts." On the one hand, despite of this, we offer a framework for experience and application research as an account of what we believe is a synthesis of current practices. On the other hand, in the spirit of Polanyi, we believe that strong cooperation among researchers within this framework will greatly benefit experience and application research.

We aim to define an EARC life cycle model to attain the goals stated in the previous section. The EARC life cycle consists of the following phases: exploration of the problem situation; *feasibility and quality in interaction* (F&Q); *demonstration and validation* (D&V); and finally the experience phase. Several life cycle research models have been suggested, including one for software research (Glass 1994). First is the informational phase, where information is gathered via reflection and literature is surveyed. Next is the propositional phase, where a model, method, hypothesis, or an algorithm is presented. Following the propositional phase is the analytical phase, where the proposition is analyzed leading to a demonstration and/or formulation of a theory. Finally, an evaluation phase where the proposal or the result of the analysis is evaluated in controlled experiments or observations. Similar to this is a cycle of exploratory research followed by confirmatory research with refinements or improvements of the artifact in between. Exploratory research is often conducted with qualitative, non-empirical techniques but confirmatory research with quantitative, empirical techniques (Straub 1989).

One of the challenges we put forward to scientists is to break away from tradition, and carefully select a research method for each of the EARC phases, depending on the goal. To guide the researcher we propose the following dimensions that characterize the phases. It is not our intention to specify what type of research methods should be applied in each of these phases, but to let the framework guide the researcher in selecting appropriate methods and tools. The second challenge is to ask scientists to abandon a strict transition in a cycle between the phases and let the research questions and the answers guide them between the phases. It should motivate researchers to bridge the gap between exploration/evaluation and creation, with the former feeding into the latter. In describing the EARC life cycle in subsequent sections, we examine how the research setting changes throughout the life cycle with respect to the following dimensions which can be used during research design:

- *Approaches.* Description, explanation, prediction, invention, and evaluation activities
- *Methods.* Qualitative or quantitative. Case study, field study, action research, lab experiment, field surveys, applied research, or normative writings

- *Artifacts.* Different types of models that provide the capability to experience the technology
- *Context.* Organizational, social and technological, cognitive, temporal, and spatial contexts
- *Interfaces.* Interfaces are provided, either explicitly or implicitly, to other systems that are computational, mechanical, electrical, biological, geophysical, and human
- *Scope and span.* The scope and the span of the artifacts and the research platform changes throughout the life cycle
- *Transformation*: The transformation, outcome, or the benefit, resulting from development of the technology intervention
- *Reliability and validity.* Expectations of degree of reliability and validity of the research
- *Resources.* Human resources, equipment, tools, and infrastructure are needed

20.3.1 Exploration of the Problem Situation

There has to be a motivation for research, or else it is not valid. We want to advance knowledge by explaining, creating, or evaluating phenomena. EARC assumes that knowledge discovery takes place in context of some problem situation. We are able to describe an existing situation that will motivate search for knowledge, which can be in the form of product, process, model, or theory. Jackson (1995) emphasizes analyzing the problem context, to include two distinct domains, i.e., both the system and the application domains that contain the real world or subject matter. Lieberman and Selker (2000) focus on the context of the system itself. Context is everything but the explicit input and output of an application. Context is the state of the user, state of the physical environment, state of the computational environment, and the history of user–computer–environment interaction. Phrased differently, the context of the system includes *who* is the actor, *when* (e.g., in time) an actor operates the system, *where* the actor operates it, and *why* the actor activates the system (Abowd and Mynatt 2002). Within human–computer interaction, situated action has been extensively discussed as the current situation or context that drives the users' actions ad hoc and not a predefined plan (Suchman 1987). The method of situated action is a detailed method analyzing activity deeply at an atomic level. Contrary to this, activity theory uses structures for situational analysis, such as to situate work and technology historically, to situate the technology in a web of activities, characterize the use, consider the support needed, identify the objects worked on, in or through the technology, consider the web of activities and the contradictions in and between them (Bertelsen and Bødker 2003). Artifacts, i.e., technologies as instruments, are the focus of analysis, how they are used in activities and how activities influence them. In situation awareness, the emphasis is on perception of the situation, comprehending its meaning and projecting future status of entities in the situation (Endsley 1995).

It will not be sufficient to learn about the current problem situation and look at current user requirements, but we need to predict future problem situations and create user requirements. This projection based on current knowledge will be fundamental in information technology research. One of the aims of this phase is to try to predict the problem situation longevity so that when researchers have found the solution, competitors will not have solved it, the problem removed by the advent of another technology or other changes in context.

A research team should focus the scope of the problem situation under inspection, but can virtually widen the scope and span by building a connection to another adjacent problem situation currently under consideration by a different team. During this phase, we expect to see descriptive and evaluative approaches used to help understand the problem situation. It is desirable to perform the work in actual context, with the expectation to produce greater validity. Reliability will be increased with the same protocol applied to more then one instance, thus increasing the span.

Models, created during descriptive or evaluative activities in this phase, can explain characteristics and relationships between phenomena. Examples are task analysis models, conceptual models, cognitive work analysis models (Vicente 1999), consolidated work models (Beyer and Holtzblatt 1998), and scenarios (Go and Carroll 2004). They are products of ethnography and contextual inquiries. Quality models (ISO 2001) should be built during this phase, to understand the constraints of the development and to prepare for validation in later phases. Furthermore, the expected reliability and degree of validity should be described.

20.3.2 Feasibility and Quality of Interaction

Once an initial concept to solve a problem situation has been created, or a first attempt of an intervention has been developed, we evaluate its feasibility. We evaluate whether it is possible to build the desired product, process, or model, taking into account the current resources, constraints, and the expected benefits. A feasibility study can address questions such as what methods, tools, and components we currently have at our disposal to build the desired solution, and whether we can design the solution with the current manpower. One of the feasibility questions may attempt to evaluate how much reliability we expect to validate the solution. Boehm (1988) describes round zero of the spiral life cycle model for software development as a feasibility study. Closely related to such a study is risk analysis.

We expect to conduct the feasibility study analytically in a laboratory, but initial validation of quality of interaction should be carried out preferably in context. The validation is initially expected to be small scale, in a limited scope. We foresee that a small contextual laboratory be created in the workplace, at home, or elsewhere in the application domain. Parts of contexts, e.g., users, may be available, but other contexts may not be obtainable, such as working environment, especially in real-time safety-critical applications. If real context is not attainable, then it can be simulated in a research laboratory. The ideas are similar to industry-as-laboratory, which was proposed by Potts (1993) to avoid to research first and then transfer the knowledge. Action research works on similar premises, where the practioner is the researcher. Development to realize Ambient Intelligence does not confine contextual laboratories to industry, but can be built in any type of application domain. Contextual laboratories are not restricted to technical research only but include social, organizational, and cognitive factors.

The vision, that people in Ambient Intelligence are in the foreground, stimulates us to validate quality of people's participation in Ambient Intelligence. Usability is a quality characteristic of interaction between people and information technology. Usability is measured by evaluating user's effectiveness, efficiency, and satisfaction while carrying out tasks in a particular context. Other related characteristics are accessibility, aesthetics, creativity, and emotions. In Ambient Intelligence, technology will be highly interconnected and therefore developers stress any type of quality. This applies, e.g., to security, performance, and interoperability. Good maintainability allows users and developers to change technology according to evolving demands and interacting components. Openness provides a technology platform that is open for reviews and changes. Transparency shields users from heterogeneity in underlying implementations. Adaptive technology adapts to other interacting systems, but adaptable technology is customizable by the actor through an interface, e.g., the user interface.

While inspecting solutions, a number of options will be evaluated and their tradeoffs and contradictions examined, documented, and weighed. Hence, iteration between invention and evaluation is frequent and the activity will be repeated after successive exploration of the problem situation.

20.3.3 Demonstration and Validation

To widen the span of a technology experience, a demonstration center shows to the general public results of their research and development. A demonstration center allows people to observe, interact, and reflect on technology. A D&V can help raise interest in research with young people and enhance people's understanding. D&V centers can be seen as marketing platforms. Pre-marketing demonstration can be a vehicle to increase future user acceptance. A second goal of a demonstration and validation center is to provide a platform for larger scale validation than in the feasibility and quality of interaction phase.

The D&V center can be physical or virtual depending on the technology researched. A demonstration center is a synthesized application domain, such as a house, an office, a school room, a manufacturing floor, a car, a control room, a hospital surgery, etc. D&V centers are for sharing resources and a platform for cooperation among different research teams working on interacting technologies from interdisciplinary fields, regardless of their similarity or dissimilarity. Another reason for building D&Vs is that when developing emerging technologies intended to be deployed in the future, say several years, it is not possible to test them in a current context.

Whereas the span of the F&Q center is small and limits the number of people having access to research results, the D&V center allows more people of different background to survey the results. The goal is to increase the scope and span of validation in order to increase the validity and reliability of the research. While in F&Q, where a laboratory is built, e.g., in one hospital, a D&Vs synthesized hospital room allows researchers and developers of technologies to bring their products together to demonstrate it in context with other new technologies. The scope of an experiment can be enlarged through interfaces to other technologies.

Controlled experiments that produce quantitative data may be a suitable choice of validation method for this phase. As an example is a questionnaire, asking about user acceptance, Davis and Venkatesh (2003) have presented and studied a model called Technology Acceptance Model (TAM) extensively, which measures perceived usefulness, ease of use, and intention to use. They formulate and validate a Unified Theory of Acceptance and Use of Technology (UTAUT) that extends the three constructs in TAM with attitude towards using the system, social influence, facilitating conditions, self-efficacy, and anxiety. The D&V center is also suitable for a range of qualitative studies such as interviews or observational studies.

D&V centers provide research methods, tools, and skills to conduct the studies. The aim is to provide researchers with shared resources, whether they are tools or contexts.

20.3.4 Assisted Reality

When new technologies are researched in assisted reality, they are placed in the application domain, but since they may be unfamiliar to interacting components or actors, assistance is provided. An example is an automatic, unmanned, check-in teller, placed in an airport. Stressed passengers may avoid the automatic teller and want to go to the familiar desk where a clerk will check them into a flight. Therefore, a human assistant helps passengers gain confidence in the new technology.

Assisted reality centers are not only used for transfer of technology or adoption, but also empirical research that can feed back into development. Experiments can either be controlled or uncontrolled. As in the case of D&V centers, assisted reality centers provide methods, tools, and environment to carry out experiments, but in actual context, whereas in the D&V it is synthesized.

If an airport is declared as an assisted reality zone, it needs to be done in close cooperation with airport operators and researchers need to have facilities to conduct their studies. Researchers must ensure that it does not interfere with the privacy or intentions of passengers, guests or employees and we hope they will be rewarded directly through improved environment.

The cycle is closed when technology has been fully embraced; it becomes a natural part of the application domain, and we move again to the problem situation, analyzing motivations for new research and development.

20.4 Facilitators of Experience Research

The following sections discuss four building blocks that we propose are useful in EAR: prototypes, experiments, experience and knowledge management, and cooperation. These are not mutually exclusive but we believe they can work well together to reach the goals of experience research.

There have been a number of examples of user experience research, e.g., in laboratories at IBM (Russel and Cousins 2004) and Philips (de Ruyter and Aarts 2004), gaining experience from ethnographical and contextual research and methods. The use of performative activities has been growing for exploration and testing of new information technology products and is potentially for the design of products for Ambient Intelligence. *Experience prototyping* enables design team members, users, and other stakeholders to gain appreciation of existing or future conditions through active engagement and role playing with prototypes.

Researchers need to pay more attention to how to interleave exploration, creativity, and successful evaluation to make up valuable innovation. Experiments will be an invaluable tool, but we need to analyze critically their methodology to see how they are suitable for AmI research.

One of the premises of success of EARC is good cooperation across research and development teams in order to increase the span and the scope of experiences and to encourage interdisciplinary research.

20.4.1 Prototypes

Prototypes have been used for designing interactive systems since early 1980s, when Mason and Carey (1983) suggested that scenarios and iterative development with user involvement help detect problems early. The motivation was that interpretive models were too complex for the user to understand. In his spiral model, Boehm (1988) advocated the development of prototypes to minimize risks.

Mock-ups are low-fidelity prototypes that can be built out of paper, cardboard boxes or anything else to simulate a finished information technology product or tool to empower the worker in workplace democracy (Bødker et al. 1987). Mock-ups were used as tools in participatory design, where the workers, as representative of the union, participated in technology development. There has been active research on methods in the tradition of participatory design, not only in Scandinavia where it originated but also in the United States, albeit with different emphasis (Muller et al. 1993).

Buchenau and Suri (2000) describe experience prototyping as a form of prototyping that enables design team members, users, and clients to gain first-hand appreciation of existing or future conditions through active engagement with prototypes. Experience prototyping can utilize mock-ups to create experiences early in the development process. Experience prototyping can be used to understand existing user experiences and their contexts, to explore and evaluate new designs, and communicate ideas to designers and stakeholders. Iacucci et al. (2002) give an overview of works that have been published describing group performances to experience ideas during early design phases. Experience prototyping is included in the overview as is their own case of situated and participative enactment of scenarios. Pering (2002) demonstrates that we do not need prototypes for software only, but are often building new hardware too and need to demonstrate the software–hardware interaction. This supports our view that interaction between components is relevant in Ambient Intelligence.

Olsson (2004) presents a participatory design process, after conducting empirical studies with skilled workers, which recommends that stakeholders need time to develop a design with user involvement, that users need to reflect on their needs and new designs, and that greater degree of user involvement is imperative. Designers and users need to get out of the box, thus freeing themselves from their current situation, and learn about new technologies, while maintaining a balance between work and technology. A D&V center can be a place that supports this participatory design process.

Cultural probes and artifact walkthroughs are examples of methods that help designers create new designs and let users or designers, as apprentices to domain experts, experience them. In artifact walkthroughs, a user is asked to apply the think-aloud protocol. Cultural probes are meant to evoke people to communicate their experiences through, e.g., diaries, photography, or videos, while mobile probes attempt to offer contextual interactive probing tools (Gaver et al. 2004; Hulkko et al. 2004). Making Tea (Schraefel et al. 2004) is a design elicitation method that aims to bridge the communication gap between designers and domain experts through analogy. Probology is meant to be playful and subjective, inspiring new design ideas. It has roots in ethnography and hence produces qualitative data.

Research in the area of prototyping has advanced from offering a prototype as an artifact to a process of using the prototype to help elicit information about the context, project future scenarios, and validate a design. Research investigating the validity of low-fidelity prototypes has been carried out to see whether users can express ease of use, usefulness, and intention as reliably in pre-prototypes as in high-fidelity prototypes (Davis and Venkatesh 2004). The study showed that usefulness measures taken before prototype development loosely approximate hands-on usefulness measures and are predictive of usage intentions up to 6 months after workplace implementation of the system. In the study, a survey is used to collect user perception. In a qualitative study,

Hall (2001) reports results of four studies that show how different prototype fidelities can provide different types of data. For Ambient Intelligence it is noteworthy that to examine visual, auditory, and tactile needs of the user higher-fidelity prototypes are needed, while lower-fidelity prototypes can be used to examine the cognitive or information processing needs. The guidelines that can be used for role playing and low-fidelity prototypes in user-centered and participatory design for the software design of future mobile systems will be useful to developers (Svanæs and Seland 2004).

Alavi (1984) has shown that one of the problems with prototypes is that they are difficult to manage and control. The three studies cited above investigate the process of working with prototypes. Given the popularity and long use of prototyping, it is disappointing that researchers have not provided better connections to software engineering models, which give developers tools to build quality software. In Ehn's (1992) discussion on prototypes, he not only discusses prototypes and mock-ups but also descriptions and representations to enable us to share knowledge. Unfortunately, relationships between other types of representations and prototypes are lacking, and models that capture results of prototype evaluation have not been visible in research.

20.4.2 Experiments

The EARC life cycle should lend itself well to experiments, both controlled and uncontrolled in various settings, ranging from out of context to full context. We hesitate to use the word laboratory here for out of context, since we see both that a laboratory can exist in the field, e.g., the F&Q centers, and that the context can be partially brought into a synthesized setting, such as the D&V centers.

Experiments can serve both exploration and evaluation purposes. There can be several reasons why experiments in context are not prevalent, e.g., in human–computer interaction research on mobile devices as reported by Kjeldskov and Graham (2003). Their review showed that the majority (61%) of the papers engineered technology in applied research and evaluation was performed in an artificial laboratory setting in majority of cases (71%), only 19% in field experiments and the remaining 10% through surveys. Fundamental fallacies can be the cause as reported by Tichy (1998): we are experimenting enough, experiments are too costly, demonstration will suffice, there is too much noise, e.g., impossible to control, experimentation will slow progress, technology changes too fast, and experiments will not get published. Some of these are reported from the researcher in academia point of view, but the practitioner in industry could give other excuses, e.g., that experiments do not produce returns on investment or that they increase product's time to market. We need to remember that experimentation in industry is conducted to fulfill the needs of specific organizational goals instead of advancing general knowledge. If we expect Ambient Intelligence to be realized, we need methods and tools to conduct experiments that researchers and developers in academia and industry can use with visible satisfactory results.

20.4.3 Experience and Knowledge Management

Experience factories attempt to capture experiences for software development organizations (Seaman et al. 2003). The concept of experience factory was introduced well before knowledge management became popular. The same concept can be used for exploiting the experiences of introducing a new system into a domain or on a meta-level to learn from the experience of creating a new system for Ambient Intelligence.

In EARC we build and learn from experiences created in the past. Experiential computing provides an environment to researchers that can help them to explore and experience events from multiple perspectives, but customized to their own research interests (Jain 2003). Digital libraries, validated web resources as well as semantic web will be important tools. Data mining should be explored as a means of searching for best practices, patterns, and guidelines of interaction, especially between man and machine. The tools of EARC will motivate intelligent processing of empirical data that we can aim to use as a basis for automatic design and validation. In addition to automatic analysis, EARC emphasizes strong visualization of experiences. The expected complex and even conflicting criteria of Ambient Intelligence require tools that can evaluate different design solutions. The scope of the design is different in EARC than we have seen before because it will enable collaborative design among groups and it will span interdisciplinary teams. This calls for collaborative intelligence that is one of the seven layers of knowledge management.

20.4.4 Cooperation

For many years, scientific programs such as the European Research Framework Programs have encouraged cooperation in research. Cooperation has taken place across different dimensions, e.g., size of partners, industry vs. academia, culture, different application domains or industries, etc. In the European Research Framework, since several years ago, demonstrators and field studies with the construction of test beds have been encouraged, if not required. In the early days of Internet applications, this was prevalent. Moreover, in many of the experiments undertaken, cooperation with industries has been made a mandatory part of research projects. Understandably, researchers have focused on reporting results of these test beds and field studies, but dissemination of experiences of project management and the research platforms themselves has been less than desired. Nonetheless, the knowledge and experience should give researchers good foundation for cooperation in EAR.

Different types of instruments have been used for this purpose. Examples are networks of excellence and integrated projects where a set of organizations or research groups carry out research independently on a shared platform under the umbrella of a larger goal. Internet technology has been a major facilitator of research cooperation with high-speed connections, audio/video communication, hyperlinks, exchange mark-up languages, meta-data, etc. Mobility of researchers has also been encouraged, realizing that sole virtual presence is not always adequate, especially if previous physical encounter has not occurred.

Studies that examine the relationships between complexity components and project performance need to be performed for large research projects. Xia and Lee (2004) present a study among 541 North-American IS development projects, in which they found that in order to improve performance, focus on technological aspects is not enough, but that organizations should pay close attention to organizational aspects such as relationship with top management and end-users.

The difference between past research projects and EARC is that it will be larger in scale and scope than hitherto. This will impose new challenges in communications of goals and results that we expect emerging technologies like semantic web and grids, which have been associated to e-science (De Roure and Hendler 2004), can contribute to. Interdisciplinary teams or communities will require shared understanding and a common language to discuss research findings. Ontologies attempt to bridge the language barrier. Interdisciplinary research is known to be difficult and a powerful mechanism is required. Several attempts have been made to bridge, e.g., the gap between human–computer interaction and software engineering.

20.5 Example Developments in Experience Research

There are a number of initiatives and projects that are examples of Experience and Application Research. A brief description follows, but further details are enlisted in an ISTAG Report on Experience and Application Research (ISTAG 2004).

20.5.1 Living Tomorrow

Living Tomorrow (LivTom, online) is a demonstration center in Brussels that encourages cooperation of companies in different sectors. The center was founded in 1995 but has been rebuilt twice to accommodate new projects and technologies. Participating companies fund the center but in return demonstrate their products and prototypes to a wide audience, thus building image and enriching knowledge of future customers.

20.5.2 Philips HomeLab

Philips has built a special observation home where multidisciplinary teams of researchers can install technologies in normal home environments and involve people in experiments. The Home is a two-story house with a living room, a kitchen, two bedrooms, a bathroom, and a study. Adjacent is a lab, consisting of two observation rooms with two-way mirrors and video cameras, which researchers can use to observe real-life experiments. HomeLab has been used to investigate a number of problems, including how to locate people, interaction with speech-based systems, inference between wireless systems, gesture recognition, and hiding systems. A crosscutting emphasis has been on capturing the quality of user experiences.

20.5.3 Telenor's House of the Future

Telenor's House of the Future was built as a laboratory for living where research included user experiences in everyday life in future technology environments. In addition to laboratory experience work, basic research has also been conducted. Technical activities have involved different areas such as infrastructure, automation, and networks for homes, user behavior, and user interfaces. Telenor's experience was that the house contributes to research by giving focus to research, providing a platform for interdisciplinary collaboration in real-life contexts, and by supporting communication of complex solutions.

20.5.4 Brazilian Sao Paulo e-Government Facility

The Electronic Government Innovation and Access project (eGOIA, online) is an activity funded by the European Union. The main target of eGOIA is the demonstration of future-oriented public administration services to a broad public. The vision of the eGOIA project is to support the interaction of citizens, independent of social status, gender, race, abilities, and age, and the public administration in a simple and cost-effective way. The established eGOIA consortium consists of partners from Latin America – Brazil and Peru – and European partners from Germany, UK, and Portugal.

20.5.5 Responsive Home

CUHTec [CUHTec, online; Monk et al. 2004] is a joint venture initiated by the Joseph Rowntree Foundation and the University of York. CUHtec has established the Responsive home, which is a fully equipped house that will be used for the development and the demonstration of new products and systems. One important area of the home is to help disabled people and the elderly who want to live independently. An interdisciplinary team of researchers from psychology, computer science, and electronic engineering works in the Responsive home.

20.5.6 Innovation Lab Katrinebjerg

Innovation Lab Katrinebjerg in Aarhus is an innovation company, a miniature group of networking companies with close contacts to university research. Innovation Lab is a place where you can see and feel things. Projects in the lab need to fulfill several requirements. They need to be innovative. They must be used for presentation, like a prototype that can be touched, displayed, and can tell a story. The projects need to involve the Katrinebjerg's researchers, the lab's permanent partner companies, and small, specialized companies.

20.5.7 Intelligent House Duisburg Innovation Center – inHaus

The inHaus project (inHaus, online) offers a networked living environment following an integral concept of product-oriented innovations. The basis of the project is the inHaus facility in Duisburg, which includes a residential home, a workshop, a networked car, and a networked garden. Seventeen prominent national and international companies, which hold 5-year contracts with the Fraunhofer Institute, are involved in this project. The project is supervised by the Fraunhofer Institute for Microelectronic Circuits and Systems.

The above examples illustrate how a real-life home or office has been built and used as a platform for development, demonstrations, and evaluations. A number of lessons have been learnt that we summarize as follows (ISTAG 2004):

- *Facilities.* Facilities that are dedicated to longer-term studies should be located in the center of town or in a residential area. Facilities that are located in campus type environments are better suited for short-term studies, experiments, or developments.
- *Research Focus.* A coherent research program needs to be established around the facilities. This will make the facilities sustainable beyond individual projects. Some facilities like the Living Tomorrow are suitable for mass data collection but facilities like HomeLab are better suited for studies with small number of people.
- *Research Data.* Collection and analysis of research data should be planned well and made accessible across partners.
- *Costs.* Facilities like Living Tomorrow can be expensive especially since they need to be rebuilt every five years. In interdisciplinary research and across company's divisions, their needs to be an agreement and commitment to the investment needed. Other examples show that it is possible to do considerable research with a limited budget.
- *Field Research.* Experience of field trials shows that users want to actively participate in trials, and can provide useful feedback to developers. Field research can both be small and large scale, and span a small or large population. Field research can provide contradictory results to those previously obtained from other instruments such as focus groups or interviews but add validity.

- *Pitfalls.* The resources invested in the observation infrastructure used for experiments and the technology that is being experienced need to be balanced. A half developed prototype can be deterrent for future users. Care should be taken to make facilities human centered and not technology centered which may be a tendency when demonstrating technologies. There should be a focus on how users can participate and be involved, how they interact with technology instead of a single-sided technology assessment.

The IST (Information Society Technology) priority within the sixth Framework Program has encouraged further development by including a call for projects taking an EAR approach in the domain of Collaborative Working Environments.

20.6 Challenges

This chapter proposes an EARCs life cycle model of four phases. Horizontal facilitators across the phases have been described in this chapter, but others are undoubtedly necessary. Several common dimensions of the phases have been presented, but the chapter has not in any detail specified how the phases differ along these dimensions. Although the work presented here is based on experience and literature analysis, it remains to be seen whether EAR and the life cycle proposed results in innovative IT research and development. We challenge researchers to propose research methods that fulfill the stated goals of experience research in the framework of the life cycle model, possibly relying on the facilitators as building blocks. In particular, we challenge researchers to consider research methods that can deal with evolving environments and that can resolve conflicts between technologies, between goals, and between stakeholders, as these are among the essential characteristics of Ambient Intelligence.

References

AAAI (2005), *Smart Rooms & Household Appliances*, Webpage, http://www.aaai.org/AITopics/html/rooms.html.

Aarts, E. (2004), Ambient Intelligence: A Multimedia Perspective, *IEEE Multimedia*, 11(1), 12–19.

Aarts, E. (2005), Ambient Intelligence Drives Open Innovation, *Interactions*, XII.4, 66–68.

Aarts, E., and L. Appelo (1999), Ambient intelligence: thuisomgevingen van de toekomst, *IT Monitor*, 9/99, 7–11 (in Dutch).

Aarts, E., and J.K. Lenstra (eds.) (1997), *Local Search in Combinatorial Optimization*, Wiley, Chichester, UK.

Aarts, E., and S. Marzano (eds.) (2003), *The New Everyday: Views on Ambient Intelligence*, 010 Publishers, Rotterdam, The Netherlands.

Aarts, E., and R. Rovers (2003), Embedded System Design Challenges in Ambient Intelligence, in: T. Basten, M. Geilen, and H. de Groot (eds.), *Ambient Intelligence: Impact on Embedded System Design*, Kluwer Academic, Dordrecht, The Netherlands, pp. 11–29.

Aarts, E., R. Collier, E. van Loenen, and B. de Ruyter (eds.) (2003), *Ambient Intelligence, First European Symposium, EUSAI03*, Springer Lecture Notes in Computing Science, 2875, Springer, Berlin Heidelberg New York.

Aarts, E., R. Harwig, and M. Schuurmans (2002), Ambient Intelligence, in: P. Denning (ed.), *The Invisible Future: The Seamless Integration of Technology in Everyday Life*, McGraw-Hill, New York, NY, USA, pp. 235–250.

Aarts, E., J.H.M. Korst, and W.F.J. Verhaegh (2005), Intelligent Algorithms, in: W. Weber, J. Rabaey, and E. Aarts (eds.), *Ambient Intelligence*, Springer, Berlin Heidelberg New York, pp. 349–373.

Abelson, H., R. Weiss, D. Allen, D. Coore, C. Hanson, G. Homsy, T.F. Knight, R. Nagpal, E. Rauch, and G.J. Sussman (2000), Amorphous Computing, *Communications of the ACM*, 43(5), 74–82.

Abowd, G.D., and E.D. Mynatt (2002), Charting Past, Present, and Future Research in Ubiquitous Computing, *ACM Transactions on Computer–Human Interaction*, 7, 29–58.

Abowd, G.D., A.K. Dey, P.J. Brown, N. Davies, M. Smith, and P. Steggles (1999), Towards a Better Understanding of Context and Context-Awareness, *Proceedings of the International Symposium on Handheld and Ubiquitous Computing*, September 27–29, Karlsruhe, Germany, pp. 304–307.

Addington, M., and D.L. Schodek (2004), *Smart Materials & Technologies in Architecture*, Architectural, Elsevier Science, Oxford, UK.

ADSX (online), Homepage, http://www.adsx.com.

Alavi, M. (1984), An Assessment of the Prototyping Approach to Information Systems Development, *Communications of the ACM*, 27, 556–563.

Aldrich, J. (2003), *Using Types to Enforce Architectural Structure*, Ph.D. Thesis, University of Washington, Washington, WA, USA.

AmI@INI (online), Ambient Intelligence@INI-GraphicsNet, INI-GraphicsNet Foundation, Homepage, http://ami.inigraphics.net.

AmI@Work (online), Ambient Intelligence@Work, Homepage, http://europa.eu.int/information_society/activities/atwork/ami_family/index_en.htm.

Andrews, A. (2003), Networks, Systems and Society, in: E. Aarts and S. Marzano (eds.), *The New Everyday: Views on Ambient Intelligence*, 010 Publishers, Rotterdam, The Netherlands, pp. 212–217.

Andrews, A., M. Bueno , and J. Cass (2003), Deep Customization, in: E. Aarts and S. Marzano (eds.), *The New Everyday: Views on Ambient Intelligence*, 010 Publishers, Rotterdam, The Netherlands, pp. 114–119.

Andrews, A., L. Geurts, and S. Kyffin (2005), *Experience Innovation – Shaping the Future Together*, Internal Report, Philips Design, Eindhoven, The Netherlands.

Antifakos, S., F. Michahelles, and B. Schiele (2002), Proactive Instructions for Furniture Assembly, in: G. Boriello and L.E. Holmquist (eds.), *Ubicomp 2002: Ubiquitous Computing, Fourth International Conference*, Springer Lecture Notes in Computing Science, 2498, Springer, Berlin Heidelberg New York, pp. 351–360.

Argyle, M., and J. Dean (1965), Eye Contact, Distance and Affiliation, *Socionometry*, 28, pp 289–304.

Ark, W., D.C. Dryer, and D.J. Lu (1999), The Emotion Mouse, *Proceedings of the International Conference on Human Computer Interaction*, August 30–September 3, Edinburgh, UK, pp. 818–823.

Arnstein, L., G. Borriello, S. Consolvo, C. Hung, and J. Su (2002), Labscape: A Smart Environment for the Cell Biology Laboratory, *IEEE Pervasive Computing*, 1(3), 13–21.

Atkins, D. (ed.) (2004), *Revolutionizing Science and Engineering Through Cyberinfrastructure*, Report of the National Science Foundation Blue-Ribbon Panel on Cyberinfrastructure, http://www.nsf.gov.

Atkinson, M., J. Crowcroft, C. Goble, J. Gurd, T. Rodden, N. Shadbolt, M. Sloman, I. Sommerville, and T. Storey (2002), *Computer Science Challenges to Emerge from e-Science*, Report, http://www.semanticgrid.org/docs/cschallenges.pdf.

Baader, F., D. Calvanese, D. McGuinness, D. Nardi, and P. Patel-Schneider (eds.) (2003), *The Description Logic Handbook*, Cambridge University, New York, NY, USA.

Bachler, M., S. Buckingham, S. Shum, J. Chen-Burger, J. Dalton, D. De Roure, M. Eisenstadt, J. Komzak, D. Michaelides, K. Page, S. Potter, N. Shadbolt, and A. Tate (2004), Collaborative Tools in the Semantic Grid, in: Proceedings Semantic Grid Applications Workshop, *Proceedings of the GGF11 – The 11th Global Grid Forum*, July 4–7, Honolulu, HI, USA, pp. 75–85.

Bailenson, J., J. Blascovich, A. Beall, and J. Loomis (2001), Equilibrium Theory Revisited: Mutual Gaze and Personal Space in Virtual Environments, *Presence*, 10(6), 583–598.

Balsamo, S., A. Di Marco, P. Inverardi, and M. Simeoni (2004), Model-Based Performance Prediction in Software Development: A Survey, *IEEE Transactions on Software Engineering*, 30(5), 295–310.

Banerjee, U., R. Eigenmann, A. Nicolau, and D. Padua (1993), Automatic Program Parallelisation, *Proceedings of the IEEE*, 81(2), 211–243.

Bar-Cohen, Y. (ed.) (2001), *Electroactive Polymer (EAP) Actuators as Artificial Muscles – Reality, Potential and Challenges*, SPIE, Bellingham, WA, USA, Volume PM98.

Baresi, L., R. Heckel, S. Thone, and D. Varro (2003), Modeling and Validation of Service-Oriented Architectures: Application vs. Style, *Proceedings of the Fourth Joint Meeting of the European Software Engineering Conference and the ACM SIGSOFT Symposium on the Foundations of Software Engineering*, September 1–5, Helsinki, Finland, pp. 68–77.

Basili, V.R., R.W. Selby, and D.H. Hutchens (1986), Experimentation in Software Engineering, *IEEE Transactions on Software Engineering*, SE-12(7), 733–743.

Bass, L., P. Clements, and R. Kazman (2003), *Software Architecture in Practice* (2nd edition), SEI Series in Software Engineering, Addison-Wesley, Reading, MA, USA.

Basten, T., M. Geilen, and H. de Groot (2003), *Ambient Intelligence: Impact on Embedded System Design*, Kluwer Academic, Dordrecht, The Netherlands.

Baudrillard, J. (1968), *Le Système des Objets*, Gallimard, Paris, France, pp. 255–283, translated by J. Mourrain as *The System of Objects*, in: M. Poster (ed.) (1988), Jean Baudrillard: *Selected Writings*, Stanford University, Palo Alto, CA, USA, pp. 10–29.

Baudrillard, J. (1981), *Simulacres et Simulation*, Galilée, Paris, translated by S.F. Glaser as *Simulacrum and Simulation* (1995), University of Michigan, Ann Arbor, MI, USA.

Becker, C., and W. Glad (2000), Light-Activated EAP Materials, Jet Propulsion Laboratory's NDEA Technologies, *World Wide – ElectroActivePolymer Newsletter*, 2(1), p. 11.

Beigl, M., T. Zimmer, A. Krohn, C. Decker, and P. Robinson (2003), *Smart-Its – Communication and Sensing Technology for UbiComp Environments*, Technical Report, ISSN 1432-7864 2003/2, TecO, Karlsruhe, Germany.

Bernardo, M., and P. Inverardi (eds.) (2003), *Formal Methods for Software Architectures, Third International School on Formal Methods for the Design of Computer, Communication and Software Systems: Software Architectures, SFM 2003*, Lecture Notes in Computer Science, 2804, Springer, Berlin Heidelberg New York.

Berners-Lee, T. (1999), *Weaving the Web*, Harper Collins , San Francisco CA, USA.

Berners-Lee, T., J. Hendler, and O. Lassila (2001), The Semantic Web, *Scientific American*, 284(5), 34–43.

Bernstein, P. (1996), Middleware: A Model for Distributed System Services, *Communications of the ACM*, 39(2), 87–98.

Bertelsen, O.W., and S. Bødker (2003), Activity Theory, in: J.M. Carroll (ed.), *HCI Models, Theories and Frameworks*, Morgan Kaufmann, San Francisco, CA, USA, pp. 291–324.

Bertozzi, D., L. Benini, and G. De Micheli (2002), Low Power Error Resilient Encoding For On-Chip Data Buses, *Proceedings of the Conference on Design, Automation and Test in Europe*, March 4–8, Paris, France, pp. 102–109.

Beslay, L., and Y. Punie (2002), *The Virtual Residence: Identity, Privacy and Security*, IPTS Report, 67, pp. 17–23.

Bettstetter, C., and C. Renner (2000), A Comparison of Service Discovery Protocols and Implementation of the Service Location Protocol, *Proceedings of the Sixth EUNICE Open European Summer School: Innovative Internet Application*, September 13–15, Enschede, The Netherlands, http://wwwtgs.cs.utwente.nl/eunice/summerschool/papers/paper5-1.pdf.

Beyer, H., and K. Holtzblatt (1998), *Contextual Design*, Academic, London, UK.

Beyer, H., and K. Holtzblatt (2002), *Contextual Design: A Customer-Centered Approach to Systems Design*, Morgan Kaufmann, San Francisco, CA, USA.

Beyne, E. (2004), 3D Interconnection and Packaging: Impending Dream or Reality, *International Solid State Circuits Conference, Digest of Technical Papers*, February 14–19, San Francisco, CA, USA, pp. 138–139.

Bhatia, K., N. Koba, D. Lioupis, A.D. Lloyd, S. Mendiola, D. Milojicic, K. Sankar, and I. Taylor (2003), *Appliance Aggregation Architecture Terminology, Survey, and Scenarios*, Report of the GGF Appliance Aggregation Architecture Group, http://www.hpl.hp.com/hosted/ggf/AppAggSurveyApr03.doc.

Blair, G., and R. Campbell (eds.) (2000), *Proceedings of the Workshop on Reflective Middleware*, April 7–8, New York, NY, USA.

Bluetooth SIG (online), Homepage, http://www.bluetooth.com.

Bobrow, D. (1977), An Overview of KRL, *Cognitive Science*, 1(1), 3–64.

Bødker, S., P. Ehn, J. Kammersgaard, M. Kyng, and Y. Sundblad (1987), A Utopian Experience: On Design of Powerful Computer-Based Tools for Skilled Graphical Workers, in: G. Bjerknes, P. Ehn, and M. Kyng (eds.), *Computers and Democracy – A Scandinavian Challenge*, Avebury, Aldershot, UK, pp. 251–278.

Boehm, B.W. (1988), A Spiral Model of Software Development and Enhancement, *IEEE Computer*, 21(5), 61–72.

Bohn, V., M. Coroama, F. Langheinrich, M. Mattern, and M. Rohs (2004), Social, Economic, and Ethical Implications of Ambient Intelligence and Ubiquitous Computing, in: W. Weber, J. Rabaey, and E. Aarts (eds.), *Ambient Intelligence*, Springer, Berlin, Germany, pp. 5–29.

Bohnenberger, T., A. Jameson, A. Krüger, and A. Butz (2002), User Acceptance of a Decision-Theoretic, Location-Aware Shopping Guide, *Proceedings of the International Conference on Intelligent User Interfaces*, January 13–16, San Francisco, CA, USA, pp. 178–179.

Bongers, A.J., J.H. Eggen, D.V. Keyson, and S.C. Pauws (1998), Multi-modal Interaction Style, *Human Computer Interaction Letters*, 1(1), 3–5.

Böttner H. (2002), *Thermoelektrische Wandler, Stand der Technik*, Talk presented at the Second GMM-Workshop Energieautarke Sensorik, June 6–7, Dresden, Germany.

Bougant F., F. Delmod, and C. Pageot-Millet (2003), The User Profile for the Virtual Home Environment, *IEEE Communications Magazine*, 41, 93–98.

Boulkenafed, M., D. Sacchetti, and V. Issarny (2004), Using Group Management to Tame Mobile Ad Hoc Networks, *Proceedings of the IFIP TC8 Working Conference on Mobile Information Systems*, September 15–17, Oslo, Norway, pp. 245–260.

Bratman, M.E., D.J. Israel, and M.E. Pollack (1988), Plans and Resource-Bounded Practical Reasoning, *Computational Intelligence*, 4, 349–355.

Braunisch, H., S.N. Towle, R.D. Emery, C. Hu, and G.J. Vandentop (2002), Electrical Performance of Bumpless Build-Up Layer Packaging, *Proceedings of the Electrical Components and Technology Conference*, May 28–31, San Diego, CA, USA, pp. 353–358.

Broer, D.J., J. Lub, and G.N. Mol (1995), Wide-Band Reflective Polarizers from Cholesteric Polymer Networks with a Pitch Gradient, *Nature*, 378(6556), 467–469.

Broer, D.J., G.N. Mol, K.D. Harris, and C.W.M. Bastiaansen (2005), Thermo-Mechanical Responses of Liquid-Crystal Networks with a Splayed Molecular Organization, *Advanced Functional Materials*, 15(7), 1155–1159.

Bromberg, Y.-D., and V. Issarny (2004), Service Discovery Protocols Interoperability in the Mobile Environment, *Proceedings of the Fourth International Workshop on Software Engineering and Middleware*, September 20–21, Linz, Austria, pp. 64–77.

Brooks, R. (1986), A Robust Layered Control System for a Mobile Robot, *IEEE Journal of Robotics and Automation*, 2(1), 14–23.

Brooks, R.A. (2002), *Flesh and Machines*, Pantheon Books, New York, NY, USA.

Brown, G. H. (1983), Liquid Crystals – The Chameleon Chemicals, *Journal of Chemical Education*, 60(10), 900–905.

Brown, P.J. (1996), The Stick-e Document: A Framework for Creating Context Aware Applications, *IFIP Proceedings of Electronic Publishing '96*, September 22–26, Palo Alto, CA, USA, pp. 259–272.

Brumitt, B., B. Meyers, J. Krumm, J. Kern, and S.A. Shafer (2000), EasyLiving: Technologies for Intelligent Environments, in: P.J. Thomas and

H.-W. Gellersen (eds.), *Handheld and Ubiquitous Computing, Second International Symposium, HUC 2000*, Lecture Notes in Computer Science, 1927, Springer, Berlin Heidelberg New York, pp. 12—29.

Buchenau, M., and J.F. Suri (2000), Experience Prototyping, in: D. Boyarski and W.A. Kellogg (eds.), *Proceedings of the ACM Conference on Designing Interactive Systems*, August 17–19, New York, NY, USA, pp. 424–433.

Bui, H. (2003), A General Model for Online Probabilistic Plan Recognition, *Proceedings of the International Joint Conference on Artificial Intelligence*, August 9–14, Acapulco, Mexico, pp. 1309–1318.

Burgelman, J.-C. (2000), Innovation of Communication Technologies: Some General Lessons for the Future from the Past, in: B. Cammaerts and J.-C. Burgelman (eds.), *Beyond Competition: Broadening the Scope of Telecommunication Policy*, VUB, Brussels, Belgium, pp. 229–237.

Burkhardt, W., T. Christmann, B.K. Meyer, W. Niessner, D. Schalch, and A. Scharman (1999), W- and F-doped VO_2 Films Studied by Photoelectron Spectrometry, *Thin Solid Films*, 345(2), 229–235.

Butz, A., M. Schneider, and M. Spassova (2004), SearchLight – A Lightweight Search Function for Pervasive Environments, in: A. Ferscha and F. Mattern (eds.), *Pervasive Computing, Second International Conference, PERVASIVE 2004*, Lecture Notes in Computer Science, 3001, Springer, Berlin Heidelberg New York, pp. 351–356.

Cabrera, M., and C. Rodríguez (2004), Sociability Versus Individualism in the Aging Society: The Role of AmI in the Social Integration of the Elderly, in: G. Riva, F. Vatalaro, F. Davide, and M. Alcañiz (eds.), *Ambient Intelligence*, Emerging Communication Series,IOS, Amsterdam, The Netherlands, pp. 267–282.

Cabrera, M., J.-C. Burgelman, M. Boden, O. da Costa, and C. Rodríguez (2004), *eHealth in 2010: Realising a Knowledge-Based Approach to Healthcare in the EU, Challenges for the Ambient Care System*, IPTS Technical Report Series, EUR 21486EN, European Commission, Luxemburg, Luxemburg.

Calabrese, A., and J.-C. Burgelman (eds.) (1999), *Communication, Citizenship, and Social Policy: Re-thinking the Limits of the Welfare State*, Rowman & Littlefield, Lanham, MD, USA.

Calhoun, B., and A. Chandrakasan (2005), Ultra-dynamic Voltage Scaling Using Sub-threshold Operation and Local Voltage Dithering in 90 nm CMOS, *International Solid State Circuits Conference, Digest of Technical Papers*, February 6–10, San Francisco, CA, USA, pp. 242–243.

Callaway, E., P. Gorday, L. Hester, J.A. Gutierrez, M. Naeve, B. Heile, and V. Bahl (2002), Home Networking with IEEE 802.15.4: A Developing Standard for Low-Rate Wireless Personal Area Networks,*IEEE Communications Magazine*, August 2002, pp. 70–77.

Calvin, W.H. (1995), How to Think What No One Has Ever Thought before, in: J. Brockman and K. Matson (eds.), *How Things Are: A Science Tool-Kit for the Mind*, William Morrow, London, UK, pp. 151–164.

Carchon, G., X. Sun, G. Posada, D. Linten, and E. Beyne (2005), Thin Film as Enabling Passive Integration Technology for RF-SoC and SiP, *International Solid State Circuits Conference, Digest of Technical Papers*, February 6–10, San Francisco, CA, USA, pp. 324–325.

CARE (online), http://europa.eu.int/comm/transport/care/index_en.htm.

Carriero, N., and D. Gelernter (1989), Linda in Context, *Communications of the ACM*, 32(4), 444–458.

Castro, P., B. Greenstein, R. Muntz, P. Kermani, C. Bisdikian, and M. Papadopouli (2001), Locating Application Data Across Service Discovery Domains, *Proceedings of the Seventh ACM SIGMOBILE Annual International Conference on Mobile Computing and Networking*, July 16–21, Rome, Italy, pp. 28–42.

Catthoor, F., K. Danckaert, C. Kulkarni, E. Brockmeyer, P.-G. Kjeldsberg, T. Van Achteren, and T. Omnes (2002), *Data Access and Storage Management for Embedded Programmable Processors*, Kluwer Academic, New York, NY, USA.

Catthoor, F., K. Danckaert, S. Wuytack, and N. Dutt (2001), Code Transformations for Data Transfer and Storage Exploration Preprocessing in Multimedia Processors, *IEEE Design & Test*, 18(3), 70–81.

Cattin, P. (2003), *Biometric Authentication System Using Human Gait*, Ph.D. Thesis, ETH Zurich, Zurich, Switzerland.

Centeno, C., R. Van Bavel, and J.-C. Burgelman (2004), *eGovernment in the EU in the Next Decade: The Vision and Key Challenges Based on the Workshop Held in Seville, 4–5 March 2004 "eGovernment in the EU in 2010: Key Policy and Research Challenges"*, IPTS Technical Report Series, EUR 21376EN, European Commission, Luxemburg, Luxemburg.

Cheng, Z.Y., V. Bharti, T.B. Xu, H. Xu, T. Mai, and Q.M. Zhang (2001), Electrostrictive Poly(vinylidene fluoride-trifluoroethylene) Copolymers, *Sensors and Actuators*, A90, 138–147.

Cheverest, K., N. Davies, and K. Mitchel (2000), Developing a Context Aware Electronic Tourist Guide: Some Issues and Experiences, *Proceedings of ACM Conference on Computer Human Interaction*, April 1–6, The Hague, The Netherlands, pp. 17–24.

Chin, J., J. Harting, S. Jha, P.V. Coveney, A.R. Porter, and S.M. Pickles (2003), Steering in Computational Science: Mesoscale Modeling and Simulation, *Contemporary Physics*, 44(5), 417–434.

Claasen, T. (1999), High Speed: Not the Only Way to Exploit the Intrinsic Computational Power of Silicon, *International Solid State Circuits Conference, Digest of Technical Papers*, February 15–17, San Francisco, CA, USA, pp. 22–25.

Clements, B., G. Comyn, K. Rouhana, and J.-C. Burgelman (2004), *Building the Information Society in Europe: the Contribution of Socio-economic Research*, IPTS Report, 85, pp. 2–4.

Clements, B., I. Maghiros, L. Beslay, C. Centeno, Y. Punie, and C. Rodriguez (2003), *Security and Privacy for the Citizen in the Post-September*

11 Digital Age: A Prospective Overview, IPTS Technical Report Series, EUR 2083EN, European Commission, Luxemburg, Luxemburg.

Coen, M. (1998), Design Principles for Intelligent Environments, *Proceedings of the 15th National/10th Conference on Artificial Intelligence/Innovative Applications of Artificial Intelligence*, July 26–30, Madison, WI, USA, pp. 547–554.

Cohen, P.R., and D.R. McGee (2004), Tangible Multimodal Interfaces for Safety-Critical Applications, *Communications of the ACM*, 47(1), 41–46.

Cohen, P.R., M. Johnston, D. McGee, S. Oviatt, J. Pittman, I. Smith, L. Chen, and J. Clow (1997), QuickSet: Multimodal *Interaction* for Distributed Applications, *Proceedings of the ACM Fifth International Multimedia Conference*, November 9–13, Seattle, WA, USA, pp. 31–40.

Collings, P. (1990), *Liquid Crystal – Nature's Delicate Phase of Matter* (1st edition), Adam Hilger, Bristol, UK.

Colwell, B., et al.(2004), Better Than Worst Case Design, *IEEE Computer*, 37(3), 40–73.

COM (2004), *757 Challenges for the European Information Society beyond 2005*, Communication from the Commission to the Council, the European Parliament, the European Economic, and Social Committee and the Committee of the Regions, November 19, European Commission, Luxemburg, Luxemburg.

CombeChem (online), Homepage, http://www.CombeChem.org.

Compañó, R., C. Pascu, and J.-C. Burgelman (eds.) (2004), *Key Factors Driving the Future Information Society in the European Research Area*, IPTS Technical Report Series, EUR 21310EN, European Commission, Luxemburg, Luxemburg.

Constantine, L., and L. Lockwood (2001), Instructive Interaction: Making Innovative Interfaces Self-teaching, *User Experience*, 1(3), 14–19.

Cook, D., and S. Das (eds.) (2005), *Smart Environments: Technology, Protocols, and Applications*, Wiley, New York, NY, USA.

Cook, D., M. Huber, K. Gopalratnam, and M. Youngblood (2003), Learning to Control a Smart Home Environment, submitted to *Innovative Applications of Artificial Intelligence*, see also Prediction Algorithms for Smart Environments, in: D. Cook and S. Das (eds.), *Smart Environments: Technology, Protocols, and applications*, Wiley, New York, NY, USA, pp. 175–192.

Cooperstock, J., S. Fels, W. Buxton, and K. Smith (1997), Reactive Environments – Throwing Away Your Keyboard and Mouse, *Communications of the ACM*, 40(9), 65–73.

Cordis Acts (online), Homepage, http://www.cordis.lu/acts.

Cordis Esprit (online), Homepage, http://www.cordis.lu/esprit.

Cordis FP5 (online), Homepage, http://www.cordis.lu/fp5.

Cordis FP6 (online), Homepage, http://www.cordis.lu/fp6.

Cordis Telematics (online), Homepage, http://www.cordis.lu/telematics.

Cottet, D., J. Grzyb, T. Kirstein, and G. Tröster (2003), Electrical Characterization of Textile Transmission Lines, *IEEE Transactions on Advanced Packaging*, 26(2), 182–190.

Crowley, J.L. (2003), Context Driven Observation of Human Activity, in: E. Aarts, R. Collier, E. van Loenen, and B. de Ruyter (eds.), *Ambient Intelligence, First European Symposium, EUSAI03*, Lecture Notes in Computing Science, 2875, Springer, Berlin Heidelberg New York, pp. 101–118.

Crowley, J.L., J. Coutaz, G. Rey, and P. Reignier (2002), Perceptual Components for Context Aware Computing, in: G. Boriello and L.E. Holmquist (eds.), *Ubicomp 2002: Ubiquitous Computing, Fourth International Conference*, Lecture Notes in Computing Science, 2498, Springer, Berlin Heidelberg New York, pp. 117–134.

Csikszentmihalyi, M. (1990), *Flow: The Psychology of Optimal Experience*, Harper and Row, New York, NY, USA.

CUHTec (online), Homepage, http://www.cuhtec.org.uk.

Dally, W., and Brian Towles, (2001), Route Packets, Not Wires: On-Chip Interconnection Networks, *Proceedings of the Design Automation Conference*, June 18–22, Las Vegas, NV, USA, pp. 684–689.

DAME (online), Homepage, http://www.cs.york.ac.uk/dame.

DAML (online), Homepage, http://www.daml.org.

Darrell, T., D. Demirdjian, N. Checka, and P. Felzenswalb (2001), Plan-View Trajectory Estimation with Dense Stereo Background Models, *Proceedings of the Eighth IEEE International Conference On Computer Vision*, July 7–14, Vancouver, British Columbia, Canada, Volume 2, pp. 628–635.

Davis, F.D., and V. Venkatesh (2004), Toward Preprototype User Acceptance Testing of New Information Systems: Implications for Software Project Management, *IEEE Transactions on Software Engineering*, 51(1), 31–46.

Davies, N., A. Friday, and O. Storz (2004), Exploring the Grid's Potential for Ubiquitous Computing, *IEEE Pervasive Computing*, 3(2), 74–75.

Davies, N., K. Mitchell, K. Cheverest, and G. Blair (1998), *Proceedings of First Workshop on Human–Computer Interaction for Mobile Devices*, May 22, Glasgow, UK, pp. 64–68.

de Ruyter, B. (2003), *365 Days of Ambient Intelligence Research in HomeLab*, Neroc, Eindhoven, The Netherlands.

de Ruyter, B., and E. Aarts (2004), Ambient Intelligence: Visualising the Future, *Proceedings of the Advanced Visual Interfaces Conference*, May 25–28, Gallipoli, Italy, pp. 203–208.

de Ruyter, B., and G. Hollemans (1997), *Towards a User Satisfaction Questionnaire for Consumer Electronics: Theoretical Basis*, Technical Note, NL-TN, 406/97, Philips Research, Eindhoven, The Netherlands.

de Ruyter, B., P. Saini, P. Markopoulos, and A. van Breemen (2005), Assessing the Effects of Building Social Intelligence in a Robotic Interface for the Home, *Interacting with Computers* (to appear).

De Gennes, P.G., and J. Prost (1993), *The Physics of Liquid Crystals* (2nd edition), Oxford University, Oxford, UK.

De Man, H., F. Catthoor, R. Marichal, C. Verdonck, J. Sevenhans, and L. Kiss (2003), Filling the Gap Between System Conception and Silicon/Software Implementation, *International Solid State Circuits Conference, Digest of Technical Papers*, February 10–12, San Francisco,CA, USA, pp. 158–159.

De Roure, D. (2003), On Self-organization and the Semantic Grid, *IEEE Intelligent Systems*, 18, 77–79.

De Roure, D., and J.A. Hendler (2004), e-Science: The Grid and the Semantic Web, *IEEE Intelligent Systems*, 19(1), 65–71.

De Roure, D., N.R. Jennings, and N.R. Shadbolt (2001), *Research Agenda for the Semantic Grid: A Future e-Science Infrastructure*, National e-Science Centre, Report UKeS-2002-02, Edinburgh, UK.

De Roure, D., N.R. Jennings, and N.R. Shadbolt (2005), The Semantic Grid: Past, Present and Future, *Proceedings of the IEEE*, 93(3), 669–681.

Delin, K.A., S.P. Jackson, D.W. Johnson, S.C. Burleigh, R.R. Woodrow, J.M. McAuley, J.M. Dohm, Felipe Ip, T.P.A. Ferré, D.F. Rucker, and V.R. Baker (2005), Environmental Studies with the Sensor Web: Principles and Practice, *Sensors*, 5, 103–117.

Denning, P., and R.M. Metcalfe (1997), *Beyond Calculation: The Next Fifty Years of Computing*, Copernicus, New York, NY, USA.

Dertouzos, M. (1999), The Future of Computing, *Scientific American*, 281(2), 52–55.

Dertouzos, M. (2001), *The Unfinished Revolution*, Harper Collins , New York, NY, USA.

Dey, A.K. (2001), Understanding and Using Context, *Personal and Ubiquitous Computing*, 5(1), 4–7.

Dey, A.K., and G.D. Abowd (2000), The Context Toolkit: Aiding the Development of Context-Aware Applications, *Proceedings of the Workshop on Software Engineering for Wearable and Pervasive Computing*, June 5–6, Limerick, Ireland, pp. 207–226.

DIAMOND (online), Homepage, http://www.diamond.ac.uk.

Dick, B. (2000), A Beginner's Guide to Action Research, in: B. Dick, R. Passfield, and P. Wildman (eds.), *Action Research and Evaluation*, Webpage, Southern Cross University, Lismore, Australia, http://www.scu.edu.au/schools/gcm/ar/arp/guide.html.

Diederiks, E., B. Eggen, and J. van Kuijk (2002), *Ambient Intelligent Lighting Concepts*, Technical Note NL-TN 2002/179, Philips Research, Eindhoven, The Netherlands.

Dirks, T. (1996), Close Encounters of the Third Kind, Website, http://www.filmsite.org/clos.html.

Don, A. (1992), Anthropomorphism: from ELIZA to Terminator 2, Panel Session, *Proceedings of the ACM SIGCHI Conference on Human Factors in Computing Systems*, May 3–7, Monterey, pp. 67–70.

Donato, G., M.S. Bartlett, J.C. Hager, P. Ekman, and T.J. Sejnowski (1999), Classifying Facial Actions, *IEEE Transactions on Pattern Analysis and Machine Intelligence*, 21, 974–989.

Dourish, P. (2002), *Where the Action Is: The Foundations of Embodied Interaction*, MIT, Cambridge, MA, USA.

Draper, B.A., R.T. Collins, J. Brolio, A.R. Hansen, and E.M. Riseman (1989), The Schema System,*International Journal of Computer Vision*, 2(3), 209–250.

Ducatel, K., J.-C. Burgelman, F. Scapolo, and M. Bogdanowicz (2000), *Baseline Scenarios for Ambient Intelligence in 2010*, IPTS Working Paper, European Commission, Luxemburg, Luxemburg.

Dunne, A., and F. Raby (2001), *Design Noir: The Secret Life of Electronic Objects*, Brinkhäuser, Basel, Switzerland.

DynAMITE (online), Homepage, http://www.dynamite-project.org.

DynAMITE (online), Dynamic Adaptive Multimodal IT Ensembles, Homepage, http://www.dynamite-project.org.

EC–IDA (2004), *Inovative m-Ticketing Solution to be Tested in Frankfurt, Germany*, eGovernment news – 04 November, http://europa.eu.int:80/ida/en/document/3450.

ECMA (2004), *Near Field Communications*, White Paper, Ecma/TC32-TG19/2004/1, http://www.ecma-international.org/activities/Communications/2004tg19-001.pdf.

Eggen, J.H., and E. Aarts (eds.) (2002), *Ambient Intelligence in HomeLab*, Neroc, Eindhoven, The Netherlands.

Eggen, J.H., G. Hollemans, R. van de Sluis (2003), Exploring and Enhancing the Home Experience, *Journal on Cognition Technology and Work*, 5, 44–54.

Eggen, J.H., M. Rozendaal, and O. Schimmel (2003), Home Radio – Extending the Home Experience beyond the Boundaries of the Physical House, *Proceedings of the Home Oriented Informatics and Telematics International Conference on "The Networked Home and the Home of the Future"*, April 6–8, Irvine, CA, USA, pp. 1–14.

eGOIA (online) , Homepage, http://www.egoia.info.

e-Grain (online), Homepage, http://www.e-Grain.org.

Ehn, P. (1992), Scandinavian Design: On Participation and Skill, in: P.S. Adler and T.A. Winograd (eds.), *Usability: Turning Technologies into Tools*, Oxford University, Oxford, UK, pp. 96–132.

EITO (online), European Information Technology Observatory, European Economic Interest Grouping, Homepage, http://www.eito.org.

Emmerich, W. (2000), *Engineering Distributed Objects*, Wiley, Chichester, UK.

Emmerich, W. (2002), Distributed Component Technologies and Their Software Engineering Implications, *Proceedings of the 24th International Conference on Software Engineering*, May 19–25, Orlando, Florida, USA, pp. 537–546.

Emory (online), Homepage, http://www.emory.edu.

Encarnação, J., and T. Kirste (2005), Ambient Intelligence: Towards Smart Appliance Ensembles, in: M. Hemmje, C. Niederee, and T. Risse (eds.), *From Integrated Publication and Information Systems to Virtual Information and Knowledge Environments, Essays Dedicated to Erich J. Neuhold on the Occasion of His 65th Birthday*, Springer Lecture Notes in Computer Science, 3379, Springer, Berlin Heidelberg New York, pp. 261–270.

Endres, Ch., A. Butz, and A. MacWilliams (2005), A Survey of Software Infrastructures and Frameworks for Ubiquitous Computing, *Mobile Information Systems Journal*, 1(1), 41–80.

Endsley, M.R. (1995), Toward a Theory of Situation Awareness in Dynamic Systems, *Human Factors*, 37(1), 85–104.

Engelbrecht, A.P. (2002), *Computational Intelligence: An Introduction*, Wiley, Chichester, UK.

EPC (online), Homepage, http://www.epcglobalinc.org.

Equator (online), Homepage, http://www.equator.com.

ETSI HIPERLAN/2 (online), Webpage, http://portal.etsi.org/bran/kta/Hiperlan/hiperlan2.asp.

Eurescom Personal Nets (online), Webpage, http://www.eurescom.de/public/projects/P1000-series/p1047/default.asp.

Eurobarometer (2002), *Internet Usage and the Public at Large*, Flash Eurobarometer 125, European Commission, Brussels, Belgium.

European Commission (2002), *Information Society Technologies: A Thematic Priority for Research and Development under the Specific Programme "Integrating and Strengthening the European Research Area" in the Community Sixth Framework Programme, IST Priority, WP 2003–2004*, Report, European Commission, Brussels, Belgium, http://www.cordis.lu/ist.

European Commission (2004a), *The Social Situation in the European Union*, Annual Report, European Commission, Luxemburg, Luxemburg, http://europa.eu.int/comm/employment_social/social_situation/socsit_en.htm.

European Commission (2004b), *Monitoring Industrial Research: The 2004 EU Industrial R&D Investment Scoreboard*, Technical Report, EUR 21399EN, European Commission, Luxemburg, Luxemburg.

Fagin, R., J.Y. Halpern, Y. Moses, and M.Y. Vardi (1995), *Reasoning about Knowledge*, MIT, Cambridge, MA, USA.

Farringdon, J., A.J. Moore, N. Tilbury, J. Church, and P.D. Biemond (1999), Wearable Sensor Badge and Sensor Jacket for Context Awareness, *Proceedings of the Third International Symposium on Wearable Computers*, October 18–19, San Francisco, CA, USA, pp. 107–113.

Fels, S. (2000), Intimacy and Embodiment: Implications for Art and Technology, *Proceeding of the ACM Conference on Multimedia*, October 22–27, Marina del Sol, CA, USA, pp. 13–16.

Fernandez, R., and R.W. Picard (2003), Modeling Drivers' Speech under Stress, *Speech Communication*, 40, 145–159.

Filion R., R. Wojnarowski, T. Gorcyzca, B. Wildi, and H. Cole (1994), Development of a Plastic Encapsulated Multichip Technology for High Volume, Low Cost Commercial Electronics, *Proceedings of the IEEE Applied Power Electronics Conference*, February 13–17, Orlando, FL, USA, pp. 805–809.

Fischer, T., K. Zoschke, K. Scherpinski, K. Buschick, O. Ehrmann, J.M. Wolf, and H. Reichl (2004), Flexible Circuit Carrier with Integrated Passives for High Density Integration, *Journal of Telecommunications*, 58, 58–64.

Flichy, P. (1995), *L'Innovation Technique: Récents Développements en Sciences Sociales. Vers une Nouvelle Théorie de l'Innovation.* Editions La Découverte, Paris, France.

FloodNet (online), Webpage, http://www.envisense.org.

Florida, R. (2002), *The Rise of the Creative Class*, Basic Books, New York, NY, USA.

Foster, I., C. Kesselman, and S. Tuecke (2001), The Anatomy of the Grid: Enabling Scalable Virtual Organizations, *International Journal of Supercomputer Applications*, 15(3), 200–222.

Fox, A., B. Johanson, P. Hanrahan, and T. Winograd (2000), Integrating Information Appliances into an Interactive Workspace, *IEEE Computer Graphics and Applications*, 20(3), 54–65.

Fox, D., J. Hightower, L. Liao, D. Schulz, and G. Borriello (2003), Bayesian Filtering for Location Estimation,*IEEE Pervasive Computing*, 2(3), 24–33.

Franklin, D. (1998), Cooperating with People: The Intelligent Classroom, *Proceedings of the15th National/10th Conference on Artificial Intelligence/Innovative Applications of Artificial Intelligence*, July 26–30, Madison, WI, USA, pp. 555–560.

Franklin, D., J. Budzik, and K. Hammond (2002), Plan-Based Interfaces: Keeping Track of User Tasks and Acting to Cooperate, *Proceedings of the Seventh International Conference on Intelligent User Interfaces*, January 13–16, San Francisco, CA, USA, pp. 79–86.

Frey, J.G., M. Bradley, J.W. Essex, M.B. Hursthouse, S.M. Lewis, M.M. Luck, L. Moreau, D. De Roure, M. Surridge, and A. Welsh (2003), Combinatorial Chemistry and the Grid, in: F. Berman, G. Fox, and T. Hey (eds.), *Grid Computing: Making the Global Infrastructure a Reality*, Wiley, Chichester, UK, pp. 151–174.

Frey, J.G., D. De Roure, and L.A. Carr (2002), Publication at Source: Scientific Communication from a Publication Web to a Data Grid, *Proceedings of the Euroweb 2002 Conference*, December 17–18, Oxford, UK, http://eprints.ecs.sutton.ac.uk/7852/.

Fromherz, P. (2005), Joining Ionics and Electronics: Semiconductor Chips with Ion Channels, Nerve Cells and Brain Tissue, *International Solid State Circuits Conference,Digest of Technical Papers*, February 6–10, San Francisco, CA, USA, pp. 48–49.

Gago Panel Report (2005), *Research and Technology Development in Information Society Technologies, Five-Year Assessment 1999–2003*, Final Panel Report, European Commission, Luxemburg, Luxemburg.

Garey, M.R., and D.S. Johnson (1979), *Computers and Intractability: A Guide to the Theory of NP-Completeness*, W.H.Freeman, San Fransisco, CA, USA.

Garlan, D. (2000), Software Architecture: A Roadmap, in: A. Finkelstein (ed.), *The Future of Software Engineering*, ACM, New York, NY, USA, pp. 91–101.

Garlan, D., D. Siewiorek, A. Smailagic, and P. Steenkiste (2002), Project Aura: Towards Distraction-Free Pervasive Computing, *IEEE Pervasive Computing*, 1(2), 22–31.

Gaver, W.W., A. Boucher, S. Pennington, and B. Walker (2004), Cultural Probes and the Value of Uncertainty, *Interactions*, 11(5), 53–56.

Gavigan, J., M. Ottish, and C. Greaves (1999), *Demographic and Social Trends*, Panel Report, IPTS Futures Report Series 02, European Commission, Luxemburg, Luxemburg.

Gaynor, M., S. Moulton, M. Welsh, E. LaCombe, A. Rowan, and J. Wynne (2004), Integrating Wireless Sensor Networks with the Grid, *IEEE Internet Computing*, 8(4), 32–39.

Geens, P., and W. Dehaene, (2005), A Small Granular Controlled Leakage Reduction System for SRAMs, *Proceedings of the First International Conference on Memory Technology and Design*, May 21–24, Giens, France, pp.73–76.

GENIE (online), Homepage, http://www.genie.ac.uk.

Georgantas, N., and V. Issarny (2004), User Activity Synthesis in Ambient Intelligence Environments, *Adjunct Proceedings of the Second European Symposium on Ambient Intelligence (EUSAI)*, November 8–11, Eindhoven, The Netherlands,
http://www-rocq.inria.fr/arles/doc/nikolaosEUSAI04.pdf.

Gerrits, J., J.R. Farserotu, and J.R. Long (2004), UWB Considerations for My Personal Global Adaptive Network Systems, *Proceedings of the European Solid State Circuits Conference 2004*, September 24, Leuven, Belgium, pp. 45–56.

Gershenfeld, N. (1999), *When Things Start to Think*, Henry Holt, New York, NY, USA.

Gershenfeld, N., R. Krikorian, and D. Cohen (2004), The Internet of Things, *Scientific American*, 291(4), 76–81.

GeWiTTS (online), Webpage, http://www.nesc.ac.uk/events/sc2004/talks.

GGF (online), Homepage, http://www.ggf.org.

Gibson, J.J. (1986), *The Ecological Approach to Visual Perception*, Lawrence Erlbaum Associates, Hillsdale, NJ, USA.

Gimpel, S., U. Möhring, A. Neudeck, and W. Scheibneret (2005), Integration of microelectronic devices in textiles, *MST News*, 2/05, 14.

GLA Economics (2002), *Creativity: London's Core Business*, Report, London, UK, http://www.creativelondon.org.uk/documents/ londons_creative_sector_2004_update.pdf.

Glass, R.L. (1994), The Software-Research Crisis, *IEEE Software*, 11(6), 42–47.

Glass, R.L., V. Ramesh, and I. Vessey (2004), An Analysis of Research in Computing Disciplines, *Communications of the ACM*, 47, 89–94.

Glynne-Jones, P., and N.M. White (2001), Self-powered Systems: A Review of Energy Sources, *Sensor Review*, 21(2), 91–97.

Go, K., and J.M. Carroll (2004), The Blind Men and the Elephant: Views of Scenario-Based Design, *Interactions*, 11(5), 45–53.

Goble, C.A., and D. De Roure (2004), The Semantic Grid: Myth Busting and Bridge Building, *Proceedings of the European Conference on Artificial Intelligence*, August 23–27, Valencia, Spain, pp. 1129–1135.

Goodenough, J.B. (1971), 2 Components of Crystallographic Transition in VO_2, *Journal of Solid State Chemistry*, 3(4), 490–500.

Gonzales, R., and M. Horowitz (1996), Energy Dissipation in General-Purpose Microprocessors, *IEEE Journal of Solid-State Circuits*, 31(9), 1277–1284.

Grace, P., G. Blair, and S. Samuel (2003), Middleware Awareness in Mobile Computing, *Proceedings of the First International ICDCS Workshop on Mobile Computing Middleware*, May 19–22, Providence, RI, USA, p. 382.

Gray J., and T. Hey (2001), *In Search of Petabyte Databases*, Talk presented at 2001 HPTS Workshop, October 14–17, Asilomar, CA, USA, http://www.research.microsoft/~gray.

Grid Security Papers (online), Webpage, http://www.princeton.edu/~jdwoskin/grid/gridsecpapers.html.

Grimnes, S., and Ørjan G. Martinsen (2000), *Bioimpedance and Bioelectricity Basics*, Academic, New York, NY, USA.

Grosser, V., and K.-D. Lang (2004), *Strategy for System Integration – 3D Integration*, Tutorial presented at the SMT/HYBRID/PACKAGING Conference, June 15–17, Nürnberg, Germany.

GSM World (online), Homepage, http://www.gsmworld.com/index.shtml.

GSM World Statistics (online), Growth of the Global Digital Mobile Market, http://www.gsmworld.com/news/statistics/pdf/gsma_stats_q4_04.pdf.

GTI (online), http://www.vw.co.uk/company/press/GTI_advert.

Gupta, R., S. Pande, K. Psarris, and V. Sarkar (1999), Compilation Techniques for Parallel Systems, *Parallel Computing*, 25(13–14), 1741–1783.

Gustafsson, E., and A. Jonsson (2003), Always Best Connected, *IEEE Wireless Communications*, 10, 49–55.

Haartsen, J.C. (2000), The Bluetooth Radio System, *IEEE Personal Communications*, 7(1), 28–36.

Hahn, R. (2002), *Autarke Systeme mit Piezogeneratoren, Mikrosolarmodulen und flexiblen Speichern*, Talk presented at the GMM Workshop Energieautarke Sensorik, June 6–7, Dresden, Germany.

Hahn, R., and J. Müller (2000), Future Power Supplies for Portable Electronics and Their Environmental Issues, *Proceedings of the International Congress Electronics Goes Green 2000*, September 11–13, Berlin, Germany, pp. 727–734.

Hall, R.R. (2001), Prototyping for Usability of New Technology, *International Journal of Human–Computer Studies*, 55, 485–501.

Halpern, J.Y., and Y. Moses (1992), A Guide to Completeness and Complexity for Modal Logics of Knowledge and Belief, *Artificial Intelligence*, 54, 319–379.

Hanson, A.R., and E. M. Riseman (1978), VISIONS: A Computer Vision System for Interpreting Scenes, in: A.R. Hanson and E.M. Riseman (eds.), *Computer Vision Systems*, Academic, New York, NY, USA, pp. 303–334.

Hanssens, N., A. Kulkarni, R. Tuchinda, and T. Horton (2002), Building Agent-Based Intelligent Workspaces, *Proceeding of the International Workshop on Agents for Business Automation*, June 24–27, Las Vegas, CA, USA, pp. 10–16.

Harrison, B.L., K.P. Fishkin, A. Gujar, C. Mochon, and R. Want (1998), Squeeze Me, Hold Me, Tilt Me, *Proceeding of the ACM CHI 98 Conference on Human Factors in Computing Systems*, April 18–23, Los Angeles, CA, pp. 17–24.

Hawkins, J., and S. Blakeslee (2004), *On Intelligence*, Times Books, New York, NY, USA.

Healey, J., and R.W. Picard (1998), Startlecam: A Cybernetic Wearable Camera, *Proceedings of the Second IEEE International Symposium on Wearable Computers*, October 19–20, Pittsburgh, PE, USA, pp. 42–49.

Heider, T., and T. Kirste (2002a), Architectural Considerations for Interoperable Multi-modal Assistant Systems, in: P. Forbrig, Q. Limbourg, B. Urban, and J. Vanderdonckt (eds.), *Interactive Systems: Design, Specification, and Verification, Ninth International Workshop,DSV-IS 2002*, Lecture Notes in Computing Science, 2545, Springer, Berlin Heidelberg New York, pp. 253–267.

Heider, T., and T. Kirste (2002b), Supporting Goal-Based Interaction with Dynamic Intelligent Environments, *Proceedings of the 15th European Conference on Artificial Intelligence*, Lyon, France, pp. 596–600.

Helander, M.G. (1987), *Handbook of Human Factors*, Wiley, New York, NY, USA.

Hellenschmidt, M., and T. Kirste (2004a), *A Generic Topology for Ambient Intelligence*, in: P. Markopoulis, B. Eggen, E. Aarts, and J. Crowley (eds.), *Ambient Intelligence: Second European Symposium, EUSAI04*, Springer Lecture Notes in Computer Science, 3295, Springer, Berlin Heidelberg New York, pp. 112–123.

Hellenschmidt, M., and T. Kirste (2004b), Software Solutions for Self-organizing Multimedia Appliances, *Computer & Graphics*, 28(5), 643-655.

Herfet, Th., T. Kirste, and M. Schneider (2001), EMBASSI – Multimodal Assistance for Infotainment and Service Infrastructures, *Computer & Graphics*, 25(4), 581–592.

Hess, C.K., M. Roman, and R.H. Campbell (2002), Building Applications for Ubiquitous Computing Environments, *Proceedings of the First International Conference on Pervasive Computing*, August 26–28, Zurich, Switzerland, pp. 16–29.

Hey, T., and A.E. Trefethen (2002), The UK e-Science Core Programme and the Grid, *Future Generation Computer Systems*, 18, 1017–1031.

Hey, T., and A.E. Trefethen (2003), The Data Deluge, in: F. Berman, G. Fox, and T. Hey (eds.), *Grid Computing: Making the Global Infrastructure a Reality*, Wiley, Chichester, UK, pp. 809–824.

Heylen, D., M. Ghijsen, A. Nijholt, and R. op den Akker (2005), Facial Signs of Affect during Tutoring Sessions, in: First International Conference on *Affective Computing and Intelligent Interaction (ACII2005)*, Beijing, China, October 22–24, Lecture Notes in Computer Science, Springer-Verlag, Berlin, Germany (to appear).

Heylen, D., A. Nijholt, and R. op den Akker (2005), Affect in Tutoring Dialogues, *Applied Artificial Intelligence*, 19(3–4), 287–311.

Heynderickx, I., and D.J. Broer (1991), The Use of Cholesterically-ordered Polymer Networks in Practical Applications, *Molecular Crystals and Liquid Crystals I*, 202, 113–126.

Heylen, D., M. Vissers, R. op den Akker, and A. Nijholt (2004), Affective Feedback in a Tutoring System for Procedural Tasks, in: E. André, L. Dybkjær, W. Minker, and P. Heisterkamp (eds.), *Affective Dialogue Systems, Tutorial and Research Workshop, ADS 2004*, Lecture Notes in Computer Science, 3068, Springer, Berlin Heidelberg New York, pp. 244–253.

Hightower, J., and G. Borriello (2001), Location Systems for Ubiquitous Computing, *Computer*, 34(8), 57–66.

Hikmet, R.A.M., and H. Kemperman (1998), Electrically Switchable Mirrors and Optical Components Made from Liquid-Crystal Gels, *Nature*, 392(6675), 476–479.

Hinckley, K., J. Pierce, M. Sinclair, and E. Horvitz (2000), Sensing Techniques for Mobile Interaction, *Proceedings of the 13th Annual ACM Symposium on User Interface Software and Technology*, San Diego, CA, USA, pp. 91–100.

Hofs, D., R. op den Akker, and A. Nijholt (2003), A Generic Architecture and Dialogue Model for Multimodal Interaction, in: P. Paggio, K. Jokinen, and A. Jönsson (eds.),*Proceedings of the First Nordic Symposium on Multimodal Communication*, CST Publication, Center for Sprokteknologi, September 25–26, Copenhagen, Denmark, pp. 79–92.

Hogan, P.M., A.R. Tajbakhsh, and E.M. Terentjev (2002), UV-Manipulation of Order and Macroscopic Shape in Nematic Elastomers, *Physical Review*, E65, 041720-1–041720-10.

Holland, J.H. (1992), *Adaptation in Natural and Artificial Systems* (reprint edition), MIT, Cambridge, MA, USA.

Holmquist, L.E., S. Antifakos, B. Schiele, F. Michaelles, M. Beigl, L. Gaye, H.-W. Gellersen, A. Schmidt, and M. Strohbach (2003), Building Intelligent Environments with Smart-Its, *Emerging Technologies Exhibition at SIGGRAPH*, July 27–29, San Diego, CA, USA.

Horrocks, I., and P.F. Patel-Schneider (2003), Reducing OWL Entailment to Description Logic Satisfiability, in: D. Fensel, K.P. Sycara, and J. Mylopoulos (eds.), *The Semantic Web – ISWC 2003, Second International*

Semantic Web Conference, Lecture Notes in Computing Science, 2870, Springer, Berlin Heidelberg New York, pp. 17–29.

Hospers, M., E. Kroezen, A. Nijholt, R. op den Akker, and D. Heylen (2003), An Agent-Based Intelligent Tutoring System for Nurse Education, in: J. Nealon and A. Moreno (eds.), *Applications of Intelligent Agents in Health Care*, Whitestein Series in Software Agent Technologies, Birkhauser, Basel, Switzerland, pp. 143–159.

Hughes, G., H. Mills, D. De Roure, J.G. Frey, L. Moreau, M.C. Schraefel, G. Smith, and E. Zaluska (2004), The Semantic Smart Laboratory: A System for Supporting the Chemical e-Scientist, *Organic and Biomolecular Chemistry*, 2(22), 3284–3293.

Hui, M.-H., P. Keller, B. Li, X. Wang, and M. Brunet (2003), Light-Driven Side-On Nematic Elastomer Actuators, *Advanced Materials*, 15(7–8), 569–572.

Huiberts, J.N., R. Griessen, J.H. Rector, R.J. Wijngaarden, J.P. Decker, D.G. De Groot, and N.J. Koeman (1996), Yttrium and Lanthanum Films with Switchable Optical Properties, *Nature*, 380(6571), 231–234.

Hulkko, S., T. Mattelmäki, K.Virtanen, T., and Keinonen (2004), Mobile Probes, *Proceedings of the Third ACM Nordic Conference on Human–Computer Interaction*, October 23–27, Tampere, Finland, pp. 43–51.

Iacucci, G., C. Iacucci, and K. Kuutti (2002), Imagining and Experiencing in Design, the Role of Performances,*Proceedings of the Third ACM Nordic Conference on Human–Computer Interaction*, October 19–23, Aarhus, Denmark, pp. 167–176.

IEEE 802.11 (online), The Working Group for WLAN Standards, http://grouper.ieee.org/groups/802/11/.

IEEE 802.15 (online), Working Group for WPAN, http://grouper.ieee.org/groups/802/15/.

IEEE 802.16 (online), Working Group on Broadband Wireless Access Standards, http://grouper.ieee.org/groups/802/16/.

Ingvar, D. (1985), Memory of the Future: An Essay on the Temporal Organization of Conscious Awareness, *Human Neurobiology*, 4(3), 127–136.

inHaus (online), *inHaus-Innovation-Center*, Fraunhofer-Gesellschaft, Homepage, http://www.inhaus-duisburg.de.

Integrative Biology (online), Homepage, http://www.integrativebiology.ox.ac.uk.

Internet World Stats (online), Internet Usage Statistics – The Big Picture World Internet Users and Population Statistics, http://www.internetworldstats.com/stats.htm.

Ireland, P.T., and T.V. Jones (1987), The Response-Time of a Surface Thermometer Employing Encapsulated Thermochromic Liquid-Crystals, *Journal of Physics E*, 20(10), 1195–1199.

Ishibashi, K., T. Yamashita, Y. Arima, I. Minematsu, and T. Fujimoto, T. (2003), A 9 mW 50 MHz 32b Adder Using a Self-adjusted Forward

Body Bias in SoCs, *International Solid State Circuits Conference, Digest of Technical Papers*, February 10–12, San Francisco, CA, USA, pp. 116–117.

Ishii, H., and B. Ullmer (1997), Tangible Bits: Towards Seamless Interfaces Between People, Bits and Atoms, *Proceedings of the CHI97 Human Factors in Computing Systems Conference*, March 22–27, Atlanta, GA, USA, pp. 234–241.

Issarny, V., C. Kloukinas, and A. Zarras (2002), Systematic Aid for Developing Middleware Architectures, *Communications of the ACM, Special Issue on Adaptive Middleware*, 45(6), 53–58.

Issarny, V., D. Sacchetti, F. Tartanoglu, F. Sailhan, R. Chibout, N. Levy, and A. Talamona (2005), Developing Ambient Intelligence Systems: A Solution Based on Web Services, *Automated Software Engineering*, 12(1), 101–137.

Issarny, V., F. Tartanoglu, J. Liu, and F. Sailhan (2004), Software Architectures for Mobile Distributed Computing, *Proceedings of the Fourth Working IEEE/IFIP Conference on Software Architecture*, June 12–15, Oslo, Norway, pp. 201–210.

ISO (2001), ISO/TR, *Software Engineering – Product Quality – Part 1: Quality Model*, Technical Report, TR 9126-1:2001.

IST (2004), IST 2004, Event, Homepage, http://europa.eu.int/information_society/istevent/2004/index_en.htm.

IST Ambient Networks (online), Homepage, http://www.ambient-networks.org/.

IST E2R (online), End-to-End Reconfigurability, http://e2r.motlabs.com/.

IST MAGNET (online), Website, http://www.ist-magnet.org/index.html.

IST RUNES (online), Reconfigurable Ubiquitous Networked Embedded Systems, http://www.ist-runes.org/.

IST VESPER (online), Virtual Home Environment for Service Personalization and Roaming Users, http://www.ee.surrey.ac.uk/CCSR/IST/Vesper/.

IST WINNER (online), Wireless World Initiative New Radio, https://www.ist-winner.org/.

ISTAG (2001), *Scenarios for Ambient Intelligence in 2010*, ISTAG Report, European Commission, Luxemburg, Luxemburg, ftp://ftp.cordis.lu/pub/ist/docs/istagscenarios2010.pdf.

ISTAG (2003), *Ambient Intelligence: from Vision to Reality*, Report, European Commission, Luxemburg, Luxemburg, ftp://ftp.cordis.lu/pub/ist/docs/istag-ist2003_consolidated_report.pdf.

ISTAG (2004a), *Experience and Application Research: Involving Users in the Development of Ambient Intelligence*, ISTAG Report, European Commission, Luxemburg, Luxemburg, ftp://ftp.cordis.lu/pub/ist/docs/2004_ear_web_en.pdf.

ISTAG (2004b), *Grand Challenges in the Evolution of the Information Society*, ISTAG Report, European Commission, Luxemburg, Luxemburg, ftp://ftp.cordis.lu/pub/ist/docs/2004_grand_challenges_web_en.pdf.

ITEA (online), Homepage, http://www.itea-office.org.

ITEA (2004), *Technology Roadmap on Software Intensive Systems*, ITEA Report, Eindhoven, The Netherlands, http://www.itea-office.org/index.htm.

ITU-D (online), *International Telecommunication Union, Information Technology Statistics*, Report, http://www.itu.int/ITU-D/ict/statistics/at glance/Internet03.pdf.

Jackson, M. (1995), *Software Requirements & Specifications*, Addison-Wesley, Reading, MA, USA.

Jain, R. (2003), Experiential Computing, *Communications of the ACM*, 46, 48–54.

Jantsch, A., and H. Tenhunen (eds.) (2003), *Networks on Chip*, Kluwer Academic,New York, NY, USA.

Jayapala, M., F. Barat, T. Van der Aa, F. Catthoor, G. Deconinck, and H. Corporaal (2005), Clustered L0 Buffer Organisation for Low Energy Embedded Processors, *IEEE Transactions on Computers*, 54(6), 672–683.

Jensen, F.V. (2001), *Bayesian Networks and Decision Graphs*, Springer, New York, NY, USA.

Jeremijenko, N. (2001), Dialogue with a Monologue: Voice Chips and the Products of Abstract Speech, *Proceedings of the 28th Annual ACM Conference on Computer Graphics and Interactive Techniques*, August 12–17, Los Angeles, CA, USA, pp. 1002–1033.

Jha, N. (2001), Low Power System Scheduling and Synthesis, *Proceedings of the IEEE International Conference on Computer Aided Design*, November 4–8, San Jose, CA, USA, pp. 259–263.

Johanson, B., A. Fox, and T. Winograd (2002), The Interactive Workspaces Project: Experiences with Ubiquitous Computing Rooms,*IEEE Pervasive Computing*, 1(2), 67–75.

Johnston, M., S. Bangalore, G. Vasireddy, A. Stent, P. Ehlen, M. Walker, S. Whittaker, and P. Maloor (2002), MATCH: An Architecture for Multimodal Dialogue Systems, *Proceedings of the Association for Computational Linguistics*, pp. 376–383.

Jurafsky, D. (1997), *Switchboard SWBD-DAMSL Shallow-Discourse-Function Annotation*, Technical Report, 97-02, University of Colorado, CO, USA.

Jurafsky, D. (2001), Pragmatics and Computational Linguistics, in: L.R. Horn and G. Ward (eds.), *Handbook of Pragmatics*, Blackwell, Oxford, UK.

Kaelbling L.P., M.L. Littman, and A.W. Moor (2004), Reinforcement Learning: A Survey, *Journal of AI Research*, 4, 237–285.

Kallmayer, C., M. Niedermayer, S. Guttowski, and H. Reichl (2005), Packaging Challenges in Miniaturization, in: W. Weber, J. Rabaey, and E. Aarts (eds.), *Ambient Intelligence*, Springer, Berlin Heidelberg New York, pp. 327–348.

Kallmayer, C., R. Pisarek, S. Cichos, A. Neudeck, and S. Gimpel (2003), New Assembly Technologies for Textile Transponder Systems, *Proceedings of the Electronic Components Technology Conference*, May 27–30, New Orleans, LA, USA.

Kearns, M., and U. Vazirani (1994), *An Introduction to Computational Learning Theory*, MIT, Cambridge, MA, USA.

Kern, N., B. Schiele, H. Junker, P. Lukowicz, and G. Tröster (2002), Wearable Sensing to Annotate Meeting Recording, *Proceedings of the Sixth IEEE International Symposium on Wearable Computers*, October 7–10, Seattle, WA, USA, pp. 186–196.

Kidd, C.D., R.J. Orr, G.D. Abowd, C.G. Atkeson, I.A. Essa, B. MacIntyre, E. Mynatt, T.E. Starner, and W. Newstetter (1999), The Aware Home: A Living Laboratory for Ubiquitous Computing Research, *Proceedings of the Second International Workshop on Cooperative Buildings*, October 1–2, Pittsburgh, PA, USA, pp. 191–198.

Kindberg, T., and A. Fox (2002), System Software for Ubiquitous Computing, *IEEE Pervasive Computing*, 1(1), 70–81.

Kjeldskov, J., and C. Graham (2003), A Review of Mobile HCI Research Methods, in: L. Chittaro (ed.), *Human–Computer Interaction with Mobile Devices and Services, Fifth International Symposium, Mobile HCI 2003*, Lecture Notes in Computer Science, 2795, Springer, Berlin Heidelberg New York, pp. 317–334.

Knapp, K. (2000), Metaphorical and Interactional Uses of Silence, *EESE: Erfurt Electronic Studies in English*, Volume 7, http://webdoc.sub.gwdg.de/edoc/ia/eese/eese.html.

Kornbluh, R., R. Pelrine, Q. Pei, and S.V. Shastri (2001), Application of Dielectric EAP Actuators, in: Y. Bar-Cohen (ed.), *Electroactive Polymer (EAP) Actuators as Artificial Muscles*, SPIE, Volume PM98, Bellingham, WA, USA, pp. 457–460.

Kort, B., R. Reilly, and R.W. Picard, An Affective Model of Interplay between Emotions and Learning: Reengineering Educational Psychology – Building a Learning Companion, *Proceedings of the IEEE International Conference on Advanced Learning Technologies*, August, Madison, WI, USA, pp. 43–48.

Kourouthanasis, P., D. Spinellis, G. Roussos, and G. Giaglis (2002), Intelligent Cokes and Diapers: MyGrocer Ubiquitous Computing Environment, *Proceedings of the First International Mobile Business Conference*, July 8–9, Athens, Greece, pp. 150–172.

Koza, J.R. (1994), *Genetic Programming II: Automatic Discovery of Reusable Programs*, MIT, Cambridge, MA, USA.

Krasner, G.E., and S.T. Pope (1988), A Description of the Model-View-Controller User Interface Paradigm in the Smalltalk-80 System, *Journal of Object-Oriented Programming*, 1(3), 26–49.

Kravets, R., C. Carter, and L. Magalhaes (2001), A Cooperative Approach to User Mobility, *ACM Computer Communications Review*, 31, 57–69.

Kröner, A., S. Baldes, A. Jameson, and M. Bauer (2004), Using an Extended Episodic Memory Within a Mobile Companion, *Proceedings of the Pervasive 2004 Workshop on Memory and Sharing of Experiences*, April 20, Vienna, Austria, pp. 59–66.

Kuenning, G., and G. Popek (1997), Automated Hoarding for Mobile Computers, *Proceedings of the 16th ACM Symposium of Operating Systems Principles*, October 5–8, St. Malo, France, pp. 264–275.

Kyffin, S., L. Feijs, and T. Djajadiningrat (2005), Exploring Expression of Form, Action and Interaction, *Proceedings of the International Conference on Home Oriented Informatics and Telematics*, April 13–15, York, UK (to appear).

Lagorce, L.K., O. Brand, and M.G. Allen (1999), Magnetic Microactuators Based on Polymer Magnets,*IEE Journal of Microelectromechanical Systems*, 8, 2–9.

LaMarca, A., Y. Chawathe, S. Consolvo, J. Hightower, I. Smith, J. Scott, T. Sohn, J. Howard, J. Hughes, F. Potter, J. Tabert, P. Powledge, G. Borriello, and B. Schilit (2005), Place Lab: Device Positioning Using Radio Beacons in the Wild, in: H.-W. Gellersen, R. Want, and A. Schmidt (eds.), *Pervasive Computing, Third International Conference, PERVASIVE 2005*, Lecture Notes in Computing Science, 3468, Springer, Berlin Heidelberg New York, pp. 116–133.

Lambrechts, A., P. Raghavan, A. Leroy, G. Talavera, T. Van der Aa, M. Jayapala, F. Catthoor, D. Verkest, G. Deconinck, H. Corporaal, F. Robert, and J. Carrabina, (2005), Power Breakdown Analysis for a Heterogeneous NoC Platform Running a Video Application, *Proceedings of the International Conference on Application Specific Array Processors*, July 18–20, Samos, Greece (to appear).

Lampe, M., M. Strassner, and E. Fleisch (2004), A Ubiquitous Computing Environment for Aircraft Maintenance, Symposium on Applied Computing, *Proceedings of the 2004 ACM Symposium on Applied Computing*, March 14–17, Nicosia, Cyprus, pp. 1586–1592.

Langheinrich, M., F. Mattern, K. Romer, and H. Vogt, (2000), First Steps Towards an Event-Based Infrastructure for Smart Things, Ubiquitous Computing Workshop at PACT, October 15–19, Philadelphia, PA, USA, http://www.vs.inf.ethz.ch/publ/papers/firststeps.pdf.

Lee, H., Y. Liu, E. Alsberg, D. Ingber, R. Westervelt, and D. Ham (2005), An IC/Microfluidic Hybrid Microsystem for 2D Magnetic Manipulation of Individual Biological Cells, *International Solid State Circuits Conference, Digest of Technical Papers*, February 6–10, San Francisco, CA, pp. 52–53.

Lehmann, W., H. Skupin, C. Tolksdorf, E. Gebhard, R. Zentel, P. Krüger, M. Lösche, and F. Kremer (2001), Giant Lateral Electrostiction in Ferroelectric Liquid Crystalline Elastomers, *Nature*, 410(6827), 447–450.

Leroi Gourhan, A., (1993), *Gesture and Speech*, translated by A. Berger , MIT, Cambridge, MA, USA.

Levis, P., S. Madden, D. Gay, J. Polastre, R. Szewczyk, A. Woo, E. Brewer, and D. Culler (2004), The Emergence of Networking Abstractions and Techniques in TinyOS, *Proceedings of the First USENIX/ACM Symposium on Networked Systems Design and Implementation*, March 29–31, San Francisco, CA, USA, pp. 1–14.

Li, M., and V. Vitanyi (1993), *An Introduction to Kolmogorov Complexity and Its Applications*, Springer-Verlag, New York, NY, USA.

Lieberman, H., and T. Selker (2000), Out of Context: Computer Systems That Adapt to, and Learn from, Context, *IBM Systems Journal*, 39(3–4), 617–632.

Line56 (2003), *Wal-Mart's RFID Mandate*, http://www.line56.com/articles/default.asp?ArticleID=4710.

Linten, D., X. Sun, G. Carchon, W. Jeamsaksiri, A. Mercha, J. Ramos, S. Jenei, L. Aspemyr, A. Scholten, P. Wambacq, S. Decoutere, S. Donnay, and W. De Raedt (2004), A 328 mW 5 GHz VCO in 90 nm CMOS with High Quality Thin Film Post-processed Inductor, *Proceedings of the Custom Integrated Circuits Conference*, October 3–6, Orlando, pp. 701–704.

Lisetti, C., and D. Schiano (2000), Facial Expression Recognition: Where Human Computer Interaction, Artificial Intelligence, and Cognitive Science Intersect, *Pragmatics and Cognition*, 8(1), 185–235.

LivTom (online), Homepage, http://www.livingtomorrow.be.

Lukowicz, P., H. Junker, M. Staeger, T.V. Bueren, and G. Tröster (2002), WearNET: A Distributed Multi-sensor System for Context Aware Wearables, in: G. Borriello and L.E. Holmquist (eds.), *UbiComp 2002: Ubiquitous Computing, Fourth International Conference*, Lecture Notes in Computer Science, 2498, Springer, Berlin Heidelberg New York, pp. 361–370.

Lundqvist, A.H., and A. van Santen (2004), *Research Profile TU/e*, Eindhoven, The Netherlands.

Lynce, I., and J. Marques-Silva (2002), Building State-of-the-Art SAT Solvers, *Proceedings of the 15th European Conference on Artificial Intelligence*, July 21–26, Lyon, France, pp. 166–170.

Maghiros, I., Y. Punie, S. Delaitre, E. Lignos, C. Rodríguez, M. Ulbrich, M. Cabrera, B. Clements, L. Beslay, and R. Van Bavel (2005), *Biometrics at the Frontiers: Assessing the Impact on Society*, Technical Report, EUR 21585EN, European Commission, Luxemburg, Luxemburg.

Markopoulis, P., B. Eggen, E. Aarts, and J. Crowley (eds.) (2004), *Ambient Intelligence, Second European Symposium,EUSAI04*, Springer Lecture Notes in Computer Science, 3295, Springer, Berlin Heidelberg New York.

Martin, D.L., A.L. Cheyer, and D.B. Moran (1999), The Open Agent Architecture: A Framework for Building Distributed Software Systems, *Applied Artificial Intelligence*, 13(1–2), 91–128.

Marvin, C. (1988), *When Old Technologies Were New: Thinking about Electric Communication in the Late Nineteenth Century*, Oxford University, Oxford, UK.

Marx, G. (2001), Murky Conceptual Waters: The Public and the Private, *Ethics and Information Technology*, 3(3), 157–169.

Marzano, S. (ed.) (1995), *Vision of the Future*, V+K, Bussum, The Netherlands.

Marzano, S. (1998), *Creating Value by Design: Thoughts*, Lund Humphries, London, UK.

Marzano, S. (ed.) (1999), *La Casa Prossima Futura*, Philips Design, Eindhoven, The Netherlands.

Marzano, S. (ed.) (2000),*New Nomads*, Philips Design, Eindhoven, The Netherlands.

Mascolo, C., L. Capra, S. Zachariadis, and W. Emmerich (2002), XMIDDLE: A Data-Sharing Middleware for Mobile Computing, *Wireless Personal Communications*, 21(1), 77–103.

Maslow, A. (1943), A Theory of Human Motivation,*Psychological Review*, 50, 370–396.

Maslow, A. (1954), *Motivation and Personality*, Harper, New York, NY, USA.

Maslow, A. (1971), *The Farther Reaches of Human Nature*, Viking, New York, NY, USA.

Mason, R.E.A., and T.T. Carey (1983), Prototyping Interactive Information Systems, *Communications of the ACM*, 26, 347–354.

Masuoka, R., B. Parsia, and Y. Labrou (2003), Task Computing – The Semantic Web Meets Pervasive Computing, in: D. Fensel, K.P. Sycara, and J. Mylopoulos (eds.), *Semantic Web – ISWC 2003, Second International Semantic Web Conference*, Lecture Notes in Computing Science, LNCS, 2870, Springer, Berlin Heidelberg New York, pp. 866–881.

Matsuoka, R., Y. Labrou, B. Parsia, and E. Sirin (2003), Ontology-Enabled Pervasive Computing Applications, *IEEE Intelligent Systems*, 18(5), 68–72.

Maybury, M., and W. Wahlster (eds.) (1998), *Readings in Intelligent User Interfaces*, Morgan Kaufmann, San Francisco, CA, USA.

McCorduck (1979), *Machines Who Think: A Personal Inquiry into the History and Prospects of Artificial Intelligence*, Freeman, New York, NY, USA.

McGee, D.R., P.R. Cohen, R.M. Wesson, and S. Horman, (2002), Comparing Paper and Tangible, Multimodal Tools, *Proceedings of the CHI 2002 Conference on Human Factors in Computing Systems*, April 20–25, Minneapolis, MN, USA, pp. 407–414.

McLuhan, M. (1964), *Understanding Media: The Extensions of Man*, MIT, Cambridge, MA, USA.

McLuhan, M., and Q. Fiore (1967), *The Medium Is The Massage: An Inventory of Effects*, Random House, New York, NY, USA.

Mei, B., S. Vernalde, D. Verkest, H. De Man, and R. Lauwereins (2003), ADRES: An Architecture with Tightly Coupled VLIW Processor and Coarse-Grained Reconfigurable Matrix, in: P.Y.K. Cheung, G.A. Constantinides, and J.T. de Sousa (eds.), *Field-Programmable Logic and Applications, 13th International Conference, FPL 2003*, Lecture Notes in Computer Science, 2778, Springer, Berlin Heidelberg New York, pp. 61–70.

Mental Images (online), Homepage, http://www.mentalimages.com.

Mentor Graphics (2001), *Board Systems Design and Verification – Redefining Systems Design for the Electronic Community*, Technical Paper 608, Mentor Graphics Corporation, San Jose, CA, USA, http://www.mentor.com/pcb/techpapers.

Michahelles, F., and B. Schiele (2003), Sensing Opportunities for Physical Interaction, *Proceedings of the Physical Interaction Workshop on Real World User Interfaces at MobileHCI03*, September 2, Udine, Italy, http://www.vision.ethz.ch/publ/sensingoppi03.pdf.

Michahelles, F., S. Antifakos, J. Boutellier, A. Schmidt, and B. Schiele, (2003), Instructions Immersed into the Real World: How Your Furniture Can Teach You, Interactive Poster, *Proceedings of the International Conference on Ubiquitous Computing*, October 12–15, Seattle, Washington, Canada.

Michahelles, F., P. Matter, A. Schmidt, and B. Schiele (2003), Applying Wearable Sensors to Avalanche Rescue, *Computers & Graphics*, 27(6), 839–847.

Miles, I., K. Flanagan, and D. Cox (2002), *Ubiquitous Computing: Toward Understanding European Strengths and Weaknesses*, European Science and Technology Observatory Report for IPTS, PREST, Manchester, UK.

Miles, S., J. Papay, V. Dialani, M. Luck, K. Decker, T. Payne, and L. Moreau (2003), Personalised Grid Service Discovery, *IEEE Proceedings – Software*, 150(4), 252–256.

Minsky, M. (1975), A Framework for Representing Knowledge, in: P. Winston (ed.), *The Psychology of Computer Vision*, McGraw-Hill, New York, NY, USA, pp. 211–277.

Minsky, M. (1986), *The Society of Mind*, Simon and Schuster, New York, NY, USA.

Minsky, M., and S. Papert (1969), *Perceptrons: An Introduction to Computational Geometry*, MIT, Cambridge, MA, USA.

Monk, A., J. Brant, P. Wright, and J. Robinson (2004), CUHTec: The Centre for Usable Home Technology, *Proceedings of the ACM Conference on Human Factors in Computing Systems*, April 24–29, Vienna, Austria, pp. 1073–1074.

Morrison, J., and J.F. George (1995), Exploring the Software Engineering Component in MIS Research, *Communications of the ACM*, 38, 80–91.

Morrison, R., G. Kirby, D. Balasubramaniam, K. Mickan, F. Oquendo, S. Cimpan, B. Warboys, B. Snowdon, and R.M. Greenwood (2004), Support for Evolving Software Architectures in the ArchWare ADL,*Proceedings of the Fourth Working IEEE/IFIP Conference on Software Architecture*, June 12–15, Oslo, Norway, pp. 69–78.

Mozer M. (1998), The Neural Network House: An Environment That Adapts to Its Inhabitants, *Proceedings of the AAAI Spring Symposium on Intelligent Environments*, March 23–23, Pal Alto, CA, USA, pp. 110–114.

Muhkerjee, S., E. Aarts, M. Ouwerkerk, R. Rovers, and F. Widdershoven (eds.) (2005), *AmIware: Hardware Drivers for Ambient Intelligence*, Springer, Berlin Heidelberg New York.

Muller, M.J., D.M. Wildman, and E. White (1993), Taxonomy of PD Practices: A Brief Practitioner's guide, *Communications of the ACM*, 36, 25–28.

Murphy, A.L., G.P. Picco, and G.-C. Roman (1999), Lime: Linda Meets Mobility, *Proceedings of the 21st International Conference on Software Engineering*, May 16–22, Los Angeles, CA, USA, pp. 368–377.

MyGrid (online), Homepage, http://www.mygrid.org.

Natural Motion (online), Homepage, http://www.naturalmotion.com.

Negroponte, N. (1995), *Being Digital*, Alfred A. Knopf, New York, NY, USA.

Negus, K.J., A.P. Stephens, and J. Lansford (2000), HomeRF: Wireless Networking for the Connected Home, *IEEE Personal Communications*, 7(1), 20–27.

NERC (online), Homepage, http://www.ndg.nerc.ac.uk.

Neudeck, A., S. Gimpel, U. Möhring, H. Müller, and W. Scheibner (2004), Textile-Based Electronic Substrate Technology, *Journal of Industrial Textiles*, 3, 179–189.

NGG Expert Group (2004), *Next Generation Grids 2 – Requirements and Options for European Grids Research 2005–2010 and Beyond*, IST Report, IST Programme Grid Technologies Unit, ftp://ftp.cordis.lu/publ/ist/docs/ngg_cq_final_pdf.

Niedermayer, M., D.-D. Polityko, G. Fotheringham, S. Guttowski, and W. John (2004), Design Aspects of Self-sufficient Distributed Microsystems, *Journal of Telecommunications*, 58, 54–57.

Nieland, S., A. Ostmann, R. Aschenbrenner, and H. Reichl (2000), Immersion Soldering – A New Way of Ultra Fine Pitch Bumping, *Proceedings of the International Congress Electronics Goes Green 2000*, September 11–13, Berlin, Germany, pp. 165–167.

Niemegeers, I.G.M.M., and S.M. Heemstra de Groot (2003), Research Issues in Ad-Hoc Distributed Personal Networking, *Wireless Personal Communications*, 26(23), 149–167.

Nijholt, A. (2003), Multimodality and Ambient Intelligence, in: W. Verhaegh, E. Aarts, and J. Korst (eds.), *Algorithms in Ambient Intelligence*, Kluwer Academic, Dordrecht, The Netherlands, pp. 23–53.

Nijholt, A., T. Rist, and K. Tuijnenbreijer (2004), Lost in Ambient Intelligence? Panel Session, *Proceedings of Conference on Human Factors in Computing Systems*, April 24–29, Vienna, Austria, pp. 1725–1726.

Nilsson, N.J. (1998), *Artificial Intelligence: A New Synthesis*, Morgan Kaufmann, San Francisco, CA, USA.

Nishida, T., D. Rosenberg, and R. Fruchter (2005), Understanding Mediated Communication: The Social Intelligence Design (SID) Approach, *AI & Society, The Journal of Human-Centred Systems*, 19(1), 1–7.

Norman, D.A. (1988), *The Psychology of Everyday Things*, Basic Books, New York, NY, USA.

Norman, D.A. (1993), *Things That Make Us Smart*, Perseus Books, Cambridge, MA, USA.

Notten, P.H.L., M. Kremers, and R. Griessen (1996), Optical Switching of Y-Hydride Thin Film Electrodes – A Remarkable Electrochromic Phenomenon, *Journal of the Electrochemical Society*, 143, 3348–3353.

Noyce, R. (1977), Microelectronics, *Scientific American*, 237(3), 63–69.

O'Brien, J., T. Rodden, M. Rouncefield, and J. Hughes (1999), At Home with the Technology: An Ethnographic Study of a Set-Top-Box Trial, *ACM Transactions on Computer–Human Interaction*, 6, 282–308.

OECD (1992), *Technology and the Economy. The Key Relationships,* Report, Organisation for Economic Co-operation and Development, OECD Publications, Paris, France.

OECD (2002), *Measuring the Information Economy 2002*, Report, Organisation for Economic Co-operation and Development, OECD Publications, Paris, France.

Olsson, E. (2004), *Designing Work Support Systems – For and With Skilled Users*, Ph.D. Thesis, Uppsala University, Uppsala, Sweden.

O'Neill, H., and J. Woodward, (2000), *Construction of a Bio-hydrogen Fuel Cell*, Talk presented at the DARPA Advanced Energy Technologies Energy Harvesting Program, April 13–14, Arlington, VA, USA, http://www.darpa.mil/dso/trans/energy/briefings/8ornl.pdf.

Osman, I.H., and J.P. Kelly (eds.) (1996), *Meta-Heuristics: Theory and Applications*, Kluwer Academic, Boston, MA, USA.

Ostmann, A., A. Neumann, J. Auersperg, C. Ghahremani, G. Sommer, R. Aschenbrenner, and H. Reichl (2002a), Integration of Passive and Active Components into Build-Up Layers, *Proceedings of the Electronics Packaging Technology Conference*, December 10–12, Singapore.

Ostmann, A., A. Neumann, S. Weser, E. Jung, L. Boettcher, and H. Reichl (2002b), Realization of a Stackable Package Using Chip in Polymer Technology, *Proceedings of the Polytronic Conference*, June 23–26, Zalaegerzeg, Hungary, pp. 160–164.

Oviatt, S.L. (2002), Breaking the Robustness Barrier: Recent Progress on the Design of Robust Multimodal Systems, in: M. Zelkowitz (ed.), *Advances in Computers*, 56, 305–341.

Oviatt, S.L, and W. Wahlster (eds.) (1997), Multimodal Interfaces, *Human–Computer Interaction Journal*, 12(1–2).

Oxygen (online), Homepage, http://oxygen.lcs.mit.edu/index.html.

Pahl, B., C. Kallmayer, R. Aschenbrenner, and H. Reichl (2002), *A Thermode Bonding Process for Fine Pitch Flip Chip Applications on Flexible Substrates*, Talk presented at the IMAPS Nordic Conference, September 29–October 2, Stockholm, Sweden.

Palmore, R., and G. Tayhas (2000), *Biofuel Cells*, DARPA Advanced Energy Technologies Energy Harvesting Program, April 13–14, Arlington, VA, USA, http://www.darpa.mil/dso/trans/energy/briefings/10PALMORE.PDF.

Papadimitriou, C.H., and K. Steiglitz (1982), *Combinatorial Optimization: Algorithms and Complexity*, Prentice-Hall, Englewood Cliffs, NJ, USA.

Papanikolaou, A., M. Miranda, H. Wang, F. Lobmaier, and F. Catthoor (2005), A System-Level Methodology for Fully Compensating Process Variability Impact of Memory Organizations in Periodic Applications, *Proceed-*

ings of the IEEE International Conference on Hardware/Software Codesign and System Synthesis, September 19–21, New York, NY, USA (to appear).

Papazoglou, M.P., and D. Georgakopoulos (eds.) (2003), Service-Oriented Computing, *Communications of the ACM*, 46(10).

Paradiso, J.A. (2005), Energy Scavenging for Mobile and Wireless Electronics, *IEEE Pervasive Computing*, 4(1), 18–27.

Pascoe, J. (1998), Adding Generic Contextual Capabilities to Wearable Computers, *Proceedings of the Second International Symposium on Wearable Computers*, October 19–20, Pittsburgh, PA, USA, pp. 92–99.

Pauws, S.C., and J.H. Eggen (2003), Realization and User Evaluation of an Automatic Playlist Generator, *Journal of New Music Research*, 32 (2), 179–192.

Pearl, J. (1988), *Probabilistic Reasoning in Intelligent Systems: Networks of Plausible Inference*, Morgan Kaufmann, San Francisco, CA, USA.

PECo (online), *Personal Environment Controller*, Project Homepage, http://www.igd.fhg.de/igd-a1/projects/peco/index.html.

Penterman, R., S.I. Klink, H. De Koning, G. Nisato, and D.J. Broer (2002a), Single Substrate Liquid Crystal Displays by Photo-Enforced Stratification, *Nature*, 417(6884), 55–58.

Penterman, R., S.I. Klink, H. De Koning, G. Nisato, and D.J. Broer (2002b), Single Substrate LCDs Poduced by Photo-Enforced Stratification, *Society of Information Displays 2002 International Symposium, Digest of Technical Papers*, pp. 1020–1024.

Perik, E., B. de Ruyter, P. Markopoulos, and B. Eggen (2004), The Sensitivities of User Profile Information in Music Recommender Systems,*Proceedings of the Second Annual Conference on Privacy, Security and Trust*, October 13–15, Fredericton, New Brunswick, Canada, http://dev.hil.unb.ca/Texts/PST/.

Pering, C. (2002), Interaction Design Prototyping of Communicator Devices: Towards Meeting the Hardware–Software Challenge, *Interactions*, 9(6), 37–46.

PerSec (online), IEEE International Workshop on Pervasive Computing and Communication Security, Homepage, http://www-lce.eng.cam.ac.uk/~fms27/persec-2005.

Picard, R.W. (1997), *Affective Computing*, MIT, Cambridge, MA, USA.

Picard, R.W., and J. Klein (2002), Computers that Recognise and Respond to User Emotion: Theoretical and Practical Implications, *Interacting with Computers*, 14(2), 141–169.

Pieper, R. (1999), *Ambient Intelligence*, Keynote presented at the Digital Living Room Conference, June 21, Dana Point, CA, USA, http://www.ambientintellgence.net/DLR_keynote.htm.

Pine, B.J., and J.H. Gilmore (1999), *The Experience Economy*, Bradford Books, New York, NY, USA.

Pinto, P., L. Bernardo, and P. Sobral (2004), UMTS–WLAN Service Integration at Core Network Level, in: M. Freire, P. Chemouil, P. Lornz, and A.

Gravrey (eds.), *Universal Multiservice Networks, Third European Conference, ECUMN 2004*, Lecture Notes in Computer Science, 3262, Springer, Berlin Heidelberg New York, pp. 29–39.

Point Topic (online), *World Broadband Statistics: H2 2004*, http://www.point-topic.com/contentDownload/dslanalysis/world%20broadband%20statistics%20q4%202004.pdf.

Poladian, V., J.P. Sousa, D. Garlan, and M. Shaw (2004), Dynamic Configuration of Resource-Aware Services, *Proceedings of the 26th International Conference on Software Engineering*, May 23–28, Edinburgh, Scotland, UK, pp. 604–613.

Polanyi, M. (1962), *Personal Knowledge Towards a Post-critical Philosophy*, University of Chicago, Chicago, IL, USA.

Popper, R., and I. Miles (2004), *The FISTERA Delphi. Future Challenges, Applications and Priorities for Socially Beneficial IST.* WP4 IST Futures Forum, PREST, Manchester, UK, http://fistera.jrc.es.

Potts, C. (1993), Software-Engineering Research Revisited, *IEEE Software*, 10, 19–28.

Powerscape (2005), Homepage, http://www.powerescape.com.

Prahalad, C.K. (2004), *The Fortune at the Bottom of the Pyramid: Eradicating Poverty Though Profits*, Wharton School Publishing/Pearson, Philadelphia, PA, USA.

Prahalad, C.K, and V. Ramaswamy (2004), *The Future of Competition*, Harvard Business School, Cambridge, MA, USA.

Punie, Y. (2005), The Future of Ambient Intelligence in Europe: The Need for More Everyday Life, *Communication & Strategies*, 57, 141–165.

Quillian, M.R. (1968), Semantic Memory, in: M. Minsky (ed.), *Semantic Information Processing*, MIT, Cambridge, MA, USA, pp. 227–270.

Rabaey, J., J. Amer, T. Karalar, S. Li, B. Otis, M. Sheets, and T. Tuan (2002), PicoRadios for Wireless Sensor Networks: The Next Challenge in Ultra-Low Power Design, *International Solid State Circuits Conference, Digest of Technical Papers*, February 3–7, San Francisco, CA, USA, pp. 200–210.

Rahnema, M. (1993), Overview of the GSM System and Protocol Architecture,*IEEE Communications Magazine*, 31(4), 92–100.

Ramm, P., A. Klumpp, R. Merkel, J. Weber, and R. Wieland (2004), Vertical System Integration by Using Interchip Vias and Solid–Liquid Interdiffusion Bonding, *Japanese Journal of Applied Physics*, 43(7A), L829–L830.

Ramm, P., A. Klumpp, R. Merkel, J. Weber, R. Wieland, A. Ostmann, J. Wolf, and H. Reichl (2003), 3D System Integration, *Proceedings of the MicroSystems International Meeting*, April 21–25, San Fransisco, CA, USA, pp. 3–14.

Reeves, B., and C. Nass (1996), *The Media Equation*, Cambridge University, New York, NY, USA.

Reidsma, D., R. op den Akker, R. Rienks, A. Nijholt, R. Poppe, D. Heylen, and J. Zwiers (2005), Virtual Meeting Rooms: from

Observation to Simulation, *Proceedings of the International Conference on Social Intelligence Design*, March 24–26, Stanford, CA, USA, http://wwwhome.cs.utwente.nl/~heylen/Publicaties/sid2005.pdf.

Rekimoto, J. (1996), Tilting Operations for Small Screen Interfaces, *Proceedings of the ACM Symposium on User Interface Software and Technology*, November 6–8, Seattle, Washington, Canada, pp. 67–168.

Rich, E., and K. Knight (1991), *Artificial Intelligence* (2nd edition), McGraw-Hill, New York, NY, USA.

Richardson, T., F. Bennett, G. Mapp, and A. Hopper (1994), Teleporting in an X Window System Environment, *IEEE Personal Communication Magazine*, 1(3), 6–12.

Rifkin, J. (2004), *The European Dream*, Tarcher, Penguin Books, New York, NY, USA.

Rijpkema, E., K. Goossens, A. Radulescu, J. Gielissen, J. Van Meerbergen, P. Wielage, and E. Waterlander (2003), Trade-Offs in the Design of a Router with Both Guaranteed and Best-Effort Services for Networks on Chip, *Proceedings of the Conference on Design, Automation and Test in Europe*, March 6–10, Munich, pp. 350–355.

Riva, G., F. Vatalaro, F. Davide, and M. Alcañiz (eds.) (2005), *Ambient Intelligence: The Evolution of Technology, Communication and Cognition Towards the Future of Human–Computer Interaction*, IOS, Amsterdam, The Netherlands.

Roberts, L. (1998), quoted in S. Segaller, *Nerds: A Brief History of the Internet*, TV Books, New York, NY, USA.

Rodden, T., K. Cheverest, K. Davies, and A. Dix (1998), Exploiting Context in HCI Design for Mobile Systems, *Proceedings of the Workshop on Human Computer Interaction with Mobile Devices*, May 21–23, Glasgow, UK, http://www.dcs.gla.uk/~johnson/papers/mobile/HCIMD1.

Rodriguez, C., C. van Wunnik, L. Delgado, J.-C. Burgelman, and P. Desruelle (eds.) (2004), *The Future Impact of ICTs on Environmental Sustainability*, IPTS Technical Report Series, EUR 21384EN, DG JRC, European Commission, Luxemburg, Luxemburg.

Roman, M., and R.H. Campbell (2003), A Middleware-based Application Framework for Active Space Applications, *Proceedings of the ACM/IFIP/USENIX International Middleware Conference*, June 16–20, Rio de Janeiro, Brazil, pp. 433–454.

Roman, M., C.K. Hess, R. Cerqueira, A. Ranganat, R.H. Campbell, and K. Nahrstedt (2002a), Gaia: A Middleware Infrastructure for Active Spaces, *IEEE Pervasive Computing*, 1(4), 74–83.

Roman, M., H. Ho, and R.H. Campbell (2002b), Application Mobility in Active Spaces, *Proceedings of the First International Conference on Mobile and Ubiquitous Multimedia*, December 11–13, Oulu, Finland, http://choices.cs.uiuc.edu/gaia/papers/mum2002.pdf.

Roman, G.-C., G.P. Picco, and A.L. Murphy (2000), Software Engineering for Mobility: A Roadmap, in: A. Finkelstein (ed.), *The Future of Software Engineering*, ACM, New York, NY, USA, pp. 241–258.

Rosenblatt, F. (1957), *The Perceptron: A perceiving and Recognizing Automaton*, Report 85-460-1, Project PARA, Cornell Aeronautical Laboratory, Ithaca, New York, NY, USA.

Roundy, S.R., *Energy Scavenging for Wireless Sensor Nodes with a Focus on Vibration to Electricity Conversion*, Ph.D. Thesis, University of California, Berkeley, CA, USA.

Roundy, S., and J. Rabaey (2004), *Energy Scavenging for Wireless Sensor Networks with Special Focus on Vibrations*, Kluwer Academic, New York, NY, USA.

Roundy, S., B. Otis, Y.-H. Chee, J. Rabaey, and P. Wright (2003), A 1.9 GHZ Transmit Beacon Using Environmentally Scavenged Energy, *Proceedings of the International Symposium on Low Power Electronics and Devices*, August 5 , Seoul, Korea, pp. 25–27.

Roush, W. (2004), Beyond the Bar Code, *MIT's Magazine of Innovation, Technology Review*, 107(10), 20–21.

Rovers, A.F., and H.A. Van Essen (2004), HIM: A Framework for Haptic Instant Messaging, *Proceedings of International Conference on Computer Human Interaction 2004*, April 24–29, Vienna, Austria, pp. 1313–1316.

Rumelhart, D.E., G.E. Hinton, and R.J. Williams (1986), Learning Representations by Backpropagating Errors, *Nature*, 323, 533–536.

Rush (online), Homepage, http://www.rush.edu.

Russel, D.M., and S.B. Cousins (2004), IBM Almaden's User Sciences & Experience Research Lab., *Proceedings of the ACM Conference on Human-Factors of Computer Systems*, April 24–29, Vienna, Austria, pp. 1079–1080.

Russell, S., and P. Norvig (2003), *Artificial Intelligence: A Modern Approach* (2nd edition), Prentice Hall, Englewood Cliffs, NJ, USA.

Sailhan, F., and V. Issarny (2003), Cooperative Caching in Ad Hoc Networks, *Proceedings of the Fourth International Conference on Mobile Data Management*, January 21–24, Melbourne, Australia, pp. 13–28.

Sakurai, T. (2003), Perspectives on Power-Aware Electronics, *International Solid State Circuits Conference, Digest of Technical Papers*, February 10–12, San Francisco, CA, USA, pp. 26–27.

Salkintzis, A.K., C. Fors, and R. Pazhyannur (2002), WLAN–GPRS Integration for Next-Generation Mobile Data Networks, *IEEE Wireless Communications*, October, 112–124.

Samjani, A.A. (2002), General Packet Radio Service (GPRS), *IEEE Potentials*, 21(2), 12–15.

Sanger (online), Homepage, http://www.sanger.ac.uk.

Schank R.C., and R.P. Abelson (1977), *Scripts, Plans, Goals and Understanding*, Lawrence Erlbaum Associates, Hillsdale, NJ, USA.

Scheirer, J., R. Fernandez, and R.W. Picard (1999), Expression Glasses: A Wearable Device for Facial Expression Recognition, Extended abstract, *In-*

ternational Conference on Human Factors in Computer Systems, May 15–20, Pittsburgh, PA, USA.

Scherer, K. (1987), *Toward a Dynamic Theory of Emotion: The Component Process Model of Affective States*, http://www.unige.ch/fapse/emotion/publications/list.html.

Schlansker, M., T. Conte, J. Dehnert, K. Ebcioglu, J. Fang, and C. Thompson (1997), Compilers for Instruction-Level Parallelism, *IEEE Computer Magazine*, 30(12), 44–50.

Schneider, M. (2004), Towards a Transparent Proactive User Interface for a Shopping Assistant, *Proceedings of the Workshop on Multi-user and Ubiquitous User Interfaces,*January 13–16, Funchal, Portugal, pp. 31–35.

Schilit, B., and M. Theimer, (1994), Disseminating Active Map Information to Mobile Hosts, *IEEE Network*, 8, 22–32.

Schmidt, A. (1999), *Implicit Human-Computer Interaction Through Context*, Paper presented at the Second Workshop on Human Computer Interaction with Mobile Devices Edinburgh, Scotland, August 31, http://www.teco.uni-karlsruhe.de/~albrecht/publication/mobile99/impl_hci.pdf.

Schmidt, A., K.A. Aidoo, A. Takaluoma, U. Tuomela, K. Van Laerhoven, and W. Van de Velde (1999), Advanced Interaction in Context, in: Hans-Werner Gellersen (ed.), *Handheld and Ubiquitous Computing, First International Symposium, HUC'99*, Lecture Notes in Computing Science, 1707, Springer, Berlin Heidelberg New York, pp. 89–101.

Schmidt, A., M. Strohbach, K. Van Laerhoven, A. Friday, and H.W. Gellersen (2002), Context Acquisition Based on Load Sensing, in: G. Borriello and L.E. Holmquist (eds.), *Ubicomp 2002: Ubiquitous Computing, Fourth International Conference*, Lecture Notes in Computer Science, 2498 Springer, Berlin Heidelberg New York, pp. 333–350.

Schmitz, A., M. Tranitz, S. Wagner, R. Hahn, and F. Hebling (2003), Planar Self-breathing Fuel Cells, *Journal of Power Sources*, 5213, 1–10.

Schraefel, M.C., G. Hughes, H. Mills, G. Smith, and J.G. Frey (2004a), Making Tea: Iterative Design Through Analogy, *Proceedings of the ACM Conference on Designing Interactive Systems, Processes, Practices, Methods, and Techniques*, August 1–4, Cambridge, MA, USA, pp. 49–58.

Schraefel, M.C., G. Hughes, H. Mills, G. Smith, T. Payne, and J. Frey (2004b), Breaking the Book: Translating the Chemistry Lab Book to a Pervasive Computing Environment, *Proceedings of the ACM SIGCHI Conference on Human Factors in Computer Systems*, April 24–29, Vienna, Austria, pp. 25–32.

SDR Forum (online), Homepage, http://www.sdrforum.org.

Seaman, C.B., M.G. Mendonca, V.R. Basili, and Y. Kim (2003), User Interface Evaluation and Empirically-Based Evolution of a Prototype Experience Management Tool, *IEEE Transactions on Software Engineering*, 29(9), 838–850.

Searle, J.R. (1980), Minds, Brains, and Programs, in: *The Behavioral and Brain Sciences*, Cambridge University, New York, NY, USA.

SEC (2005), *The Local Dimension of the Information Society*, Commission Staff Working Document, SEC(2005)/206, European Commission, Luxemburg, Luxemburg.

Servat, D., and A. Drogoul (2002), Combining Amorphous Computing and Reactive Agent-Based Systems: A Paradigm for Pervasive Intelligence? *Proceedings of the First International Joint Conference on Autonomous Agents and Multiagent Systems*, July 15–19, Bologna, Italy, pp. 441–448.

Shadbolt, N. (2003), Ambient Intelligence, *IEEE Intelligent Systems*, 18(4), 2–3.

Shahinpoor, M., and K.J. Kim (2001), Ionic Polymer–Metal Composites, I. Fundamentals, *Smart Materials Structures*, 10, 819–833.

Shaw, M., and D. Garlan (1996), *Software Architecture: Perspectives on an Emerging Discipline*, Prentice-Hall, Englewood Cliffs, NJ, USA.

Shazam (online), Homepage, http://www.shazam.com.

Siegemund, F., and C. Flöer (2003), Interaction in Pervasive Computing Settings Using Bluetooth Enabled Active Tags and Passive RFID Technology Together with Mobile Phones, *Proceedings of the First IEEE International Conference on Pervasive Computing and Communications*, March 23–26, Fort Worth, TX, USA, pp. 378–387.

Sim, I., P. Gorman, R.A. Greenes, R.B. Haynes, B. Kaplan, H. Lehmann, and P.C. Tang (2001), Clinical Decision-Support Systems for the Practice of Evidence-Based Medicine,*Journal of the American Medical Informatics Association*, 8, 527–534.Singh , H., M-H. Lee, L. Guangming, F.J. Kurdahi, N. Bagherzadeh, F. Chaves (2000), MorphoSys: An Integrated Reconfigurable System for Data-Parallel and Computation-Intensive Applications, *IEEE Transactions on Computers*, 49(5), 465–481.

Smart-Its (online), Homepage, http://www.smart-its.org.

SmartTea (online), Homepage, http://www.SmartTea.org.

Smela, E., O. Inganänas, and I. Lundström (1995), Controlled Folding of Micrometer-Size Structures, *Science*, 268, 1735–1738.

Smits, R., and S. Kuhlmann (2004), The Rise of Systemic Instruments in Innovation Policy, *International Journal of Foresight and Innovation Policy*, 1(1–2), 4–32.

SOA (online), http://www.service-architecture.com/web-services/articles/service-oriented_architecture_soa_definition.html.

Sokolowski, W.M., A.B. Chiemelewski, and S. Hayashi (1999), Cold Hibernated Elastic Memory (CHEM) Self-deployable Structures, in: Y. Bar-Cohen (ed.), *Smart Structures and Materials 1999: Electroactive Polymer Actuators and Devices, Proceedings of SPIE, 3669*, pp. 179–185.

Sousa, J.P., and D. Garlan (2002), Aura: An Architectural Framework for User Mobility in Ubiquitous Computing Environments, *Proceedings of the Third IEEE/IFIP Conference on Software Architecture*, August 25–30, Montreal, Canada, pp. 29–43.

Sousa, J.P., and D. Garlan (2003), *The Aura Software Architecture: An Infrastructure for Ubiquitous Computing*, Carnegie Mellon University, Technical Report CMU-CS-03-183.

SPC-Conf (online), *International Conference on Security in Pervasive Computing*, April 6–8, 2005, Boppard, Germany, Homepage, http://www.spc-conf.org.

Spilling, P., and Y. Lundh (2004), Features of the Internet History, *Telektronikk*, 100(3), 113–133.

SPPC (online), *Workshop on Security and Privacy in Pervasive Computing*, April 20, Vienna, Austria, Homepage, http://www.vs.inf.ethz.ch/events/sppc04.

Stefik, M. (1995), *Introduction to Knowledge Systems*, Morgan Kaufmann, San Francisco, CA, USA.

Stevens, R., H.J. Tipney, C. Wroe, T. Oinn, M. Senger, P. Lord, C. Goble, A. Brass, and M. Tassabehji (2004), Exploring Williams–Beuren Syndrome Using myGrid, *Proceedings of the 12th Conference on Intelligent Systems for Molecular Biology*, July 31–August 4, Glasgow, UK, pp. 303–310.

Stoop, E. (2003), Mobility: Freedom of Body and Mind, in: E. Aarts and S. Marzano (eds.), *The New Everyday: Views on Ambient Intelligence*, 010 Publishers, Rotterdam, The Netherlands, pp. 140–145.

Stork, D.G. (1997), *HAL's Legacy: 2001's Computer as Dream and Reality*, MIT, Cambridge, MA, USA.

Storz, O., A. Friday, and N. Davies (2003), Towards "Ubiquitous" Ubiquitous Computing: An alliance with "the Grid", *Proceedings of the Fifth International Conference on Ubiquitous Computing*, October 12, Seattle, WA, USA.

Strang, T., and C. Linnhoff-Popien (2004), A Context Modeling Survey, *Proceedings of the Sixth International Conference on Ubiquitous Computing*, September 7, Nottingham, UK.

Straub, D.W. (1989), Validating Instruments in MIS Research, *MIS Quarterly*, 13(2), 147–169.

Strese, H. (2005), Technologies for Smart Textiles, *MST News*, 2/05, 6.

Stuart, D. (2002), Talk presented at the National e-Science Centre Workshop, June 2, Edinburgh, UK.

Su, J., J.S. Harrison, T.L. St. Clair, Y. Bar-Cohen, and S. Leary (1991), Electrostrictive Graft Elastomers and Applications, *Proceedings of Materials ResearchSocietyFall Meeting*, December 2–6, Boston, MA, USA, pp. 131–135.

Suchman, L. (1987), *Plans and Situated Actions: The Problem of Human–Machine Communication*, Cambridge University, Cambridge, UK.

Sutton, R.S., and A.G. Barto (1998), *Reinforcement Learning: An Introduction*, MIT, Cambridge, MA, USA.

Svanæs, D., and G. Seland (2004), Putting the Users Center Stage: Role Playing and Low-fi Prototyping Enable End Users to Design Mobile Systems,

Proceedings of the ACM Conference on Computer Human Interaction, April 24–29, Vienna, Austria, pp. 479–486.

Szyperski, C., D. Gruntz, and S. Murer (2002), *Component Software – Beyond Object-Oriented Programming* (2nd edition), Addison-Wesley, Reading, MA, USA.

ter Horst, H.J. (2004), Extending the RDFS Entailment Lemma, in: Sheila A. McIlraith, D. Plexousakis, and F. van Harmelen (eds.), *The Semantic Web – ISWC 2004, Third International Semantic Web Conference*, Lecture Notes in Computing Science, 3298, Springer, Berlin Heidelberg New York, pp. 77–91.

Talkingpresents (online), Homepage, http://www.talkingpresents.com.

Talkingproducts (online), Homepage, http://www.talkingproducts.co.uk.

Tanaka, T., I. Nishio, S.-T. Sun, and S. Ueno-Nishio (1982), Collapse of Gels in an Electric Field, *Science*, 218, 467–469.

Tang, A., C. Owen, F. Biocca, and W. Mou (2003), Comparative Effectiveness of Augmented Reality in Object Assembly, *Proceedings of the Conference on Human Factors in Computing Systems, Proceedings of the 2003 Conference on Human Factors in Computing Systems*, April 5–10, Ft. Lauderdale, Fl, USA, pp. 73–80.

Tarasewich, P., C. S. Campbell, T. Xia, and M. Dideles (2003), Evaluation of Visual Notification Cues for Ubiquitous Computing, *Proceedings of the International Conference on Ubiquitous Computing*, in: A.K. Dey, A. Schmidt, and J.F. McCarthy (eds.), *UbiComp 2003: Ubiquitous Computing, Fifth International Conference*, Lecture Notes in Computer Science, 2864, Springer, Berlin Heidelberg New York, pp. 349–366.

Taylor, J.M. (online), http://www.e-science.clrc.ac.uk.

Taylor, S.J., and R. Bogdan (1998), *Introduction to Qualitative Research Methods*, John Wiley, Chichester, UK.

Thomson D.L., P. Keller, J. Naciri, R. Pink, H. Jeon, D. Shenoy, and B.R. Ratna (2001), Liquid Crystal Elastomers with Mechanical Properties of a Muscle, *Macromolecules*, 34, 5868–5875.

Thorstensen, B., T. Syversen, T.A. Bjørnvold, and T. Walseth (2004), The Electronic Shepherd: A low Cost, Low Bandwidth, Wireless Network System, *Proceedings of the International Conference on Mobile Systems, Applications and Services*, June 6–9, Boston, MA, USA, pp. 245–255.

Tichy, W.F. (1998), Should Computer Scientists Experiment More? *IEEE Computer*, 31(5), 32–40.

Tinyos (2005), Homepage, http://www.tinyos.net.

Todd, J. (2002), The Hopi Environmental Ethos, Videorecording produced by Pat Ferrero, *New Day Films, E99.H7 H675 1990z VHS*, New Day Film Library, San Francisco, CA, USA.

Tröster, G., T. Kirstein, and P. Lukowicz (2005), Wearable Computing: Packaging in Textiles and Clothes, *Proceedings of the 14th European Microelectronics and Packaging Conference & Exhibition*, June 23–25, Friedrichshafen, Germany, pp. 23–25.

Tsang, E.P.K. (1993), *Foundations of Constraint Satisfaction*, Academic, London, UK.

Tuomi, I. (2003), *Beyond User-Centric Models of Product Creation*, Paper presented at the COST A269 Conference: The Good, The Bad and The Irrelevant, September 3–5, Helsinki, Finland.

Tuominen, R., and P. Palm (2002), Development of Industrial Scale Manufacturing Line for Integrated Module Board technology, Talk presented at the Sixth VLSI Packaging Workshop of Japan, November 12–14, Kyoto, Japan.

Turing, A.M. (1936) On Computable Numbers, with an Application to the Entscheidungs-Problem, *Proceedings of the London Mathematical Society*, 42, 230–265 and 43, 544–546.

Turing, A.M. (1950), Computing Machinery and Intelligence, *Mind*,59(236), 433–460.

UK e-Science (online), Webpage, http://www.nesc.ac.uk/events/townmeeting0405.

Ullmer, B., and H. Ishii (2001), Emerging Frameworks for Tangible User Interfaces, in: J. M. Carroll (ed.),*Human–Computer Interaction in the New Millenium,* Addison-Wesley, Reading, MA, USA, pp. 579–601.

UMTS Forum (2005), *3G/UMTS Subscribers Hit 16 Million*, http://www.umts-forum.org/servlet/dycon/ztumts/umts/Live/en/umts/News_PR_Article050105.

UMTS Forum (online), Webpage, http://www.umts-forum.org/servlet/dycon/ztumts/umts/Live/en/umts/Home.

van de Sluis, R., J.H. Eggen, J. Jansen, and H. Kohar (2001), User Interface for an In-Home Environment, *Proceedings of the International Conference on Human Computer Interaction*, July 9–13, Tokyo, Japan, pp. 383–390.

van den Hoven, E., and J.H. Eggen (2003), The Design of a Recollection Supporting Device – A Study into Triggering Personal Recollections, *Proceedings of the International Conference on Human Computer Interaction*, June 22–27, Crete, Greece, Part II, pp. 1034–1038.

Van Bavel, R., Y. Punie, and I. Tuomi (2004), *ICT-Enabled Changes in Social Capital*, IPTS Report, 85, pp. 28–32.

Van Essen, H.A., and A.F. Rovers (2005), Foot IO – Design and Evaluation of a Device to Enable Foot Interaction Over a Computer Network, *Proceedings of WorldHaptics 2005*, March 18–20, Pisa, Italy, pp. 521–522.

Van Laerhoven, K., K. Aidoo, and S. Lowette (2001), Real-Time Analysis of Data from Many Sensors with Neural Networks, *Proceedings of the Fifth IEEE International Symposium on Wearable Computers*, October 8–9, Zurich, Switzerland, pp. 115–122.

Van der Sluis, P., and V.M.M. Mercier (2001), Solid State Gd–Mg Electrochromic Devices with ZrO2Hx Electrolyte, *Electrochimica Acta*, 46, 2167—2171.

Van der Sluis, P., M. Ouwerkerk, and. P.A. Duine (1997), Optical Switches Based on Magnesium Lanthanide Alloy Hydrides, *Applied Physics Letters*, 70, 3356–3358.

Van Turnhout, K., J. Terken, I. Bakx, and J.H. Eggen (2005), Identifying the Intended Addressee in Mixed Human–Human and Human–Computer Interaction from Non-verbal Features, *Proceedings of the Seventh International Conference on Multimodal Interfaces*, October 4–6, Trento, Italy (to appear).

Van Welie, M., G.C. van der Veer, and A. Elins (1999), Breaking Down Usability,*Proceedings of the International Conference on Human–Computer Interaction*, August 30–September 3, Edinburgh, UK, pp. 613–620.

Venkatesh, V., M.G. Morris, G.B. Davis, and F.D. Davis (2003), User Acceptance of Information Technology: Towards a Unified View, *MIS Quaterly*, 27(3), 425–478.

Verhaegh, W., E. Aarts, and J. Korst (eds.) (2004), *Algorithms in Ambient Intelligence*, Kluwer Academic, Dordrecht, The Netherlands.

Vicente, K.J. (1999), *Cognitive Work Analysis*, Lawrence Erlbaum Associates, Mahwah, NJ, USA.

Vildjiounaite, J., E.J. Malm, J. Kaartinen, and P. Alahuhta (2002), Location Estimation Indoors by Means of Small Computing Power Devices, Accelerometers, Magnetic Sensors and Map Knowledge, in: F. Mattern and M. Naghshineh (eds.), *Pervasive Computing, First International Conference, Pervasive 2002*, Lecture Notes in Computer Science, 2414, Springer, Berlin Heidelberg New York, pp. 211–224.

Vogels, J.P.A., S.I. Klink, R. Penterman, H. De Koning, H.E.A. Huitema, and D.J. Broer (2004), Robust Flexible LCDs with Paintable Technology, *Society of Information Displays 2004 International Symposium, Digest of Technical Papers*, May 22–27, Boston, MA, USA, pp. 767–769.

Wagner, S., R. Hahn, J. Grillmayer, H. Gaul, and H. Reichl (2003), Development of Thin Film and Micropatterning Technologies for Miniaturized Planar Fuel Cells, *Proceedings of the International Conference on Micro System Technologies*, October 7–8, Munich, Germany, pp. 303–311.

Wagner, S., R. Hahn, and H. Reichl, (2005), Foil Type Microfuel Cell, Talk presented at the Seventh Small Fuel Cells Conference, April 27–29, Washington DC, WA, USA.

Wahlster, W. (2003) Towards Symmetric Multimodality: Fusion and Fission of Speech, Gesture, and Facial Expression, in: A. Günter, R. Kruse, and B. Neumann (eds.), *KI 2003: Advances in Artificial Intelligence, 26th Annual German Conference on AI*, Lecture Notes in Artificial Intelligence, 2821, Springer, Berlin Heidelberg New York, pp. 1–18.

Wahlster, W., E. André, W. Finkler, H.J. Profitlich, and T. Rist (1993), Plan-Based Integration of Natural Language and Graphics Generation, *Artificial Intelligence*, 63, 387–427.

Wal-Mart (2004), Annual Report 2004, http://www.walmartstores.com/Files/annualreport_2004.pdf.

Wang, A., and A. Chandrakasan (2004), 180 mV FFT Processor Using Subthreshold Circuit Techniques, *International Solid State Circuits Conference, Digest of Technical Papers*, February 14–19, San Francisco, CA, USA,pp. 292–301.Want, R., A. Hopper, V. Falcao, and J. Gibbons (1992), *The Active Badge Location System, ACM Transactions on Information Systems*, 10(1), 91–102.

Ward, A., A. Jones, and A. Hopper (1997), A New Location Technique for the Active Office, *IEEE Personal Comunications*, 4, 42–47.

Waris, T., R. Tuominen, J. Kivilahti (2001), Panel-Sized Integrated Module Board Manufacturing, *Proceedings of the Polymers and Adhesives in Microelectronics and Photonics, Polytronic Conference*, October 21–24, Potsdam, Germany, pp. 218–223.

Warwick, K. (2004), Homepage, http://www.kevinwarwick.com.

Wasinger, R., and A. Krüger (2004), Multi-modal Interaction with Mobile Navigation Systems, *Information Technology*, 46(6), 322–331.

Watkins, C.J. (1989), *Models of Delayed Reinforcement Learning*, Ph.D. Thesis, Psychology Department, Cambridge University, UK.

Wayman, J., A.K. Jain, D. Maltoni, and D. Maio (2003), *Biometric Systems: Technology, Design and Performance Evaluation*, Springer, Berlin Heidelberg New York.

Wayt Gibbs, W. (2005), Considerate Computing, *Scientific American*, 291(1), 243–250.

Weber, W., R. Glaser, S. Jung, C. Lauterbach, G. Stromberg, and T. Sturm (2003), Electronics in Textiles – The Next Stage in Man Machine Interaction, *Proceedings of the Second CREST Workshop in Advanced Communicating Techniques for Wearable Information Playing*, May 23–24, Nara, Japan, pp. 35–41.

Weber, W., J. Rabaey, and E. Aarts (eds.) (2005), *Ambient Intelligence*, Springer, Berlin Heidelberg New York.

Wehrle, T., S. Kaiser, S. Schmidt, and K. Scherer (2000), Studying the Dynamics of Emotional Expression Using Synthesized Facial Muscle Movements, *Journal of Personality and Social Psychology*, 78(1), 105–119.

Weiser, M. (1991), The Computer of the 21st Century, *Scientific American*, 265, 94–101.

Weiser, M., and J.S. Brown (1996), The Coming Age of Calm Technology, in: P.J. Denning and R.M. Metcalfe (eds.), *Beyond Calculation*, Copernicus, Springer, New York, pp. 75–85.

Wermter H., and H. Finkelmann (2001), Liquid Crystalline Elastomers as Artificial Muscles, *e-Polymers*, 13, 1–13.

Williams, S. (2000), IrDA: Past, Present and Future,*IEEE Personal Communications*, 7(1), 11–19.

Winograd, T. (ed.) (1996), *Bringing Design to Software*, Addison-Wesley, Reading, MA, USA.

Winograd, T. (2000), Architecture for Context, *Human Computer Interaction*, 16, 401–419.

Winston, P.H. (1992), *Artificial Intelligence* (3rd edition), Addison-Wesley, Reading, MA, USA.

Wolf, J. (2005), *The e-Grain Concept: Technologies for Wireless Sensor Networks*, Talk presented at the Fraunhofer Technology Summit, German Innovation meets Japanese Business, April 5–8, Japan.

Wooldridge, M. (1999), Intelligent Agents, in: G. Weiss (ed.), *Multiagent Systems: A Modern Approach to Distributed Artifical Intelligence*, MIT, Cambridge, MA, USA.

Wroe, C., C. Goble, M. Greenwood, P. Lord, S. Miles, J. Papay, T. Payne, and L. Moreau (2004), Automating Experiments Using Semantic Data on a Bioinformatics Grid, *IEEE Intelligent Systems*, 19(1), 48–55.

Wüst, C.C., and W.F.J. Verhaegh (2004), Dynamic Control of Scalable Media Processing Applications, in: W. Verhaegh, E. Aarts, and J. Korst (eds.), *Algorithms in Ambient Intelligence*, Kluwer Academic, 259–276.

WWI (online), Wireless World Initiative, Homepage, http://www.wireless-world-initiative.org/.

Wynekoop, J.L., and S.A. Conger (1990), A Review of Computer Aided Software Engineering Research Methods, *Proceedings of the IFIP TC8 WG 8.2 Working Conference on The Information Systems Research Arena of The 90s*, December 14–16, Copenhagen, Denmark, Elsevier Science, New York, NY. USA, pp. 129–154.

W3C (online), http://www.w3.org/TR/2004/NOTE-ws-arch-20040211.

W3C OWL (online), http://www.w3.org/TR/owl-features.

Xia, W., and G. Lee (2004), Grasping the Complexity of IS Development Projects, *Communications of the ACM*, 47, 69–74.

Yamaoka, M., K. Osada, R. Tsuchia, M. Horiutchi, S. Kimura, and T. Kawahara (2004), Low Power SRAM Menu for SoC Application Using Ying-Yang Feedback Memory Cell,*2004 VLSI Circuits Symposium, Digest of Technical Papers*, June 17–19, Honolulu, pp. 288– 291.

Yang, P., C. Wong, P. Marchal, F. Catthoor, D. Desmet, D. Verkest, and R. Lauwereins, (2001), Energy-Aware Runtime Scheduling for Embedded Multiprocessor SoCs, *IEEE Design and Test*, 18(3), 70–82.

Zauner, J., M. Haller, A. Brandl, and W. Hartmann (2003), Authoring of a Mixed Reality Assembly Instructor for Hierarchical Structures, *Proceedings of the IEEE/ACM International Symposium on Mixed and Augmented Reality*, October 7–10, Tokyo, Japan, pp. 237–246.

Zigbee Alliance (online), Homepage, http://www.zigbee.org/en/index.asp.

Zoschke, K., J. Wolf, M. Toepper, O. Ehrmann, T. Fritzsch, K. Scherpinski, H. Reichl, and F.J. Schmückle (2004), Thin Film Integration of Passives – Single Components, Filters, Integrated Passive Devices,*Proceedings of the Electronic Components and Technology Conference*, June 1–4, Las Vegas, NV, USA, pp. 294–301.

Zoschke, K., J. Wolf, M. Toepper, O. Ehrmann, T. Fritzsch, K. Scherpinski, H. Reichl, and F.J. Schmückle (2005), Fabrication of Application Specific Integrated Passive Devices Using Wafer Level Packaging Technologies, *Proceedings of the Electronic Components and Technology Conference*, May 31–June 3, Orlando, FL, USA.

Zrínyi, M. (online), Department of Physical Chemistry, Budapest University of Technology and Economics Laboratory of Soft Matters, Homepage, http://web.fkt.bme.hu/zrinyi/ism.htm.

Index

E

F

G

H

I

K

L

M

N

O

P

Q

R